T0215235

# FOOD CRIME

This book addresses the various forms of deviance and criminality found within the conventional food system. This system—made up of numerous producers, processors, distributors, and retailers of food—has significant, far-reaching consequences bearing upon the environment and society.

*Food Crime* broadly outlines the processes and impacts of this food system most relevant for the academic discipline of criminology, with a focus on the negative health outcomes of the US diet (e.g., obesity and diabetes) and negative outcomes associated with the system itself (e.g., environmental degradation). The author introduces the concept of "food criminology," a new branch of criminology dedicated to the study of deviance in the food industry. Demonstrating the deviance and criminality involved in many parts of the conventional food system, this book is the first to provide exhaustive coverage of the major issues related to what can be considered food crime. Embedded in the context of state-corporate criminality, the concepts and practices exposed in this book bring attention to harms associated with the conventional food system and illustrate the degree of culpability of food companies and government agencies for these harms.

This book is of interest to students, scholars, and practitioners seeking a more just and healthy food system and encourages further future research into food crimes in the disciplines of criminology, criminal justice, and sociology.

**Matthew Robinson** is Professor of Government and Justice Studies at Appalachian State University. He is the author of 22 books, including *Social Justice, Criminal Justice: The Role of American Law in Effecting and Preventing Social Change* with Routledge, and more than 100 other publications. He is a past president of both the North Carolina Criminal Justice Association and the Southern Criminal Justice Association. He was recently ranked the 19th most influential criminologist in the world today. Robinson has been writing about the subject of "food crime" for several years.

# FOOD CRIME

## An Introduction to Deviance in the Food Industry

*Matthew Robinson*

Routledge
Taylor & Francis Group

NEW YORK AND LONDON

Designed cover image: © Getty Images

First published 2024
by Routledge
605 Third Avenue, New York, NY 10158

and by Routledge
4 Park Square, Milton Park, Abingdon, Oxon, OX14 4RN

*Routledge is an imprint of the Taylor & Francis Group, an informa business*

ISBN: 978-1-032-28353-1 (hbk)
ISBN: 978-1-032-28045-5 (pbk)
ISBN: 978-1-003-29645-4 (ebk)

DOI: 10.4324/9781003296454

Typeset in Sabon
by Apex CoVantage, LLC

# CONTENTS

# PREFACE

As suggested by the title, this is a book about food crime. Stated simply, food crime is any deviant and/or criminal acts committed by actors in the conventional food system—from growers to producers to distributors, and so on. As will be shown in this book, much food crime is not illegal at all but instead is deviant or immoral in the sense that it is wrong, and it causes harm to humans, animals, or both. As one example, much of the food produced and sold in the United States (and even around the world) is ultra-processed, has no nutritional value whatsoever, and is actually very dangerous to our health and well-being. Another example is the enormous harms produced by the system itself, such as environmental damage and contribution to climate change.

In this book, you will clearly see many significant problems associated with food. Many of these make up the focus of the food crime literature, the subject of this book. But, this book is different from others in at least two ways. First, this book explicitly focuses on harms associated with both food and the conventional food system, showing specifically what is wrong with what we eat and how our food is produced. Second, this book identifies the culpable harms that comprise food crimes—that is, those behaviors that make up food crimes. The main argument of the book is that we must hold food corporations responsible for the culpable, harmful acts they commit. We must also hold governments responsible for these harms as well since it is their obligation to regulate food corporations in the public interest, something they are generally failing to do.

The book is organized into ten chapters, each with a unique focus. Chapter 1 pertains to food crime itself. I offer definitions and examples. You will come to understand the important context of food crime, where profit-seeking underlies virtually every behavior discussed in this book. You will see, incredibly, that contemporary dietary patterns now kill more people each year than murder, as well as all illicit drugs combined. Given the involvement of key government agencies in the production and sale of food, plus its (lack of) regulation for safety and health, I introduce the concept of state-corporate crime to help us frame the issue of food crime as an act of elite deviance (behaviors committed by the powerful, some illegal, some not), conducted by corporations for profit, with the assistance, approval, reward, and/or complicity of government. I will argue in the book that there is culpability among both food companies and government agencies when it comes to matters of food crime; even in cases where there is no intent to harm people or animals through food crime, there is often negligence, recklessness, and knowing behaviors involved in food crime.

Chapter 2 deals with the conventional food system, the term used to describe all of the actors and organizations involved in the growth, production, and sales of food products. You will see that the system is highly industrialized (Chickenized and McDonaldized), corporatized, monopolized, and globalized. You will see that the major implication of these realities is that the great bulk of foods being produced are largely unhealthy and the same nearly everywhere around the world. They are grown, produced, marketed, and sold by companies that are far less concerned with quality or healthiness than they are with turning a profit for Wall Street and shareholders. This has enormous health consequences for Americans as well as people around the world.

Chapter 3, co-authored with my wife Briana Robinson, shows what a healthy diet is comprised of, based on government guidelines for food intake. I compare that with what people eat and show you that Americans eat too much fat, saturated fat, sugar, and salt, plus lots of additives, and far too few vegetables, fruits, vitamins, and minerals. I argue in the chapter that this reality exists for at least two major reasons. First, food companies historically played a large role in hindering the release of adequate food guidance for consumers. Second, the foods being produced by major food corporations are largely unhealthy. In particular, the cheapest and most widely available foods are the ones most consumed by Americans, as well as people around the world.

Chapter 4 is a version of an article that previously appeared in the *Journal of International Criminal Justice Studies*. It illustrates that it is the food itself that is the first crime. That is, the food itself is largely unhealthy, with high levels of fat, saturated fat, sugar, and salt, plus potentially unhealthy additives. The use of antimicrobials in animal feed is also a significant problem within the conventional food system as it is leading to increasing cases of antibiotic resistance in humans. In the chapter and in the book, I do not argue that any food should be made criminal and thus am not actually calling on lawmakers to make unhealthy foods food "crimes," but I am instead calling the attention of regulators to better regulate the production of these foods and to greatly restrict their marketing, especially to vulnerable children.

Chapter 5 is a version of an article that appeared in the journal, *Drug Science, Policy and Law*. It demonstrates that many of these foods can be considered drugs in that they produce changes in the brain, lead to cravings, produce tolerance and withdrawal, and are habit-forming and lead to compulsive use even in the face of serious deleterious outcomes associated with their consumption. You will see that ultra-processed and hyperpalatable foods seem to be the most addictive, and these are the foods being most produced in the United States and around the world. I argue that food companies are culpable for the harmful outcomes related to food addiction produced by the foods they create and sell.

Chapter 6 turns to the issue of specific harms associated with the standard American diet (or SAD). I examine major illnesses including obesity, diabetes, mental health, and other conditions. The section on diabetes was originally published in the journal, *State Crime*. I show that nearly 400,000 Americans die every year from the foods they eat (and don't eat enough of), and health-care costs associated with poor diet now dwarf those direct harms associated with street crime every year. The chapter also illustrates what a healthier diet looks like and I confront you with the issue of whether people actually have a full choice when it comes to the foods they eat. I argue that the notion of free will is a myth, particularly when it comes to what we eat; we eat what we like but we don't choose to like it, at least not freely.

Chapter 7 covers some of the harms associated with the conventional food system itself. These include harms associated with killing and eating animals, including animal welfare concerns stemming from factory farming and concentrated animal feeding operations (CAFOs), pathogens in our food and environment from food production, hazardous working conditions, significant environmental damage, and the confusing dual reality of food insecurity and food waste. The issue of environmental damage may stand out the most, given the enormous and serious consequences of issues such as climate change.

Chapter 8 deals with the important issue of the culpability of food companies for specific behaviors. Behaviors examined include producing excess calories than needed by consumers to be and stay healthy, putting food in non-food environments so that we are surrounded by junk food nearly everywhere we go, funding research to create one-sided studies, using front groups to confuse consumers as part of the industry playbook to keep regulators and lawmakers at bay, advertising unhealthy products, deceptive advertising, food fraud, and product shrinkage, not to mention all the deleterious health outcomes documented in Chapter 4 and other harms of the system documented in Chapter 5. I argue that there is great recklessness in food companies when it comes to creating more food than is needed by the population, especially when so much of it ends up going to waste. I also argue that it is reckless to surround us with especially unhealthy foods essentially wherever we go. There is likely specific intent to mislead when it comes to funding research and using front groups to make claims and defeat honest efforts to make food healthier and less harmful. There is also likely intent involved in the misleading claims made by food companies like those discussed in the chapter, advertising unhealthy products just to boost sales, and shrinking the size of packages over time to create the false impression that people are getting more food than they really are when they make purchases.

Chapter 9 shows that there are significant economic benefits of the conventional food system, including access to food across the nation and globe, expansive contribution to gross domestic product (GDP), significant employment, access to health care, as well as enormous multiplier effects. The contribution to GDP is more than $1 trillion per year, and tens of millions of people are employed in some part of the system. Still, you will see that wages associated with food system work tend to be quite a low relative to other jobs and that benefits are often lacking completely in the case of part-time and seasonal workers. The benefits of the system may very well outweigh its costs, although the costs of the system are still enormous and need to be reduced.

Chapter 10 summarizes the book and then offers some modest proposals both to make food healthier and to hold key actors in the conventional food system (and government) responsible for the harms they inflict on society. I make 25 proposals to achieve these goals. Overall, the book makes a valuable contribution to the food crime literature by drawing attention to specific problematic aspects of food, the conventional food system, as well as specific culpable and harmful behaviors committed in the context of the system.

Keep in mind that the focus of this book is on both deviance (legal yet immoral) acts of food companies and actual illegal (i.e., criminal) acts. Yet, the book touches upon issues that have been and are being handled in civil law (i.e., lawsuits against companies). Consider the following cases, as examples.

First, the number of class action lawsuits against food companies is growing over time, from just 19 in 2008 to 325 in 2021. The following are some of the most recent as well as most known suits. You will see that many of these suits may come off to you as frivolous in nature (e.g., Starbucks was sued for putting too much ice in their drinks), whereas others are deadly serious (i.e., many companies were sued over deadly outbreaks of bacterial contamination).

Starting with claims about potentially misleading products, of which there are more and more every day. Anheuser-Busch was sued because its "Made with Agave" statement on its Bud Light hard seltzer drinks is potentially misleading; the product is flavored with agave syrup rather than agave spirits. Similarly, A&W Root Beer was sued because of claims it is crafted with aged vanilla when it contains no real vanilla whatsoever. Barilla pasta was sued over its slogan "Italy's No. 1 brand of pasta"; the pasta is actually made in the United States. A similar lawsuit against Texas Pete hot sauce alleged that the name and imagery of the product are misleading since it is actually made in North Carolina. Blue Diamond was sued because its "Smokehouse" almonds were not actually flavored in a smokehouse but instead were flavored with natural and artificial flavors (i.e., chemicals). Kellogg's was sued because its frosted strawberry Pop-Tarts contain no real strawberries. Similarly, Trader Joe's was sued because its strawberry pastry is mostly made from apples. Krispy Kreme was sued because fruit-flavored desserts contained no actual fruit.

Canada Dry was sued because it claims its ginger ale is "Made with Real Ginger." Coca-Cola was sued over allegations that consumers were misled over some products that claimed to be "100% naturally flavored" even though they contained artificial flavoring. Coors was sued over its claims that some of its hard seltzers were sources of Vitamin C that were "nutritionally-equivalent" to some fruits. Dunkin Donuts was sued over the contents of its Angus Steak & Egg sandwich because the meat in the patty was not steak but instead just beef. Yum Brands/Taco Bell was sued over meat in its products that allegedly contain binders and further do not meet the minimum requirements set by the US Department of Agriculture to be labeled as "beef." And Subway was sued over claims related to its tuna sandwiches over allegations that it is not 100% tuna. There have also been allegations alleging Subway's chicken sandwiches contain large amounts of soy protein.

Frito-Lay was sued for allegedly not using enough real lime juice in one of its chips products claiming a "hint of lime." KFC was sued for false advertising over the fact that its "fill-up" chicken bucket depicted on television as overflowing with chicken was only about half full. Snack Pack pudding was sued over its claims that its products were advertised as being "made with real milk" when it is actually made with fat-free skim milk. The issue of misleading advertisements is discussed in Chapter 9.

Then there what is known as product shrinkage, where less and less of a product is sold over time in the same size container, is reviewed in Chapter 9. As a potential example of this, Hershey was sued for selling packages of candy that are not full but instead only about half full.

More seriously, General Mills was sued over the allegation that the potentially cancer-causing ingredient glyphosate is in its Cheerios products. LaCroix was sued over its label claiming its water was "natural" even though there are non-natural or synthetic components in the product such as ethyl butanoate, limonene, linalool, and linalool propionate. Mars was sued over its Skittles candy because the product contains titanium oxide, a substance that may be "unfit for human consumption"; the International Agency for Research on

Cancer claims the substance is a potential carcinogen. Further, Beech-Nut Nutrition Co., Gerber, and other companies have been sued by customers alleging claims that baby foods are safe for consumption when in fact they contain potentially harmful heavy metals including mercury, arsenic, lead, and cadmium.

Robert's American Gourmet settled a class action that claimed that the firm misstated the calorie and fat content of its Pirate's Booty snack. Nestlé and Ferrero were sued over *Escherichia coli* and *Salmonella* outbreaks.

ConAgra was sued over a major outbreak of *E. coli* in a batch of peanut butter, which led to at least 750 victims. Years ago, Foodmaker/Jack in the Box was sued over an *E. coli* outbreak that killed four children and infected hundreds more in several states. Maple Leaf Foods was sued over a listeria outbreak that killed 20 people and made ill thousands of people. And Nestlé and Ferrero were sued over *E. coli* and *Salmonella* outbreaks. The issue of harmful substances in our food is also discussed in Chapter 9.

Kellogg's and Post Food were sued over the amount of sugar in their cereals and over false advertising that the products were claimed to be healthy, wholesome, and nutritious. The issue of what is in our food is analyzed in Chapter 4.

McDonald's has faced many lawsuits over the years. It was famously sued over a cup of coffee that was too hot and caused extremely serious injury after being spilled on the leg of a customer. Hundreds of customers had complained about the temperature of coffee prior to the incident, yet McDonald's failed to ever take corrective action. McDonald's was also sued over the fact that it claimed its French fries are cooked in vegetable oil even though they are cooked in beef fat. McDonald's and Burger King are being sued for fraud, misrepresentation, and false advertising for claiming the food they serve is safe when it contains highly toxic fluorinated chemicals called per- and poly-fluoroalkyl substances in their product packaging. McDonald's was also sued for alleged misleading information on the nutritional information in their products and claims their food is addictive (a subject discussed in Chapter 5).

Finally, numerous companies (Agri Beef Company, American Foods Group, Cargill, Hormel Foods, Iowa Premium, JBS, National Beef Packing Company, Perdue Farms, Seaboard Foods, Smithfield Foods, Triumph Foods, and USA Food Company) are all being sued over allegations that the companies have conspired for years to keep worker pay lower than the market allows, in violation of the Sherman Antitrust Act.

Again, many of these kinds of issues are reviewed in future chapters. Keep an open mind as you read, understanding that I am not necessarily recommended making such behaviors crimes.

# 1

# AN INTRODUCTION TO "FOOD CRIMES"

## Introduction

Have you ever heard of the term "food crime?" Until you picked up this book, it is unlikely that you ever have. Yet, if you eat, you should become aware of it. Even in academic disciplines that study crime, such as criminology and criminal justice, there is very little focus on the subject matter (Robinson, 2017). Though it is shame, given the wide range of harms produced by the behaviors included in food crimes, the lack of focus is not surprising given that academic disciplines such as criminology and criminal justice are almost exclusively focused on street crimes, things such as murder, rape, robbery, assault, theft, motor vehicle theft, burglary, arson, and other crimes such as drug abuse and terrorism. Such acts are unquestionably serious, as they produce a significant loss of life and loss of property and also diminish the quality of life (Robinson, 2020).

What if I told you that the behaviors of focus in this book cause harms that easily dwarf those caused by the street crimes that academic criminologists tend to study? It is true: Food crimes cause more death, injury, and loss of property than all street crimes combined, and it is not even close. That reality is illustrated clearly throughout this book.

In the chapter, I provide a definition of food crime and illustrate the types of behaviors that are included in it. Since many food crimes are not actually illegal but instead are deviant or morally wrong, I differentiate the terms crime and deviance. Then, I delve into the topic of state-corporate criminality, which gives us an important context in which to discuss food crimes. Here, the issue of culpability is defined and discussed; it is important to understand that people can be held morally and legally responsible for their behaviors even when they are not committed intentionally. Finally, I end the chapter with a discussion of why food crime matters in the first place and should matter to us all.

## What Is "Food Crime"?

A growing body of literature on crimes of and within the food industry has emerged (e.g., see Gray & Hinch, 2018). The term "Food Crime" was first used by Hazel Croall (2007,

DOI: 10.4324/9781003296454-1

p. 206) to describe the "many crimes that are involved in the production, distribution, and selling of basic foodstuffs." Later, Croall (2013, p. 167) defined food crime as the behaviors in the food industry that are "morally dubious" and also sometimes illegal, which involves "manipulation and exploitation of the planet." Gray and Hinch (2015, p. 97) add that food crimes involve "serious harms . . . addressed beyond the traditional definitions of crime" that have "negative consequence for a variety of both human & non-human victims." Gray (2019a, p. 12) defines food crime as "illegal, criminal, harmful, unjust, unethical or immoral food-related issues." De Rosa et al. (2019, p. 44) define it as "dishonesty relating to the production or supply of food, that is either complex or likely to be seriously detrimental to consumers, businesses or the overall public interest."

According to Hinch and Gray (2019, p. 1), food crime includes

> a wide range of offences . . . involving economic and physical harms, issues of personal safety and health, and many different kinds of frauds, from the evasion of subsidies and quotas and the avoidance of revenue, to food adulteration and misrepresention through written and pictorial indications, the quality and contents of food.
>
> *(Croall, 2007, p. 207)*

Gray (2019a, p. 20) adds that food crime includes acts that

> criminally defined legal events (the use of agricultural slave labour, or the intentional adulteration of food) . . . directly or indirectly harmful (targeted food marketing, or the influence on climate change) . . . unethical, immoral or unjust (food deserts in vulnerable communities, or the use of humane slaughter techniques to justify livestock murder).

Food crime is widespread across the globe, including in the European Union which has had notable cases of food fraud (De Rosa et al., 2019). Gray and Hinch (2018, p. 12) suggest that food crime includes "illegal, criminal, harmful, unjust, unethical or immoral food-related" behaviors and omissions of behavior (such as poor regulation of harmful foods by the government). So, it is important to note at the outset that food crime includes many legal behaviors that are "lawful but awful" (Passas, 2005). Some will question the logic of studying legal behaviors in the context of criminology or criminal justice since the academic disciplines are traditionally focused on criminal acts (i.e., acts that violate the criminal law). Yet, a social harm perspective along the lines of Zemiology would essentially mandate examining these behaviors given the enormous harms they impose on society (Canning & Tombs, 2021; Dorling et al., 2008). The nature and extent of those harms are the subject of this book.

## Examples of Food Crimes

As examples of food crimes, Croall (2013) discusses food poisoning caused by negligence or recklessness of food producers (also see Leighton, 2019), fraud, misleading advertisements, inequality of food availability, illegal trade and pricing in the food industry, labor exploitation, and animal cruelty (also see Beirne, 1999). Monopolization and unfair trade practices can also logically be seen as unethical and even illegal behaviors, as can wage deflation, bullying of workers, and control of prices by suppliers.

With regard to fraud, Croall examines foods that are watered down as well as companies that sell branded foods that are no better or different than generic foods. Croall also examines foods that are actually deadly based on additives or inadequate regulation for safety. Poor regulation also produces widespread food poisoning, which produces illness and occasional death. Smith et al. (2017, p. np) define food fraud as "the deliberate and intentional substitution, addition, tampering, or misrepresentation of food, food ingredients, or food packaging; or false or misleading statements made about a product for economic gain." These behaviors are "carried out intentionally to avoid detection by regulatory bodies or consumers." Many of these behaviors are discussed later in this book.

Other unethical and potentially illegal behaviors by food companies include exploitation of labor, such as in the case of seasonal and migrant workers—described as the "new slavery" by some (Lawrence, 2008, p. 111)—and forcing employees to work in hazardous conditions. Deaths produced by hazardous conditions amount to about 55,000 a year just in the United States, far more than are murdered every year in the country (Reiman, 2020). Factory farming is not only horribly cruel to animals but is also often harmful to employees as well as the environment, as in the case where animal waste and chemicals pollute waterways as well as the air (e.g., see Mugni et al., 2012). These issues are examined in this book, as well.

Gray and Hinch (2015) also discuss different forms of potential deviant and criminal behavior in the food industry. These include low pay for farm workers, child labor and slavery in the production of food (e.g., see Coe & Coe, 2013; Schrage & Ewing, 2005), the dominant power held by agribusiness over traditional farmers (e.g., see Culp, 2005; Fairley, 1999; Fatka, 2013), pesticide use on crops (e.g., see Del Prado-Lu, 2019; Goodman, 2011), patented seeds and unintentional violations of those patents, laws that tend to better protect agribusiness than farmers and consumers, and weak enforcement of existing regulations. Later, the same authors (Gray & Hinch, 2018) add a focus on acts such as forced labor and slavery in the food industry (e.g., see Hinch, 2018; Satre, 2005), the use of hazardous chemicals in farming, food poisoning through corporate negligence, animal slaughter, unethical food labels and trade policy (e.g., see Lawrence, 2004), unhealthy school lunches, the impact of food production on climate change (e.g., see White & Yeates, 2018), and more. Again, many of these behaviors are examined later in the book.

Other scholars have written about the monopolization of numerous parts of the conventional food system, including biotechnology involved in genetically modified foods (GMOs) (Walters, 2006), potential harms associated with GMOs themselves (Boone, 2013; Morgan & Goh, 2004; Nottingham, 2003; Walters, 2008), environmental harms such as threats to the quality of air, soil, and water (White, 2014), buying up or seizing land for food and biofuel production and extracting valuable natural resources, referred to as "land grabbing" (White, 2012), food waste (Long & Lynch, 2018), and numerous other crimes such as bribery and corruption that stem from activities of organized crime syndicates involved in the food industry (White, 2014). Some of these behaviors are examined in later chapters of this book.

As of the time of this writing, an interesting story has appeared in the news that gives us a final example of potential food crime. Major food companies (such as McDonald's, In-N-Out, and Chipotle) are spending millions to try to block a California law that would raise salaries to $22 an hour for fast-food workers (Meyersohn, 2023). A law passed last year in 2022 that was to go into effect on January 1 of this year is now being challenged by a voter petition that has enough votes to appear on the ballot next year in 2024. Given that California's fast-food industry employs more than 500,000 workers, the cost of the law would be significant. Major fast-food companies are spending enormous amounts of money to fight this law; one could argue such behavior is unethical, especially when housed in the context of ads claiming the law would put local restaurants out of business.

Finally, White (2014, p. 835) describes the extensive government and corporate security mechanisms that have grown to protect "a platform of state, corporate, organized group wrongdoing and injustice" related to food crime. This close connection between food corporations and the state is explained more later in the book.

## The Important Context of Food Crime

All of these behaviors are confined in the context of "cheap capitalism," which is "characterized by degraded business morality, low prices and/or unsafe goods or services, and low-waged labour to maximize profits" (Asomah & Cheng, 2019, p. 194). This makes sense given that practices such as "industrialization, corporatization, and neoliberalization have drastically reformed the modern practices of the food industry, as well as the regulations which govern them" (Gray & Hinch, 2015, p. 97). In this system, corporations are designed to be greedy, and the result is much crime and deviance (Robinson & Murphy, 2008). Stated simply, the evidence presented in this book will show that global food companies, the main drivers of what is often called the "conventional food system," care about profit above and beyond all other things, including most especially assuring the food products they produce and sell are healthy, as well as not the source of harms.

In this book, I expand on this context by illustrating the companies involved in the production and sale of our food, the government agencies that both enable and regulate these companies, and the broader global economy in which all operate. Though many harms discussed in the book may be understood at the individual level, a more productive way to comprehend the issues examined within is to house them in the context of global capitalism that has taken over much of the world (Frieden, 2020; Robinson, 2014). So, for example, whereas poor diet and nutritional deficiencies will be found among certain segments of the population, the reasons this is so are owing to the nature of the economy and food economy itself. That is, the system is built to assure certain outcomes, even if the resulting harms are not intentionally conceptualized or planned. Evidence in this book illustrates that major food companies are designed to produce and profit from unhealthy food. As such, they are culpable for harms produced by it.

One simple example makes the point: it is widely known and understood that certain foods are linked to higher rates of illness and death. Specifically, diets high in fats, saturated fats, salts, and sugars, as well as low in vegetables and fruits, whole grains, nuts and seeds, and vitamins and minerals are associated with increased rates of heart disease, obesity, and other deleterious health outcomes (Micha et al., 2017). Major food companies are in the business of providing all sorts of foods to consumers, including both healthy and unhealthy foods. Yet, for reasons identified later in this book, the great bulk of foods produced and consumed are in the unhealthy variety, such as in the case of ultra-processed foods (Elizabeth et al., 2020; Monteiro et al., 2019). So, the system is set up to assure unhealthy outcomes for people. That the negative health consequences in human populations are not intended is irrelevant; the companies remain culpable—morally if not legally responsible—for the outcomes.

One unhealthy outcome is widespread death. A study examining data from 2012 found that just over 700,000 people in the United States died that year from a "cardiometabolic disease" (CMS). This included more than 500,000 from heart disease (including almost 375,000 from coronary heart disease, nearly 130,000 from stroke, and just under 70,000 from diabetes (Micha et al., 2017). Cardiometabolic disease is defined as "a combination of metabolic dysfunctions mainly characterized by insulin resistance, impaired glucose tolerance, dyslipidemia [elevated plasma cholesterol and/or triglycerides], hypertension, and central adiposity [high accumulation of fat in the lower torso]" (Saljoughian, 2016). According to Saljoughian (2016), both the American Society of Endocrinology and the World Health Organization (WHO) recognize CMS as a disease.

Of the 700,000 CMS deaths, about 395,000 (about 56%) were attributed to the foods that people ate and did not eat (Micha et al., 2017). Roughly 66,500 of the deaths identified in the study were due to high sodium intake. Another 58,000 deaths were caused by eating too much processed meat. Many other deaths were attributable to the absence of certain foods in the diet, including 59,000 deaths due to a lack of nuts and seeds in the diet, 55,000 deaths due to too little omega-3 fats, 53,000 deaths from too few vegetables, and 52,500 deaths from too few fruits. Finally, about 52,000 deaths were from too many sugar-sweetened beverages (SSBs) (Micha et al., 2017). These data are illustrated in Table 1.1.

The 395,000 deaths are very close to that reported by the US Centers for Disease Control and Prevention (CDC) with regard to poor lifestyle patterns among Americans: "estimates are that some 300,000 deaths each year in the U.S. likely are the results of physical inactivity and poor eating habits" (US Centers for Disease Control and Prevention, 2018b). For perspective, only about 42,000 people died from illicit drugs in 2012 (National Institute on Drug Abuse, 2018). And just under 15,000 people were murdered in the United States in 2012 (Federal Bureau of Investigation, 2013).

Stated simply, this means that, in 2012, the foods we ate (and did not eat) killed more than nine times more people than all illegal drugs combined and more than 26 times as many people who were murdered. Further, SSBs alone are more deadly than all illegal drugs and murder combined! Incredibly, evidence reviewed later in the book will show that SSBs are essentially drugs, yet, of course, they are legal and widely available. Extrapolate such data across so much of the world, at least in places where people eat much of the same foods produced by the same global food companies, and you get a strong sense that this is a social problem worthy of serious consideration.

**TABLE 1.1** Death Attributable to Food

| | |
|---|---|
| High sodium | 66,508 |
| Lack of nuts, seeds | 59,375 |
| Processed meat | 57,766 |
| Lack of omega-3s | 54,626 |
| Too few vegetables | 53,410 |
| Too few fruits | 52,547 |
| Sugar-sweetened beverages | 51,694 |

*Source:* Micha, R., Penalvo, J., Cudhea, F., et al. (2017). Association between dietary factors and mortality from heart disease, stroke, and type 2 diabetes in the United States. JAMA, 317(9), 912–924.

## Harms Caused by Food Crimes

In terms of other harms caused by food crimes, many will be discussed in later chapters of this book. Croall (2013) identifies the following as harms caused by food companies and the conventional food system: environmental degradation, increased greenhouse gases, food waste, airborne pollution, less soil fertility, erosion of land, increased flooding, crop resistance to antibiotics, species decline, reduced farm labor, monopolization, wage deflation, hunger, as well as various health conditions such as obesity, illness, and death. Where data are available to document these harms, they will be provided in this book.

There are, in fact, many other forms of harms caused by the conventional food system itself. For example, an examination of the animals that make up such a huge part of the system shows that there are billions of animals killed every year (Johns Hopkins Center for a Livable Future, 2021) by Americans seeking their average of about 300 pounds a meat a year (Christen, 2021). The degree of suffering these animals endure before slaughter is also astounding. Even the lives of many animals in the conventional food system who are not killed for food, such as dairy cows, are miserable and cut far short compared with free-range cattle not subjected to milking. These issues are examined later in the book.

This book will illustrate many harms associated with food crime and show that they dwarf the harms caused by both street crimes and illicit drugs typically studied by criminologists. That is the primary reason why the argument of the book is that scholars within criminology and criminal justice must study food crime. Before moving on to those, it is important to first define some key terms.

## Crime versus Deviance

*Deviance* refers to deviant behavior, acts that depart from or violate acceptable social norms (Anderson, 2017; Brown & Sefiha, 2017). *Norms* are expectations for how one is supposed to behave, forming the basis for what is considered normal in society (Best, 2020). So, deviance is synonymous with abnormal behavior, acts that are viewed as immoral or wrong, or at the very least, unusual or strange (Inderbitzin et al., 2020). As an example, walking into an elevator and facing other people in a tight, enclosed space, then striking up a conversation, would be considered abnormal. But such a behavior would not be illegal or even likely viewed as threatening to the social order. And it is not dangerous (other than during a pandemic, when people are not masked and socially distancing). Normally, people walk into an elevator and then turn around and face the door, largely without even talking. Yet, there are forms of deviance that are considered more serious, dangerous, and oftentimes even criminal. For example, having sexual relations with a minor is not considered normal in American society and is largely viewed as immoral. Typically, this behavior is illegal—statutory rape—depending on the age of the minor and adult involved. Firing a gun on your own property is generally legal in many places, but doing so in a city or into an occupied dwelling is dangerous and would be seen as abnormal, deviant, and typically even illegal.

*Crime* involves acts that actually violate the criminal law (Samaha, 2016). This includes behaviors people commit (e.g., killing another person forms the basis of the crime of murder), attempt to commit (e.g., trying to assault someone can lead to the charge of attempted battery), agree to commit (e.g., two or more people enter into an agreement to kill or hurt someone, which is a conspiracy), and even in some cases, fail to commit (e.g., *not* paying taxes or required child support) (Robinson, 2020). The major difference between deviance and crime is that deviance is not generally illegal (although in some cases, it can be), whereas crime is always illegal, since it actually requires the violation of a criminal law (Robinson, 2020).

When people think of crime, they tend to picture acts of street crime, the very behaviors the federal government views as most serious (Reiman, 2020). For example, the Part 1 Index Offenses of the Uniform Crime Reports include murder, rape, robbery, aggravated assault, theft, burglary, motor vehicle theft, and arson (Federal Bureau of Investigation, 2013). Not included in the list of serious crimes are acts that actually cause more harm and occur more frequently than these, acts such as fraud, embezzlement, defective products, hazardous workplaces, and so on (Robinson, 2020). To these, I would obviously add food crimes!

The American criminal justice system, as well as those in virtually every other nation, focuses almost exclusively on the street crimes listed earlier, plus other behaviors considered serious by the federal and state governments, things like terrorism and illicit drug use and sales (Robinson, 2020). All of this is due to one simple reason—the criminal law determines what is illegal in the first place, what is most serious in the second place, and essentially on which acts agencies of criminal justice will focus their attention. As noted by Robinson (2020), the law is not neutral, but rather it serves the interests of some more than others.

As a quick example, consider the argument of Jeffrey Reiman of the American criminal justice system. Jeffrey Reiman (2020) suggests that the criminal justice system is biased against the poor at all stages of the process, from policing to courts and all the way through corrections. Reiman suggests that criminal justice processes are actually aimed at population control and serving limited interests—the interests of those in power are served when we focus almost exclusively on street crimes rather than other types of harmful behaviors, including white-collar crimes, corporate crimes, and even governmental deviance. At the same time, those people whom we fear most (e.g., young and minority males) are routinely rounded up by the police and sent off to some form of government-controlled institution (e.g., jail or prison) or community alternative (e.g., halfway house or boot camp). This amounts to a form of population control so that the enemies in the war on crime can never win and can never achieve the types of success that can be enjoyed by those with the power to achieve.

All of this starts with the criminal law. Reiman shows that the label of "crime" (particularly "serious crime") is not used for the most harmful and frequently occurring acts that threaten us. Meanwhile, extremely dangerous acts committed by the wealthy and powerful (some he discusses include unsafe workplaces, environmental pollution, unnecessary surgery, and unnecessary prescriptions; to that, we can add unhealthy foods) are either not illegal or are but are not vigorously pursued by criminal justice agencies even though they kill and injure far more people than a crime every year.

Indeed, studies of the legal system in the United States show that the people who make it, vote for it, and pay for the political campaigns that make it possible, are not demographically representative of the population (Robinson, 2020). It is a safe bet that this is the case across the globe, but to a lesser degree, especially in parliamentary-type styles of governments where there are more political parties and thus a greater chance of representation.

When it comes to comparing campaign contributions across countries, the United Nations Office of Drugs and Crime (2019) explains: "Many countries outlawed donations by state-owned companies and donations with foreign interests [and] 60% of countries worldwide required political parties and candidates to disclose financial records" (p. 4). Yet, there is "a remarkable lack of information on the cost of elections. . . . Governments tend to not publish the details. Even if they do, they talk about very different things. It is very difficult to obtain comparable data" (p. 7). Still, there are analyses of various kinds on which we can rely to make informed judgments about how powerful interests can shape the law through funding political campaigns.

For example, compared with at least some other countries, the United States has a more stringent system of regulation when it comes to campaign finance. In the United States, there are limits on how much candidates can raise individually and through Political Action Committees (PACs), but not Super-PACs, which are allowed to raise and donate unlimited funds for candidates and political parties but cannot work with individual candidates. Take the United Kingdom, for example, where there are limits for individual candidate spending, but not for political parties. At the same time, there is less need for raising large amounts of money for things such as political advertising, since broadcast media time is free in the United Kingdom. One might assume that large corporations would have less impact on the law in the United Kingdom since they cannot directly fund individual political campaigns.

In other countries such as France, political campaigns are funded through individual donations, which are limited not only by law but also by government sources. Government funding of campaigns also ought to reduce the influence of powerful interests on the law. In other countries such as Mexico and Portugal, individual candidates do not raise funds but rather get money from political parties (Waldman, 2014).

The majority of campaign finances come from public sources in countries such as Spain, as well as the Scandinavian countries (Pinto-Duschinsky et al., 1999). In places such as Norway, the majority of funding comes from the government, as political ads are banned from television and radio (Thompson, 2012). To the degree that political campaigns are funded by public (i.e., government) sources, one might expect less influence on the law by powerful corporate interests such as major food companies.

Meanwhile, in the United States, corporate interests can and typically do have their will enacted into law by making large campaign contributions both to political parties as well as individual politicians. Another means to impact the law is through organizations such as the American Legislative Exchange Council (ALEC), a group of huge corporations that have formed partnerships with state legislators and executive employees working for governors' offices; through ALEC, corporations can and often do write laws directly through legislative staff working with state legislatures (Robinson, 2020). Keep in mind as you read this book—the criminal law does not necessarily serve the interests of the common person—so many "food crimes" will be legal in spite of the enormous harms they produce.

Places where even more corporate influence is seen in the law include India and Russia, where direct, "under the table" contributions lead to the possibility of real corruption. In Russia, media broadcast time is also distributed by the government based, in part, on the outcome of previous elections (Thompson, 2012). Private finance of political campaigns is thought, according to experts, to be one of the largest sources of corruption in the world (United Nations Office of Drugs and Crime, 2019).

In other countries, such as Brazil, an examination of winning politicians' donations reveals that nearly all of it comes from large corporations. There, nearly no one makes political contributions to candidates or parties (Thompson, 2012). The same is true in the

United States, where less than 1% of people give any sizeable amount to individual candidates running for office or political parties (Robinson, 2020). In still other counties, such as Nigeria, there are no limits in terms of how much can be spent on political campaigns (Thompson, 2012).

One analysis of 34 member nations of the Organization for Economic Cooperation and Development (OECD) showed that there are no limits on the money that can be raised and spent on political campaigns in Australia, the Czech Republic, Denmark, Estonia, Germany, Luxembourg, the Netherlands, Norway, Spain, Sweden, Switzerland, and Turkey (Waldman, 2014). Yet, Waldman (2014) notes that

> TV advertising is the single largest expense for most American congressional candidates, while in many other countries candidates are either forbidden from advertising on television or given free TV time. In most places there's substantial public funding of campaigns, and candidates are often forbidden from campaigning until a relatively short period before election day. Put all that together, and you have elections where, even if it would technically be legal to rain huge amounts of money down on candidates, nobody considers it worth their while.

Presumably, major corporations, including food companies, would be less likely to make efforts to impact the law through direct campaign contributions.

Countries with strict limits on fundraising and spending include Belgium, Canada, Chile, France, Greece, Iceland, Ireland, Israel, Japan, South Korea, Poland, and Slovenia. In these countries, one would assume major food corporations have limited access to lawmakers. Countries with limits on spending but not fundraising include Austria, Hungary, Italy, New Zealand, Slovakia, and the United Kingdom. Here, food companies are able to donate funds, but candidates could only spend a small fraction of the money. Finally, countries with fundraising limits but no spending include Finland and the United States (Waldman, 2014). Waldman asserts that

> for American candidates, it's the worst of both worlds. The lack of spending limits means they're always at risk of being outspent, which means they can never stop raising money. But the lack of contribution limits means they have to get that money in $2,600 increments, meaning they have to keep asking and asking and asking.

As American politics basically boils down to a fundraising contest, this makes the risk of corporate influence on the law very high.

The main problem with the criminal law, according to critical, conflict, and radical criminologists, is that it thus ignores the most harmful acts to us, largely because of who tends to commit them (Lynch & Michalowski, 2006; Ross, 1998). Since the people who tend to commit white-collar and corporate crimes tend to come from the same classes of people as those who make and fund the law (i.e., wealthy, older, and white males), their acts tend to either not be illegal or are but are not enforced to the degree as street crime legislation (Robinson, 2020). This issue is of primary importance for this book because, though some are violations of the law, nearly every behavior identified in this book is legal (i.e., not a crime). An example of legal yet possibly deviant behaviors in the food crime realm is producing, advertising, and selling demonstrably harmful foods. An example of an illegal behavior within the context of food crime is selling foods known to be defective. An example of a legal but possibly deviant behavior housed within food crime is employing people in known hazardous conditions. An example of an illegal behavior in food crime is knowingly violating workplace safety regulations. Others are regulatory violations subject to administrative law. There are, for example, regulations for workplace safety, as noted earlier, but so too there are regulations for food safety itself. These issues are addressed later in the book.

Still, other behaviors are legal and may not even be considered immoral or abnormal and therefore not deviant. Take manufacturing, advertising, and selling unhealthy fast foods as one example. Not only do people *not* find this deviant, but they also generally seek out these foods and even celebrate them, with them, or through them. Who did not take their young child to McDonald's for a Happy Meal with a toy in it, or to greet Ronald McDonald, their clown mascot, or even celebrate a birthday party there? Yet, when one studies fast food—what is in it, how it is manipulated to make it more appealing, how it is advertised, and so on—one might very well come to believe there is significant deviance involved. Wouldn't being in the business of selling foods that you know kill and sicken people be considered deviant by most people? If so, such behavior would be included in the realm of food crime.

## State-Corporate Crime

State-corporate crime refers to acts of elite deviance (behaviors committed by the powerful, some illegal, some not), conducted by corporations for profit, with the assistance, approval, reward, and/or complicity of government(s) (Griffin & Spillane, 2016; Kramer, 1990; Kramer & Michalowski, 1991; Michalowski & Kramer, 2006; Kramer et al., 2002; Ross, 2017; Whyte, 2018). One fundamental purpose of scholarship on state-corporate crime is to illustrate that such crimes would not be possible without the assistance of the state. As an example, a study by Friedrichs and Rothe (2014, p. 146) illustrated how large-scale fraud committed by banks and investment banks is committed with the assistance of "government-sponsored enterprises and international financial institutions." Other scholarship supports this conclusion, implicating, at a minimum, individual actors, corporate actors and enablers, and failing regulatory agencies (Hansen, 2009; Pontell, 2005). A recent analysis of the collapse of the US economy in 2008 examined reports by the US Senate Permanent Subcommittee on Investigations and the Financial Crisis Inquiry Commission and clearly illustrated that both corporate and state organizations were responsible for the "financial crisis" (Robinson & Rogers, 2018).

As globalization continues and the influence of capital continues to spread across the globe, it is increasingly clear that state-corporate crime is not limited to a nation-state such as the United States but is instead international in nature (Tombs, 2012). In this book, rising rates of obesity and diabetes in the United States and across the world are examined. The corporate and state entities that play a role in it are identified and discussed. One major reason this is happening is that large-scale companies, with international reach—assisted by states and international organizations—produce, manufacture, market, and sell their foods to citizens of nearly every nation.

One could even argue that the behaviors of powerful entities that assure the continuation of obesity and diabetes across the globe constitute a violent state-corporate crime (Kramer, 1994; Robinson & Turner, 2019), for the outcomes are widespread suffering, illness, and death. Although some might take issue with the term *crime*—because most of what is involved in the food industry is not explicitly illegal (Tappan, 1947)—it is nevertheless dangerous and often deviant in nature (e.g., when companies lie about the contents of their food products or adulterate them in ways that are dangerous to human life). Further, these acts tend to be committed with culpability, discussed next.

## Culpability Defined

Culpability is generally understood to mean moral blameworthiness (Alexander & Ferzan, 2018; Simester, 2021). From the criminal justice perspective, it refers particularly to legal blameworthiness. It is often referred to as *mens rea* or the guilty mind that is part of a legal definition of crime (Samaha, 2016). But, culpability does not just include intentional criminal acts done on purpose but also reckless, negligent, and knowing behaviors. These terms are defined with the following examples.

According to the Legal Information Institute (2021), *intentional* or purposeful action means a person does something with intent or on purpose, that the behavior was done in order to achieve the harm. An example would be picking up a loaded gun and pointing at another person and then pulling the trigger.

*Recklessly* means a person does something without regard to "a substantial and unjustified risk"; that is, acting without regard for human life or property. An example is an injury or death or property damage caused by a drunk driver; a person under the influence of alcohol does not generally intend to crash their car, but driving under the influence of alcohol is considered reckless. Many might agree with using selling harmful fast and ultra-processed foods as an example of reckless behavior since they result in such widespread harm to humans (as well as animals and the environment).

*Knowingly* means a person acted with the knowledge that a particular result will occur or is highly likely to occur. An example here is selling tobacco; tobacco companies know their products are addictive and dangerous, and they even know roughly how many people are likely to die each year from their products. Again, I would suspect that many people would agree with using selling harmful fast and ultra-processed foods as an example of knowing behavior since the outcomes of poor diet and nutrition are known to everyone involved in the food industry.

Finally, *negligently* refers to failing to do something expected of you that results in harm. For example, it is expected (and required) that companies will recall products they know to be defective; failing to do so is negligent behavior. Could one argue that many of the food products that are produced, advertised, and sold are defective? If so, failing to stop selling them might be viewed as negligent, as would be the case where companies were selling dangerous food products without warning customers of the potential harms they face when consuming them.

In the United States, the criminal law in most states follows the aforementioned approach, following the American Law Institutes' Model Penal Code. Yet, in other states, they establish culpability through the concept of malice, roughly equivalent to ill-willed. Here, there are two types of malice. The first is *express malice*, engaging in a behavior with the specific intent to harm someone. The second is *implied malice*, carrying out a behavior with indifference to harm that someone may suffer as a result of "carelessness or inattentiveness" (Legal Information Institute, 2021). When it comes to food crime, most of the behaviors that will be examined in this book relate more to implied malice then express malice, for the simple reason that they are not intended to produce harms but rather to feed people. The indifference to the suffering of humans and animals that result from food crimes is actually quite stunning when one considers it all.

The law is obviously different in every country around the world, particularly with regard to which specific behaviors are legislated as crimes. But an analysis of the US system of laws helps us understand how corporations can be held accountable for their behaviors even when harmful outcomes of their behaviors are not intended.

It is very important to understand these concepts because the great bulk of actions in the context of "food crimes" are not done intentionally. For example, companies that produce our food do not intend the deleterious health outcomes associated with their products. Nor do they intend for workers to be injured on the job. They do, of course, intentionally kill animals and farm lands and dispose of waste products in sometimes very harmful ways. That there is no intent in such cases does not mean the companies are not culpable for harms suffered by consumers and workers. Throughout the book, the issue of culpability is discussed in the context of each type of behavior examined.

To illustrate some different ways corporate and government agencies share culpability for outcomes such as obesity and diabetes, I offer Table 1.2. As shown in Table 1.2, the companies that make food are culpable for producing unhealthy food, manipulating it to make it more addictive (e.g., adding large amounts of sugar and salt), adding chemicals to food to make it taste better (e.g., artificial flavoring), and advertising food to make it more appealing. Distributors of food are culpable for advertising unhealthy foods to make them more appealing, as well as determining where food will be placed in stores (e.g., candy bars at checkout lines) and in which stores they will be sold (including stores where food was not traditionally sold, such as hardware stores). Finally, the state is culpable for working with companies to produce unhealthy food, subsidizing the least healthy foods, ineffective regulation of foods for safety, regulating in the interests of corporations, accepting donations from industries that produce unhealthy foods, and setting school lunch policy to assure large amounts of fats, sugars, and salts are served to young people. Chapter 9 examines, in much greater detail, the issue of culpability for food crimes.

**TABLE 1.2** Culpability for Outcomes Such as Obesity and Diabetes

**Food Producers**
Produce unhealthy food
Manipulate food to make it more addictive
Add chemicals to food
Advertise food to make it more appealing
**Food Distribution**
Advertise food to make it more appealing
Determine food placement in stores
**Government**
Work with companies to produce unhealthy food
Subsidize unhealthy foods
Ineffective regulation for safety
Regulate in the interests of corporations
Accept donations from industries that produce unhealthy foods
Set school lunch policy

Interestingly, the idea of shared culpability in diet-related medical conditions has already been asserted by at least one other set of scholars. For example, Schrempf-Stirling and Phillips (2019), when writing about obesity, argue that there is shared responsibility in society for the epidemic of obesity in the United States and abroad (also see Stanish, 2010). These scholars identify responsibility in (1) consumers (who choose the foods they consume, but with limited knowledge and scientific understanding, following marketing and impacted by access, convenience, affordability, taste, and even addiction), (2) corporations (who make and market the foods, both healthy and unhealthy, and who fund government science and advocacy in their own interests), and (3) governments (who not only regulate food for safety but limit advertising and marketing, as well as subsidize the production of some foods and determine school lunch policy) (Schrempf-Stirling & Phillips, 2019, pp. 111–119). That is, there is undeniably culpability in the corporations who make and market our foods, as well as the government agencies we rely on to regulate our foods for safety (but who clearly fail). This issue is discussed in much greater detail later in the book.

## The Importance of the Law

A review of lobbying efforts by Center for Responsive Politics (2019) shows that some of the top lobbying firms in America include beer, wine, and liquor interests ($30.6 million in 2018), food and beverage ($29.1 million in 2018), and restaurants and drinking establishments ($9.2 million in 2018). Additionally, agribusiness companies donated $134.4 million in 2018 (this includes agricultural products and services, tobacco, crop production and basic processing, food processing and sales, forestry and forest products, dairy, livestock, miscellaneous agriculture, and poultry and eggs). Further, another $3 million came from fisheries and wildlife, part of donations from energy and natural resource donors. These donations assure that the major actors in the conventional food system will continue to be able to produce and market their products as they see fit. Research into the influence of money in US politics shows that the law largely reflects the interests of those in power, largely because they make the law, vote for it, and fund it (Robinson, 2015).

Open Secrets (2022) is another organization tracking spending my powerful interests on the law. An analysis of spending by major food-related sectors and specific companies is shown in Table 1.3. As you can see, the Agribusiness sector spent more than $150 million in 2021. More than 1,000 (1,248) lobbyists gave this money to 527 clients. Importantly, more than half of the lobbyists (57%) were former federal employees, demonstrating the concept of the "revolving door" whereby those who lobbied for different interests at one time worked for federal agencies involved often in the regulation of the businesses they now represent. The revolving door suggests a cozy relationship between lobbyists and regulators, raising the possibility of regulation in the corporate interests over the public interest. Note that these data only refer to donations to federal lawmakers and parties, and do not thus include such donations to state politicians and parties.

The agribusiness sector includes organizations involved in food processing and sales (who donated more than $29 million in 2021), crop production and basic processing (more than $25 million in 2021), dairy (nearly $7 million in 2021), livestock (almost $4 million in 2021), miscellaneous agriculture (about $2.5 million in 2021), and poultry and eggs (more than $1 million in 2021).

TABLE 1.3 Spending by Major Food Companies on Federal Politics, 2021

| Sector | Amount | Clients | Lobbyists | Former Federal Employees |
|---|---|---|---|---|
| **Agribusiness** | $150,180,980 | 527 | 1,238 | 57% |
| Food Processing Sales | $29,062,235 | | | |
| Crop Production Basic Processing | $25,616,964 | | | |
| Dairy | $6,854,383 | | | |
| Livestock | $3,810,000 | | | |
| Misc Agriculture | $2,493,000 | | | |
| Poultry and Eggs | $1,208,125 | | | |
| **Bars and Restaurants** | $10,517,505 | 26 | 158 | 74% |
| National Restaurant Assn | $2,950,000 | | | |
| McDonald's | $1,970,000 | | | |
| Starbucks | $1,130,000 | | | |
| YUM! Brands | $1,046,000 | | | |
| Darden Restaurants | $440,000 | | | |
| Chipotle Mexican Grill | $365,168 | | | |
| Restaurant Brands Intl | $320,000 | | | |
| Domino's Pizza | $280,000 | | | |
| Independent Restaurant Coalition | $250,000 | | | |
| Dine Brands Global | $240,000 | | | |
| Landry's Inc | $220,000 | | | |
| Yum China Holdings | $200,000 | | | |
| Yum Restaurant Srvcs Group | $200,000 | | | |
| Association of KFC Franchisees | $120,000 | | | |
| Roark Capital Group | $120,000 | | | |
| OTG Management | $110,000 | | | |
| Coalition to Stop Restaurant Tariffs | $100,000 | | | |
| Roark Capital Group | $100,000 | | | |
| Taco Bell Franchise Management Advisory Council | $90,000 | | | |
| Intl Pizza Hut Franchise Holders Assn | $80,000 | | | |
| Bloomin' Brands | $66,337 | | | |
| Silver Diner | $50,000 | | | |
| Villa Restaurant Group | $40,000 | | | |
| Alicart Restaurant Group | $30,000 | | | |
| **Cattle Ranchers/Livestock** | $3,810,000 | 23 | 52 | 50% |
| National Pork Producers Council | $2,210,000 | | | |
| National Cattlemen's Beef Assn | $400,000 | | | |

*(Continued)*

**TABLE 1.3** (Continued)

| Sector | Amount | Clients | Lobbyists | Former Federal Employees |
|---|---|---|---|---|
| Ranchers-Cattlemen Action Legal Fund | $200,000 | | | |
| Texas Cattle Feeders Assn | $160,000 | | | |
| American Sheep Industry Assn | $140,000 | | | |
| Tennessee Walking Show Horse Org | $120,000 | | | |
| North American Deer Farmers Assn | $80,000 | | | |
| Agri Beef | $80,000 | | | |
| American Horse Council | $80,000 | | | |
| National Horsemen's Benevolent Protective Assn | $80,000 | | | |
| Concerned Livestock Dealers | $80,000 | | | |
| Livestock Marketing Assn | $40,000 | | | |
| Canadian Cattlemens Assn | $40,000 | | | |
| US Cattlemen's Assn | $40,000 | | | |
| National Beef Packing | $30,000 | | | |
| Woodside Ranch | $20,000 | | | |
| Ruby Vista Ranch | $10,000 | | | |
| **Farming** | **$25,616,964** | **138** | **276** | **56%** |
| American Sugar Alliance | $2,430,000 | | | |
| American Crystal Sugar | $2,338,273 | | | |
| US Beet Sugar Assn | $1,800,000 | | | |
| Florida Sugar Cane League | $1,360,000 | | | |
| Fanjul Corp | $1,205,000 | | | |
| National Cotton Council | $1,200,000 | | | |
| US Sugar | $1,160,000 | | | |
| American Soybean Assn | $915,849 | | | |
| Corn Refiners Assn | $610,000 | | | |
| Associations of Mexican Tomato Growers | $560,000 | | | |
| American Sugarbeet Growers Assn | $520,000 | | | |
| Sime Darby Plantation Berhad | $385,000 | | | |
| Mastronardi Produce | $342,000 | | | |

*(Continued)*

**TABLE 1.3** (Continued)

| Sector | Amount | Clients | Lobbyists | Former Federal Employees |
|---|---|---|---|---|
| US Hemp Roundtable | $320,000 | | | |
| Taylor Farms | $320,000 | | | |
| Western Growers Assn | $320,000 | | | |
| Global Banana Sustainability Alliance | $300,000 | | | |
| American Honey Producers Assn | $290,000 | | | |
| National Corn Growers Assn | $280,000 | | | |
| National Sorghum Producers | $280,000 | | | |
| Southern Minn Beet Sugar Co-op | $270,000 | | | |
| National Potato Council | $260,000 | | | |
| Produce Marketing Assn | $240,000 | | | |
| Dominican Sugar Industry Coalition | $240,000 | | | |
| California Rice Commission | $240,000 | | | |
| Snake River Sugar | $220,000 | | | |
| USA Rice Federation | $220,000 | | | |
| American Peanut Shellers Assn | $210,152 | | | |
| Blue Diamond Growers | $210,000 | | | |
| Illinois Corn Growers Assn | $210,000 | | | |
| Georgia Peanut Commission | $200,000 | | | |
| Ardent Mills | $200,000 | | | |
| Supima | $200,000 | | | |
| Ocean Spray Cranberries | $200,000 | | | |
| North American Millers' Assn | $200,000 | | | |
| Minnesota Corn Growers Assn | $160,000 | | | |
| Roark Farms | $155,000 | | | |
| Florida Citrus Mutual | $150,350 | | | |
| Great Lakes Sugar Beet Growers Assn | $150,000 | | | |
| California Citrus Mutual | $150,000 | | | |
| National Oilseed Processors Assn | $145,000 | | | |

(*Continued*)

**TABLE 1.3**  (Continued)

| Sector | Amount | Clients | Lobbyists | Former Federal Employees |
|---|---|---|---|---|
| Bowery Farming | $140,000 | | | |
| American Pistachio Growers | $140,000 | | | |
| Specialty Crop Farm Bill Alliance | $135,000 | | | |
| Minn-Dak Farmers Co-op | $130,000 | | | |
| Florida Fruit & Vegetable Assn | $130,000 | | | |
| Family Farm Alliance | $120,000 | | | |
| Sugar Cane Growers Co-op of Florida | $120,000 | | | |
| Sunrise Foods Intl | $120,000 | | | |
| Washington State Potato Commission | $120,000 | | | |
| Plains Cotton Growers | $114,340 | | | |
| Military Produce Group | $110,000 | | | |
| TKO Farms | $107,000 | | | |
| Prince Edward Island Potato Board | $100,000 | | | |
| American Fruit & Vegetable Processors/Growers Cltn | $100,000 | | | |
| American Cotton Shippers Assn | $80,000 | | | |
| California Dried Plum Board | $80,000 | | | |
| California Fig Advisory Board | $80,000 | | | |
| California Strawberry Commission | $80,000 | | | |
| AppHarvest Inc. | $80,000 | | | |
| California Walnut Commission | $80,000 | | | |
| National Pecan Federation | $80,000 | | | |
| National Assn of Wheat Growers | $80,000 | | | |
| Driscoll's Inc. | $80,000 | | | |
| Georgia Fruit & Vegetable Growers Assn | $80,000 | | | |
| North Dakota Grain Growers Assn | $80,000 | | | |
| NatureSweet Ltd. | $80,000 | | | |
| North American Blueberry Council | $80,000 | | | |

*(Continued)*

**TABLE 1.3** (Continued)

| Sector | Amount | Clients | Lobbyists | Former Federal Employees |
|---|---|---|---|---|
| Sun-Diamond Growers | $80,000 | | | |
| Sun-Maid Growers of California | $80,000 | | | |
| US Canola Assn | $80,000 | | | |
| San Francisco Wholesale Produce Market | $77,000 | | | |
| Mitsui & Co. | $75,000 | | | |
| Munger Companies | $70,000 | | | |
| American Beekeepers Federation | $65,000 | | | |
| Central American Sugar Assn | $60,000 | | | |
| National Sunflower Assn | $60,000 | | | |
| Produce Alliance | $60,000 | | | |
| Washington Assn of Wheat Growers | $60,000 | | | |
| Wholestone Farms | $60,000 | | | |
| California Table Grape Commission | $55,000 | | | |
| California Fresh Fruit Assn | $55,000 | | | |
| Canola Council of Canada | $50,000 | | | |
| Costa Farms | $40,000 | | | |
| Cotton Growers Warehouse Assn | $40,000 | | | |
| American Sugar Cane League | $40,000 | | | |
| National Barley Growers Assn | $40,000 | | | |
| Didion Milling | $40,000 | | | |
| Western Peanut Growers Assn | $40,000 | | | |
| USA Dry Pea & Lentil Council | $40,000 | | | |
| US Apple Assn | $40,000 | | | |
| North Carolina Sweet Potato Commission | $40,000 | | | |
| North Dakota Corn Growers Assn | $40,000 | | | |
| South Carolina Peach Council | $40,000 | | | |
| Sow Good Farms | $30,000 | | | |
| Desert Grape Growers League | $30,000 | | | |
| Family Tree Farms | $30,000 | | | |

*(Continued)*

**TABLE 1.3** (Continued)

| Sector | Amount | Clients | Lobbyists | Former Federal Employees |
|---|---|---|---|---|
| US Durum Growers Assn | $26,000 | | | |
| Humane Farming Assn | $24,000 | | | |
| Eastern Alliance of Farmworker Advocates | $20,000 | | | |
| Green Point Farms | $20,000 | | | |
| Almond Alliance of California | $20,000 | | | |
| American Malting Barley Assn | $20,000 | | | |
| US Rice Producers Assn | $20,000 | | | |
| Wisconsin State Cranberry Growers Assn | $10,000 | | | |
| Rolling Plains Cotton Growers | $10,000 | | | |
| American Olive Oil Producers Assn | $10,000 | | | |
| Green Farms Co | $10,000 | | | |
| Fresh Garlic Producers Assn | $10,000 | | | |
| Rockdale Farmers | $7,000 | | | |
| **Misc. Business, Food and Beverage** | **$27,760,216** | n/a | n/a | n/a |
| Coca-Cola Co | $5,620,000 | | | |
| National Restaurant Assn | $2,950,000 | | | |
| McDonald's Corp. | $1,970,000 | | | |
| American Beverage Assn | $1,830,000 | | | |
| Mars Inc. | $1,480,000 | | | |
| Starbucks Corp. | $1,130,000 | | | |
| YUM! Brands | $1,046,000 | | | |
| Intl Foodservice Distributors Assn | $746,411 | | | |
| Aramark Corp. | $730,000 | | | |
| Keurig Dr Pepper | $580,000 | | | |
| American Shrimp Processors Assn | $540,000 | | | |
| Delaware North Companies | $520,000 | | | |
| National Confectioners Assn | $479,000 | | | |
| Darden Restaurants | $440,000 | | | |
| Hershey Co. | $398,000 | | | |
| Ferrero Group | $390,000 | | | |
| International Bottled Water Assn | $380,000 | | | |

(*Continued*)

TABLE 1.3 (Continued)

| Sector | Amount | Clients | Lobbyists | Former Federal Employees |
|---|---|---|---|---|
| Chipotle Mexican Grill | $365,168 | | | |
| Thai Union Group | $360,000 | | | |
| Restaurant Brands Intl | $320,000 | | | |
| Domino's Pizza | $280,000 | | | |
| Independent Restaurant Coalition | $250,000 | | | |
| Dongwon Industries | $240,000 | | | |
| Drink Recess Inc. | $240,000 | | | |
| Dine Brands Global | $240,000 | | | |
| Red Bull GmbH | $240,000 | | | |
| Landry's Inc. | $220,000 | | | |
| Roark Capital Group | $220,000 | | | |
| Mazzetta Co. | $200,000 | | | |
| Yum China Holdings | $200,000 | | | |
| Yum Restaurant Services Group | $200,000 | | | |
| DO & CO New York Catering | $180,000 | | | |
| Pacific Seafood Group | $170,000 | | | |
| BlueTriton Brands | $160,000 | | | |
| Sweetener User Assn | $150,000 | | | |
| Tampa Bay Fisheries | $120,000 | | | |
| Virginia Seafood Council | $120,000 | | | |
| OATLY Inc. | $120,000 | | | |
| Association of KFC Franchisees | $120,000 | | | |
| Atlantic Sapphire | $110,000 | | | |
| OTG Management | $110,000 | | | |
| Coalition to Stop Restaurant Tariffs | $100,000 | | | |
| Trident Seafoods | $96,000 | | | |
| Niagara Bottling | $92,300 | | | |
| Metz Culinary Management | $90,000 | | | |
| Taco Bell Franchise Management Advisory Council | $90,000 | | | |
| RK Group | $90,000 | | | |
| Kombucha Brewers Intl | $90,000 | | | |
| Intl Pizza Hut Franchise Holders Assn | $80,000 | | | |
| FCF Co. | $80,000 | | | |
| Coalition of Seafood Processors | $80,000 | | | |
| Madelaine Chocolate | $80,000 | | | |
| Quest Food Management Services | $80,000 | | | |

(*Continued*)

TABLE 1.3  (Continued)

| Sector | Amount | Clients | Lobbyists | Former Federal Employees |
|---|---|---|---|---|
| Southern Shrimp Alliance | $80,000 | | | |
| JAB Holding Co. | $70,000 | | | |
| Bloomin' Brands | $66,337 | | | |
| West Coast Seafood Processors Assn | $60,000 | | | |
| Stronger America Through Seafood | $51,000 | | | |
| Silver Diner | $50,000 | | | |
| Villa Restaurant Group | $40,000 | | | |
| Pacific Seafood Processors Assn | $40,000 | | | |
| Alicart Restaurant Group | $30,000 | | | |
| Fisheries Survival Fund | $20,000 | | | |
| Haribo | $20,000 | | | |
| American Seafood Jobs Alliance | $10,000 | | | |
| Sodexo | $10,000 | | | |

*Source*: Open Secrets (2022). List of industries, 2021 data. Downloaded from: https://www.opensecrets.org/federal-lobbying/alphabetical-list?type=s

Next up is the Bar and Restaurants category, which spent $105 million on federal politics in 2021, through 158 lobbyists, to 26 clients. About three of four (74%) of lobbyists here were former federal employees. Within this category, you see some pretty big name companies, from McDonald's and Starbucks to YUM! Brands (KFC, Pizza Hut, Taco Bell, and more), Chipotle, Domino's, and Darden Restaurants (Olive Garden, LongHorn Steakhouse, Yard House, and more), and Restaurant Brands International (Burger King, Tim Hortons, Popeyes, and Firehouse Subs), but also the National Restaurant Association (NRA), which represents about 380,000 restaurant locations. Again, these data do not reflect donations to state politicians.

Next is the Cattle ranchers and livestock category, which spend nearly $4 million in 2021 on federal politics. More than 50 lobbyists (52) targeted 23 clients, and 50% of lobbyists were former federal employees. This included the National Pork Producers Council, National Cattlemen's Beef Association, Ranchers-Cattlemen Action Legal Fund, Texas Cattle Feeders Association, the American Sheep Industry Association, the North American Deer Farmers Association, Agri Beef, Concerned Livestock Dealers, the Livestock Marketing Association, the Canadian Cattlemens Association, the US Cattlemen's Association, National Beef Packing, specific ranches, as well as a few horse-related organizations. Data here only pertain to donations to federal politics.

The farming industry spent more than $25.5 million on federal politics in 2021. More than 275 lobbyists (276) targeted 138 clients, and 56% of the lobbyists were former federal employees. Notice that a large portion of groups represented here serve some aspect of the sugar industry, including the American Sugar Alliance, American Crystal Sugar, US Beet Sugar Association, Florida Sugar Cane League, the Fanjul Corporation (the top producer of refined sugar in the United States), US Sugar, American Sugarbeet Growers Association, American Honey Producers Association, National Sorghum Producers, Southern Minn Beet Sugar Co-op, Dominican Sugar, Snake River Sugar, and more. Also included in this category are organizations involved with corn (which is used not only for human food but also animal feed and sweeteners), cotton, soybeans, tomatoes, hemp, bananas, plums, figs, strawberries, blueberries, grapes, apples, peaches, potatoes, rice, peanuts and other nuts, cranberries, sunflowers, various oils, wheat, barley, garlic, and other fruits and vegetables.

The miscellaneous food and beverage category spent more than $27 million on federal politics in 2021. This included spending by the Coca-Cola Company, plus more by groups listed earlier, the American Beverage Association (ABA, which represents soda companies, water companies, and others who manufacture non-alcoholic beverages), chocolate companies (Mars Incorporated, Hershey's), International Foodservice Distributors Association, Aramark Corporation, the American Shrimp Processors Association and other producers of seafood, another sugar organization (the National Confectioners Association, NCA), and many more.

The point of all this is to show that there are powerful interests working behind the scene through lobbying lawmakers to protect the interests of a wide variety of food industry actors and organizations. Keep this in mind as you read through this book. You may find yourself asking yourself, as you examine specific "food crimes," how can this happen, or how can this be legal? The answer may simply be owing to the fact that the law often serves the interests of the powerful—including the conventional food system—over even the interests of the common person.

### Conclusion: Why Food Crime Matters (and Should Matter to Criminology and Criminal Justice and the Rest of Us)

Food crimes—even the legal ones!—are certainly relevant for criminology and criminal justice, although only a handful of scholars have examined these behaviors (Croall, 2013; Fitzgerald, 2010; Gray & Hinch, 2015, 2018; Nally, 2011; Walters, 2006; White, 2012, 2014). To the degree criminology and criminal justice are the study of harmful behaviors, especially those committed by others against us, the behaviors encompassed by food crime should be a serious area of study in the discipline. Whereas individual consumers have a large role in determining the foods they eat, most people eat foods about which they typically have no idea of their sources or means of production, nor much else about the behaviors that are engaged in by actors in the conventional food system that lead to the production, distribution, advertising, and sales of the foods we consume. Indeed, consumers have little to no knowledge of where their food comes from (Constance et al., 2019). For example, Silbergeld (2016, p. 62) claims:

> All in all, the industry is remarkably invisible to most Americans. The companies use persuasive imagery in their advertising and websites portraying themselves as small family farmers, with animals raised in natural landscapes illuminated by a kindly sun in pictures swathed in a glow of earthly perfection not seen outside of the paintings of Thomas Kinkade. The iconography of food animal production is seductive. Chickens and cows genially josh each other to promote consumption of the other in the 'Eat More Chikin' campaign of Chick-Fil-A, for example.

Even the images we associate with the conventional food system tend to be inaccurate; this will be shown in Chapter 2.

Within this system, a wide range of deviant, unethical, and even criminal behaviors sustain it. Criminology and criminal justice must turn its attention to these behaviors. The behaviors discussed in this book are meant to be studied by the new "Food Criminology" (Robinson, 2017). Food Criminology has a "social harm" focus, thereby not being constrained by the criminal law, which typically benefits the powerful people who write and fund it through political campaigns (Robinson, 2015). This focus encourages social scientists to focus on acts (and failures to act) in the food industry that are "threatening to public health and safety, or have negative consequences on either human or non-human victims, including environmental harms," whether legal or illegal (Gray & Hinch, 2018, p. 16). These outcomes can be traced back to the corporations and government agencies that produce and assist in the production of the foods we consume, as will be demonstrated in this book.

As for the rest of us—consumers and citizens—we have a real opportunity to change the path of the conventional food system by first, learning about its realities, and second, demanding changes where needed. My argument is that, ultimately, it is up to us what foods are produced and sold. We see this in many areas of food production, where institutions in the conventional food system are making efforts to bring forth and provide healthier food products in response to consumer demand. Learning about food crime will empower us all to demand better products that lead to less harm not only to humans but also to animals, the physical environment, as well as the planet itself. Speaking for myself, I want to know where my food comes from, how it is produced, and who and what suffers in order to bring it to my home, kitchen, and dining room table. I would like companies to produce healthier foods that better serve my nutritional needs as well as far more people across the globe. But I would also support measures to make sure that all actors in the conventional food system are better cared and provided for, that less animals die and suffer so that people can eat, and that food is produced in healthier and more sustainable ways. Learning about the material in this book will help us all make reasonable demands on food producers to assure these goals.

# 2

# THE CONVENTIONAL FOOD SYSTEM

## Introduction

In this chapter, I present an extensive review of the "conventional food system." This is one of the main terms scholars use to describe the network of individuals, groups, and organizations who get food to the stores and restaurants where we shop for sustenance.

In this chapter, I outline three major eras of food, from the hunter-gatherer days to today's interconnected global food system. In the chapter, I show how, over time, food has become industrialized, corporatized, monopolized, and globalized. A section on what is called the Chickenization of food is included, along with a section on McDonaldization of food. Though these terms each mean something different, each is related and adds to our understanding of the industrialization process.

One major reality of the conventional food system identified in the chapter is that, whereas all of the aforementioned developments have undeniable benefits (such as being able to produce more than enough food for every human on Earth), the major cost associated with the system is that much of what it produces amounts to junk food; it is ultra-processed and thus high in salt, sugar, fat, and additives but absent key nutrients needed to survive and thrive in this world. One major challenge for nations going forward is how to hold on to the system we have in terms of being able to feed people across the globe, while simultaneously making food more healthy. Since the great bulk of food is produced by global corporations, it is up to them to make sure this occurs, but it will not happen without consumer demand.

## Three Eras of Food

In writing this book, it is fair to say I have become fascinated with food. Before turning to the issue of food crime (introduced in Chapter 1) in more detail, it is first important to discuss a brief history of food in the United States and around the world. Reviewing this history allows us to see how we arrived at the food system we have today, a system that is industrialized, corporatized, and globalized, one that is characterized by large farms,

DOI: 10.4324/9781003296454-2

transnational corporations, heavy use of technology, and global supply chains (Constance et al., 2019). Each of those issues is discussed later in this chapter.

Walker (2019) lays out three different major eras in the human relationship to food. The first era was the hunter-gatherer phase where humans looked for and harvested both plants and animals to survive. This period started two million years ago with our ancient ancestors and lasted until about 10,000 BC. Hunter-gatherers existed across much of the world. During this time, humans and their ancestors merely searched for and collected food rather than changing the environment to grow food for nurturance.

The second era featured the first major time when humans began to modify the earth in order to grow food, through means such as propagating plants and raising farm animals. The Neolithic revolution—from about 10,000 BC until about 2,000 years ago—refers to this period of domestication of both plant and animal species for food (Winson & Choi, 2019). The neolithic revolution began in Mesopotamia, or the area now known as Iraq. It includes "horticulturalism," or the domestication of plants for food, as well as "pastoralism," or the domestication of animals for food and other purposes (Constance, 2019). According to Constance (2019, p. 81):

> Horticulturalism and pastoralism generated food surplus, which supported settlements and the accumulation of possessions. Settlements generated a more complex division of labor that allowed for distinct political, economic, and religious institutions, including war, slavery, and a steady transition from matriarchal to patriarchal societies.

Increased crop yields from cereal cropping through animal power and manure fertilizers allowed empires to be built and spread. And through the Columbian exchange, this model of food production was ultimately imposed on the New World through colonialism (Constance, 2019). In other words, all of civilization is really tied back to food. And it was technological advancements that made it all possible—mowing, plowing, reaping, threshing, using natural and then synthetic fertilizers, plus herbicides and pesticides, and more.

The third era began with widespread farming and agricultural education. Walker (2019, p. 7) writes that, during this era, in the United States, in the late 1700s,

> land was made freely available to anyone willing to homestead and farm it; states and the federal government established a platform for agriculture education, science, and paving the way to later distribute food across the country, a transcontinental railroad was built.

Given this reality, many people no longer had to grow their own food for survival, so, by 1880, less than half of the people working were laboring on farms (versus 90% in 1790). One could argue that humans' separation from food—where we don't know where our food comes from—started during this time.

What is considered to be the first wave of the Industrial Revolution took place here, in the second half of the 18th century in the north of England; this was enabled by the development and centralized location of machinery powered by water turbines, as well as labor in water mills and then factories powered by steam and then electricity (Colas et al., 2018, p. 84). This was made possible by social and economic changes including in agricultural approaches. As one example, more potatoes were grown, increasing caloric production over grains such as wheat and rye, and a surplus of root vegetables became food for livestock, allowing for greater and more consistent meat production. The amount of time and labor needed to tend fields was also reduced with the invention of reaping machines. So, more food was produced through more efficient processes. This would continue across the other eras of industrialization.

Walker (2019, p. 8) reasons that at the core of the third food era was what he calls the "grand food bargain" with two parties:

> One was a rapidly growing society of consumer who wanted more food with less effort. The other was a rapidly growing industry of food providers whose profit depended on volume. The vehicle that kept the grand food bargain on track was the modern food system.

To survive, the system had to produce more and more food each year, and the major benefit would be that food scarcity would no longer threaten the nation. Greater production of food would be incentivized by the US government starting after World War II when food rationing ended. The government "underwrote loans, funded research, created new markets, and provided insurance" to achieve more production (p. 9). Here we see early example of connections between the government and food companies, something that would continue all the way to this current day. It is unlikely that anyone saw what would come of this—more food being produced each year than is actually needed to sustain life.

By the way, the impact of the war on food was undeniable:

> The technology that had been used to build bombs and chemicals was channeled into producing food. Synthetic fertilizers and pesticides, more-powerful farm equipment, animals packed together (some on top of each other), and raised on antibiotics—all contributed to the modern industrialized food system.
>
> *(Walker, 2019, p. 10)*

And, so, some of the major benefits of the food system were born here: "Chemicals opened new doors to planting massive fields in a single crop. Antibiotics and hormones did the same for raising meat animals" (p. 11). Here is another connection between the state and food corporations.

As the military needed means to feed troops in faraway locations, techniques such as canning were ultimately created. Canning was created in 1810 and in 1937, the Hormel Company introduced Spam (spiced ham, in a metal tin). According to Colas et al. (2018, p. 76) Spam "helped sustain America, Britain, and Russia through World War II and has brought cheap, nutritious, and easily prepared food to millions of the poorest people in the world." One example of food created by the military was processed cheese, the kind now found in food products such as goldfish crackers and Cheetos.

Fuhrman (2017, p. 118) notes that "World War II fundamentally altered how Americans ate." Chemicals originally used in war, such as ammonium nitrate, were found to make crops grow faster, and thus, they began to be used as fertilizers, while nerve gases became pesticides. And an increase in corn production after World War II created new possibilities such as using surplus corn for animal feed. This lowered the cost of beef but also had the effect of requiring farmers to give cattle antibiotics to help fight off infections that could be caused by reduced immune function in cattle. According to Fuhr-man (2017, p. 121):

> Modern commercial foods were created by the military so we could effectively wage war. The technology to make these foods was then handed over to the food industry, and the basic ingredients were made cheaply so these foods became widely available and consumed in large quantities.

Yet, a drawback is that the "primary purpose of fast food was first developed to energize military personnel on the battlefield, not to maximize long-term health." Today, meat, dairy, and processed foods are widely available, and we suffer from the deleterious health outcomes of eating too many of them. This issue is revisited in Chapters 4 and 6.

Glenna and Tobin (2019, p. 97) argue that it was industrial agriculture that "led to dramatic increases in farm productivity in the US, Japan, and western Europe during the middle of the 20th Century." The issue of industrialization of food is discussed later in the chapter. And the lesson for small farmers is their "chances of rising out of poverty depend directly on their ability to increase the productivity of their crop and livestock husbandry activities" (Glenna and Tobin (2019, p. 97). The issue of poverty among farmers is discussed in Chapter 7.

Some of the problems that characterize the contemporary food system were born in the context of the grand food bargain. One of the most significant problems is the creation of far more calories per person than are needed. This issue is revisited in Chapter 9. I will argue in Chapter 9 that it is food companies who are culpable for this reality. Another was profit over environmental health, such as in the case of farmers adding too much nitrogen to their fields that would run off into streams and pollute waterways. Walker (2019, p. 13) notes: "A by-product of synthetic nitrogen is nitrous oxide, which is three hundred times more damaging to an already warming planet than carbon dioxide. Seventy-nine percent of nitrous oxide in the atmosphere comes from agriculture." He continues, writing: "Too much nitrogen or phosphorous (another fertilizer) spawns toxic algae blooms and so-called dead zones, habitats where oxygen levels in water are so low that marine life cannot be sustained." This issue is also revisited in Chapter 7. I will also argue that food companies are culpable for the environmental damage they cause.

## Sub-eras of Food Production

Winson and Choi (2019, p. 51) document three major sub-eras of food production in contemporary history or "industrial dietary regimes." All of these occurred during the third era of food identified by Walker (2019). Table 2.1 shows these sub-eras or regimes, with key occurrences for each.

As you can see, from the first regime to the third, there was a major growth in industrialization, as well as the production and consumption of processed foods. Rising rates of obesity and diabetes would follow, making food companies partially culpable for such deleterious health outcomes. These are discussed in Chapter 6.

The first sub-era is called "normalization" and occurred from 1850 to 1939. During this period, industrial processing of food began, and mass advertising in the form of print and radio ads allowed food producers to reach consumers across the country. The second wave

**TABLE 2.1** Industrial Dietary Regimes

| First Regime: Normalization (1850–1939) | Second Regime: Intensification (1950–1980) | Third Regime: Globalization (1980 to present) |
|---|---|---|
| Beginning of industrial processing (canning, milling, meatpacking) world | More consumption of processed food products | Transnational food corporations spread industrial diet around |
| Mass advertising in news and radio | Normalization of industrial diet | Poor nutritional products to developing world |
| Processed food products | Away from home fast foods | Even more consumption of processed foods |
| Degradation of whole foods (e.g., refined flour) | Mass advertising on TV | First signs of resistance to industrial diets |
| | American wheat exports as food aid, spreading of industrial diet around the world | Emergence of healthy alternatives |

of the Industrial Revolution—from the late 19th to the early 20th centuries—occurred here, and it saw "the industrialization of the food system and the reconfiguration of urban space and a sit for the preparation, distribution, and consumption of food" (Colas et al., 2018, p. 87). The application of steam power to transport with the development of the locomotive engine and the steamboat allowed goods to be transported to greater distances at quicker times. Further, an international telegraph technology allowed for greater communication across space and time to assist with planning and ultimately production. Colas and colleagues (2018) claim that agricultural space was essentially shifted closer to main arterial routes and large farms were located close to major rail hubs or in the center of towns near seaports. As cities grew, farmers had greater access to agricultural technologies that would make their jobs easier, and processed timber and advancement in tools allowed for the construction of greater and larger buildings that would allow for greater production. Ultimately, cities became major production centers for food.

Colas et al. (2018, p. 88) note:

Inland and sea ports processed grain continuously, with ships offloading their cargo into large mechanized grain elevators, where it would be graded and packed before being stored or loaded back on to ships for export. Port cities lead to an acceleration of imported and exported foods, drinks, and even styles of cooking.

Practices including pasteurization, refrigeration, canning, and tinning were created in this second wave of the Industrial Revolution. Additionally, "the mass extension of gas and electric lighting, cookers and ovens, food processors, toasters, washing machines, and internal plumbing" all assisted with the expansion of food into our lives (p. 6).

According to Colas et al. (2018, p. 6),

the Industrial Revolution increased average calorific intake across most human societies, raising life expectancy and thereby contributing to our exponential population growth. It also improved average land yields and agricultural productivity through mechanization, the use of synthetic fertilizers, and artificial irrigation.

Constance (2019, p. 79) explains:

> Three hundred years ago the Industrial Revolution replaced animal power with machines generating even more food surplus, which made possible rapid population growth and urbanization. Trade and colonialism diffused industrial agriculture around the world, the agricultural and industrial revolution transformed human existence for a world where all of us were food gatherers to a world where, in the industrialized countries, less than 1 percent of us are farmers.

Industrialization relied on "extensification," or "the process of extending farming to new lands," as well "intensification," or "the process of increasing yields per acre on those lands" (p. 79). Without all of this, Constance argues that civilization would not be possible, as noted earlier.

Technology and invention obviously have played a huge role in food. Consider the impact of the flour mill, the cotton gin, equipment such as the reaper and steel plow, threshing machines, corn planters, plus canning, and refrigeration! Walker (2019, p. 64) provides us with the following additional examples:

> Prior to farming, energy embedded in food was outside human control. Breeding new plants, enslaving other humans, deploying draft animals, building water wheels, forging moldboard plows, and combining fossilized plants (coal) with water to produce steam (and a new measure for energy called horsepower) were all incremental steps to channeling energy into making more food.

Then, there is oil:

> From crude oil came refined liquid energy that, when combined with fire, changed civilization by orders of magnitude never seen in human history. Petroleum products carried exponentially more energy punch than wood or coal, without the bulk or weight, and could be transported and stored relatively easily.
>
> *(p. 65)*

Consider the impact of diesel- and gasoline-powered tractors and harvesters, the production of nitrogen fertilizer using natural gas, the liquid energy needed to make pesticides and chemicals, as well as power the drilling of wells needed for water: "Liquid energy built the highways, railways, and waterways, then dueled the trucks, trains, and ships that became the backbone of the modern food system" (p. 66).

Further, the invention of flight allowed for the development of crop dusting planes to spray chemicals on crops to kill weeds and pests. The discovery and invention of flash freezing and preservatives allowed food to be stored longer, and the invention of the microwave allowed it to be cooked and served faster. Today, growing, cooking, and even cleaning to a large degree (when food can be microwaved and then thrown away) are largely obsolete. This is one of the ways in which problem eating is structural in nature, where attentive and thoughtful creation and consumption of food are not necessary due to technological advances emerging out of a capitalist economy. Unconscious eating—where we eat while engaged in other activities—is a potential source of overeating and deleterious health outcomes including obesity.

The third wave of the Industrial Revolution began in the early 20th century, and it saw "the application of new mass manufacturing techniques to food production." Empowered by the automobile (which eventually included mass trucking capabilities), as well as assembly line technology and the management of labor time, "convenience" food would first become possible during this time. According to Lorr (2020), packaged food made up one-fifth of all manufacturing in the United States by 1900. Further, "highly processed products that could be relatively easily prepared and served" would be born (Colas et al., 2018, p. 89). This would create an era of food that would later be associated with an epidemic of health outcomes like obesity and diabetes. Advances in food preservation and kitchen technology (e.g., the refrigerator and freezer) would allow for greater storage of food as well as the "life of perishable foods, mainly meat, fish, and fresh dairy products" (Colas et al., 2018, p. 91). Refined sugar from sugar beets and margarine were also invented; both of these products would also later be associated with negative health outcomes.

The second sub-era is called "intensification" and occurred from 1950 to 1980. In this period, industrialization became normalized, advertising picked up on television, and people increasingly began eating away from home (FAFH). FAFH is associated with worse dietary outcomes given the high caloric content of most meals as well as larger portion sizes. By 1962, when about 90% of Americans had televisions, TV commercials became the predominant way to reach American consumers (Chandler, 2019). During this time, Kentucky Fried Chicken's spokesperson, Colonel Sanders was more recognizable than President Richard Nixon and trailed only Santa Claus in popularity. It was late in this era when rates of obesity would begin to rise, as will be shown in Chapter 6.

The final sub-era is called "globalization" and exists from the 1980s to today. The molecular revolution complete with GMO crops came in this wave. As will be shown later in the book, GMO technology allows for the development of pest-resistant crops as well as faster developing animals for food production. In this period, food became globalized and companies with a national focus grew into multi-national companies, allowing them to spread industrialized ultra-processed food across the globe.

*Globalization* is generally understood to mean the internationalization of businesses or other organizations or the interdependence of the world's economies and cultures over time. The conventional food system is undeniably globalized. For example, the United States is both a major exporter and an importer of food. Food exports go from the United States to all over the world, and we are exporting a lot of junk. According to Moss (2021, p. xxii), "no part of the world's food economy was left untouched by the promotion of products that were cheap and easy." Walker (2019) says that about 90% of corn is exported, made into animal feed, or turned into biofuel. Much of the rest becomes sweeteners, oils, starches, or beverages.

When discussing agricultural exports, the Committee for Economic Development (CED) claims that they account for $70 billion each year; processed food products comprise a full 50% of all the food and agricultural outcomes exported by the United States (CED, 2017, p. 6). As you will see later in this book, this means we are exporting health problems to the rest of the world. Incredibly, "food is the *only* major export [from the US] that shows *positive* net exports today (and has for most of the past decade)" (CED, 2017, p. 18). However, much of what we are exporting to the world is food that either lacks nutritional value or is actually unhealthy. Logical outcomes would include rising rates of health conditions such as obesity around the globe.

According to Silbergeld (2016, p. 222), OECD countries and the Americas produce more than 75% of the world's food. And food trade among 164 countries is controlled by 30 agreements between various countries, facilitated by the World Trade Organization (WTO). According to Walker (2019, p. 181): "America's once localized food system is now global. Food arrives from at least two hundred different cultures and food systems around the world."

According to Ross (2022), four of the largest food-producing countries—China, India, the United States, and Brazil—rank in the top ten countries in the world for total geographic land area. Ross notes that the United States is one of the largest food exporters in the world, whereas China and India tend to consume most of the food they produce. The following is a brief summary of food production in some of these countries, as noted by Ross (2022). Brazil produces a large amount of sugarcane, soybeans, and beef, as well as fruits such as coconuts, papaya, and pineapples. Roughly one-third of its land is used as cropland. China is a major producer of food along the eastern and southern regions of the country, of rice, wheat, sheep milk, fruits such as apples, grapes, peaches, pears, plums, and watermelons, vegetables and tubers such as broccoli, cabbage, carrots, cucumbers, eggplant, green beans, lettuce, onions, potatoes, pumpkins, spinach, and tomatoes, meat including chicken, fish, goat, lamb, and pork, as well as peanuts, eggs, and honey. In the United States, the major food-producing states include California, Illinois, Indiana, Iowa, Kansas, Minnesota, Nebraska, North Carolina, Texas, and Wisconsin. Major export destinations of food produced in the United States include Canada, China, Germany, Japan, and Mexico.

According to Silbergeld (2016, p. 202), "the US consumer's market basket is increasingly filled with imported food, a total of over 16 percent overall and more than 85 percent of fish and shellfish." According to Walker (2019, p. 161), food imports are popular because of "cheaper offshore labor, ingredients, and other production factors. Over 90 percent of the seafood consumed in the United States is now imported, though most is caught by American fishermen, exported to Asia for processing, then shipped back to the United States."

According to Clausen et al. (2019, p. 174):

Seafood represents one of the most-traded products of the world food sector, with about 78 percent of seafood products contributing to international trade. Fish can be caught in the waters of one country, filleted and processed in a second country, and show up on the menu in a third country.

Clausen et al. note:

By 2013, over thirty-six million tons of seafood were exported throughout the global market, and the value of those commodities reached over $139 billion. . . . Trade in seafood is largely driven by demand from wealthy countries. For example, the top exporters of seafood are China, Norway, Vietnam, and Thailand. The top importers of fish are the United States, Japan, China, and Spain . . . as a whole, the European Union is the largest importer of seafood products in the world.

So, our food is global in the sense that it comes from all over the globe. Our food, however, is also global in another way. Consider how it is often processed across the globe. One example of the global nature of our food is the production of tomato paste, which can be processed in China, sent to Italy to be cooked with beans from Mexico, and then canned and sent to the United Kingdom where it is purchased and eaten (Bloom, 2019). Another is animal slaughtering and processing. According to Silbergeld (2016, p. 48), after some animals are slaughtered (e.g., chickens), they are then sent to places including Mexico and China, for processing, in a practice called "added value protection." The industry was able to successfully convince the US Department of Agriculture (USDA) to allow the re-importation of these animal products back into the United States.

Now that we have considered some of the history of food, we can turn our attention to the system of food that provides us with the food products we eat today. This system is often referred to as the "conventional food system." That is discussed next.

### The Conventional Food System

The great bulk of food consumed in the United States comes from the activities of the "conventional food system." The *conventional food system* refers to widespread and commercialized production, distribution, and sales of the nutritious (and non-nutritious) substances that we consume to function, grow, and maintain life. It can be differentiated with "alternative food systems" such as local food production, small organic farming, food cooperatives, fair trade systems, and so on (Eames-Sheavly & Wilkins, 2006). The conventional food system is the outcome of all the history just reviewed. It evolved into existence over time. A large majority of Americans, a large majority of the time, rely on the conventional food system to bring them nourishment. As suggested earlier, most of us have no idea about this system, and the more you know about it, the easier it becomes to see its problems. This system—made up of numerous producers, processors, distributors, and retailers of food—provides us "with a varied, relatively inexpensive, and widely available supply of food" (Nesheim et al., 2015), but not without significant costs. Figure 2.1 illustrates a conception of this system.

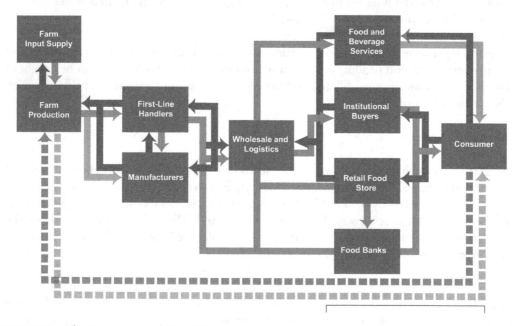

**FIGURE 2.1** The Conventional Food System

*Source:* https://www.ncbi.nlm.nih.gov/books/NBK305173/figure/fig_2-1/?report=objectonly

As with any system—although each component organization, group, and individual is vital to the success of the system as a whole—each part also has its own goals and objectives. Some aim simply to grow food to make a living, others to sell food for profit, and still others to improve public health or help protect the environment. The great food corporations of today seek profit above and beyond all else. This wide-ranging food system has significant, far-reaching consequences, impacting

> the environment (e.g., effects on biodiversity, water, soil, air, and climate), human health (e.g., direct effects on diet-related chronic disease risk, and indirect effects associated with soil, air, and water pollution), and society (e.g., effects on food accessibility and affordability, land use, employment, labor conditions, and local economies).
>
> *(Nesheim et al., 2015)*

The conventional food system is important to capitalistic economies, but it also poses an existential threat to all of human existence, due to its effects on human health and the physical environment. The impact of the system on human health is examined in Chapter 6 and on the environment in Chapter 7.

Starting with the basics, the US conventional food system comprises a "food supply chain" (Kinsey, 2001; Oskam et al., 2010; Senauer & Venturini, 2005). This food supply chain comprises every actor and institution involved in growing (e.g., grains, vegetables, and fruits), catching (e.g., fish and other seafood), and raising (e.g., cows and other livestock) our food, as well as those who provide the necessary capital and machinery to do so. It also includes all the people and organizations who turn the raw ingredients into edible food and get it to places where it can be purchased as food (Nesheim et al., 2015). This chain comprises of actors who process foods from their sources (e.g., washing, wrapping, and packing food, aggregating and storing food, and providing and utilizing other raw materials needed to finalize food products), ship to wholesalers and then to stores, stock shelves, set up advertisements, and so on. A massive transportation infrastructure is needed to achieve all of this and includes, obviously, the nation's interstate highways system as well as an enormous system of warehouses. The "state"—composed of numerous agencies and institutions, including lawmakers, regulatory agencies, and more—is charged with overseeing the entire system or food chain and, as such, is part of the system. This is an important reminder of the close connections between the government and major food corporations, first noted in Chapter 1.

Nestle (2013, p. 2) writes:

> The food industry is vast. It encompasses everyone who owns or works in agriculture (animal and plant), product manufacture, restaurants, institutional food service, retail stores, and factories that make farm machines and fertilizers, as well as people engaged in the transportation, storage, and insurance businesses that support such enterprises.

It includes CAFOs and enormous farms—and supermarket aisles overflowing with snacks, candies, cookies, sodas, and sugary foods that bear little resemblance to the plants, crops, or animals from which they were derived (p. 2). This is an important reality to confront right at the outset—most of the food products produced by the conventional food system are actually unhealthy. This issue is addressed more in Chapter 3.

After growing and/or producing food products, the next part of the conventional food system is called "first line handlers or primary processors" and includes "for-profit commodity trading companies and farmer cooperatives that aggregate the output of individual farms to gain economies of scale and market access to the rest of the food supply chain." They also include companies that wash, wax, wrap, and pack fruits and vegetables, as well as flour millers, oilseed processors, and other firms that prepare raw materials for use in the processing and manufacturing of finished food products (Nesheim et al., 2015, p. 32).

According to Nesheim and colleagues (2015, p. 39): "Most of the field crop production in the United States that is not exported or fed to livestock (roughly half the total) goes through some type of food processing and manufacturing before being consumed by people." Though processing is necessary for most foods, keep in mind that over-processing (or ultra-processing) is to blame for removing nutritional value from our foods. More on this is in Chapter 3.

For now, here is an example using Kellogg's, which "takes whole corn, strips it of almost all naturally occurring nutrients, adds sugar, salt, and chemical additives to maximize flavor, stability, and shelf life, and puts the ingredients through a complex manufacturing process" to produce its food products (Simon, 2006, p. 92). Yet, it is the whole, unprocessed grains that "are rich and fiber and nutrients. And can contribute significantly to good health" (p. 93). Of course, companies such as General Mills also offer whole grain versions of many of their products, but given the high sugar content of some such as their cereals, they remain largely unhealthy and even quite dangerous. Incredibly, some companies have created their own stamps of approval, saying things in print on boxes such as "smart choices" and "sensible solutions" to create the illusion of healthier eating. These seals and logos are often green in color, something that customers automatically associate with sustainability. This type of activity is suggestive of misleading advertising, reviewed later in Chapter 9.

The next part of the conventional food system includes all the people and organizations who turn the raw ingredients into edible food and get it to places where it can be purchased as food (Nesheim et al., 2015). They comprise what is known as the "food processing and manufacturing" sector which "includes meat packers, bakeries, and consumer product good companies that turn raw ingredients into higher-value packaged and processed food products" (Nesheim et al., 2015, pp. 32–33).

Even fresh fruits and vegetables tend to be "subjected to washing, sorting, waxing, storing, and transportation through the commercial supply chain." Most of this processing is done by large, multinational corporations (Nesheim et al., 2015, p. 186). This reality assures the ability of companies to produce and deliver foods to large populations and maintain profitability, but it also imposes its own costs, as will be identified in this book.

It may be the case that the processing sector of the conventional food system has received the most criticism since it is processing that "removes and destroys the fragile micronutrients and phytochemicals we need for cellular normalcy, and also adds toxins" (Fuhrman, 2017, p. 16). These toxins "include artificial colors, artificial flavors, preservatives, pesticides, antifoaming agents, emulsifiers, stabilizers, and thickeners." Phytochemicals are "noncaloric compounds present in plants that have health-promoting and disease-preventing properties" (p. 44). They include substances such as lycopene in tomatoes, isoflavones in soy, and flavonoids in fruit. Moss (2013, p. xvii) adds that processing strips away the nutritional value of food and converts grains into starches, which are then more easily converted into sugar by the body. This can lead to outcomes such as obesity and type 2 diabetes, to be addressed later in Chapter 6.

These food products are often then passed onto the "wholesale and logistics sector" which comprises those companies that buy these products and ship them to stores. These stores then stock shelves, set up advertisements, sell food products, and so on. One part of the wholesale and logistics sector is referred to as the "retail food sector," which comprises "grocery stores, convenience stores, vending machines," and so on that provide food mostly for food consumed at home (FAH). Another part of this sector is the "food service sector," which is made up of "restaurants, fast-food outlets, eating and drinking establishments, and institutional cafeterias" meant to provide food to be consumed away from home (FAFH). It is here, in the FAFH arena, where a greater bulk of unhealthy eating occurs, as noted earlier.

A massive transportation infrastructure is needed to achieve all of this and includes, obviously, the nation's interstate highway system as well as an enormous system of warehouses (Nesheim et al., 2015, p. 33). The "state"—composed of numerous agencies and institutions, including lawmakers, regulatory agencies, and more—is charged with overseeing the entire system or food chain, as well as the nation's highway systems. This makes the government part of the conventional food system. This reality also makes the harms that result from the conventional food system, as well as any and all deviance committed within it, relevant for state-corporate criminology, as issue introduced in Chapter 1.

Taken as a whole, the food system provides nourishment at a relatively affordable price for the US population, not to mention much of the rest of the world. It also comprises a significant portion of the economic value in the country (approximately $776 billion of GDP in 2012, or 5% of the total) (Nesheim et al., 2015)—up to more than $1 trillion in 2020. Yet, the system, as it is currently organized, also assures that a large portion of the population consumes unhealthy foods at rates that are not tolerable and that lead to terrible health outcomes including serious illness and death from conditions such as obesity and type 2 diabetes. Stated simply, the damage caused to humans by the foods we consume vastly outweighs that caused by street crime, as shown in Chapter 1. That there is culpability by economic elites in these outcomes—both in the industry and government—makes them relevant for criminologists, and in particular, state-corporate criminologists, who study crimes resulting from the interactive actions of states and corporations (Kramer & Michalowski, 1991).

In Chapter 6, I discuss medical conditions, including obesity and type 2 diabetes. I examine their nature, incidence and prevalence, trends, and causes. Throughout, you will see the culpability of key actors in the conventional food system, placing the discussion in the context of state-corporate criminality.

## Industrialization of Food

The conventional food system is industrial in nature (Winson, 2013). *Industrialization* is defined as "the act or process of industrializing" or "the widespread development of industries in a region, country, culture, etc.," and *industry* is understood to mean "manufacturing activity as a whole, "a distinct group of productive or profit-making enterprises," "a department or branch of a craft, art, business, or manufacture especially: one that employs a large personnel and capital especially in manufacturing," and "systematic labor especially for some useful purpose or the creation of something of value" (Merriam-Webster, 2022a).

Gray (2019b, p. 404) refers to the current state of food as the "industrial dietary regime." Indeed, much of the food produced, sold, and consumed is made by industry, and it does not resemble in any way food substances found in nature.

Silbergeld (2016, p. 3) notes that the term industrial "refers to the technological, economic, and social structures that have characterized other areas of human economic activity since the seventeenth century, starting in the West." And she claims that "the industrialization of food animal production is a revolution as profound as anything that has happened in the history of agriculture since our hunter-gatherer ancestors became cultivators of plants and domesticators of animals."

Earlier in the chapter, you saw how industrialization began to take hold in the conventional food system in the normalization regime from 1850 to 1939. According to Winson and Choi (2019, pp. 43–44):

> Industrialization came to the realm of food in several different ways. In the meat processing sphere, the mechanized *disassembly* line in large factory establishments emerged. In the fruit and vegetable sphere, the evolution of mechanized canning technology made possible the preservation of fruits and vegetables on a large scale so they could be consumed out of season.

Further, roller milling extended the shelf life of flours "by extracting and eliminating the wheat germ from the flour entirely, along with the outer bran. Removing the bran and germ of the wheat removed most of the valuable nutrients as well." Winson and Choi (2019) claim that white flour has less than 40% of some important nutrients including folic acid, iron, niacin, riboflavin, thiamin, and vitamin E and less than 20% of fiber, potassium, and zinc. Through these early examples, we see evidence of food production being concerned more with things like the longevity of products rather than their degree of health.

Colas and colleagues (2018, p. 91) point out that "An industrialized food system both facilitates and is further advanced by the emergence of the international food corporation." And, so, during this time, early food corporations became realities and "linked together capitalism, colonialism, and global trade. Companies like Kellogg's were instrumental in developing a marketing and advertising industry that by the mid-twentieth century formed an integral part of the food system in contemporary capitalist societies" (p. 92). With regard to colonialization, Colas et al. (2018, p. 171) write that the United States first exported "affordable grain, sugar, coffee, tea, cacao, and meat" to the Americas and Australasia, and second, launched a "postwar food regime driven by US hegemony over an inter-state system that reproduced the American model of agribusiness across the capitalist world." They explain:

> After World War II, Washington oversaw a global rule-governed food regime that combined market integration within and between allied states with the protection of domestic food prices through subsidies, quotas, foreign aid, tariffs, and fixed exchange rates, all buttressed by the newly founded Bretton Woods system and the associated U.N. specialized agencies (most notably the Food and Agriculture Organization).

It was thus in the 20th century that the conventional food system of the United States began to exert influence around the world.

Colas and colleagues (2018, p. 171) explain how this became possible:

During the mid-to late-20th Century, agriculture became concentrated and corporatized, spurred on by the accentuated mechanization of capitalist agriculture and its gradual consolidation into large-scale, intensive, specialized, monocultural farms. With the growing commodification of inputs (sowing and harvesting, pesticides, herbicides, fertilizers, feed, pharmaceuticals, and transport) and the increasing dependence on wholesale of outputs to powerful supermarket chains, vertical integration within the food system yielded the (in)famous hourglass patterns—hundreds of millions of consumers purchase food from a very narrow band of retailers, wholesalers, manufacturers, and processors, who, in turn, buy produce from millions of farmers reliant on literally a handful of input suppliers.

Vertical integration, by the way, is when a company controls more than one stage of production.

One example of this concentration is that Smithfield Farms has about 31% of the market share of American pork and 20% of its hogs. Another is that three firms (Cargill, Archer Daniels Midland, and Bunge) control 90% of all the grain in the world. Further, about 10 companies own about 65% of the seed market, and 10 companies control about 90% of the pesticide market. Consolidation of food is discussed more later in this chapter.

Today, industry dominates the conventional food system. According to Barnhill et al. (2018, p. 4): "Industrial agriculture is typified by large-scale, highly mechanized farms that grow a single crop on large areas of land and use liberal amounts of synthetic fertilizers, synthetic pesticides, synthetic herbicides, and genetically modified seeds." It was made possible by many advances in plant and soil science, selective breeding, and "subsequent technological advances and investments in machinery . . . irrigation . . . improved drilling . . . and by GMOs that produced more food and promised a reduced need for herbicides and pesticides." According to Colas et al. (2018, p. 172): "Downstream, at the consumption end of the process, agribusiness has introduced a manufactured diet whose main components are fats and sweeteners, supplemented with starches, thickeners, proteins, and synthesized flavors." The result, again, is poor health for a large portion of Americans, as well as people around the world, in both developed and developing countries.

Glenna and Tobin (2019, p. 102) identify several problems endemic to large-scale industrialization of agriculture, including labor exploitation, pest problems, and water pollution. Further, Winson and Choi (2019, p. 52) identify three interrelated processes of industrialized food that have had the effect of degrading food and endangering our health. The first is called "simplification," which has two dimensions:

The first process entails a radical reduction in the complexity of whole foods that occurs with industrial processing. The second involves a dramatic reduction in the varietal diversity of foods in food environments because of the economic imperatives of the contemporary food system.

Similarly, Pollan (2008) lays out the net effects of the industrialization of food, noting that it has resulted in a shift from whole foods to refined or processed foods, from the complexity of ingredients and foods to simplicity, from quality to quantity, and from more healthy leaves to less healthy seeds.

The second process is referred to as "speed-up of food production," which means producing products as quickly and efficiently as possible so as to maximize profits (p. 52). Winson and Choi give the example of cattle being selected for their ability to rapidly put on fat and assert that efforts to put more fat on cattle result in less healthy, more fatty meat products. Finally, the third process—"adulteration"—is defined as "a set of systematic processed by food processors to increase the palatability of highly processed foods but also to increase shelf life, make processed foods cosmetically more appealing, and lower the manufacturer's cost to produce them." This results in diminished nutritional value in food products, as does the addition of salts, sugars, fats, and artificial chemicals. These issues are discussed more in Chapter 3. For now, keep in mind that it is the industrialization of food that is responsible for a reduction in the availability of whole foods, as well as an efficient production of adulterated or unhealthy foods.

## Chickenization

A related word to industrialization, or a specific form of it, is "Chickenization." This term reportedly comes from the USDA. It captures the industrialization of meat, starting with chickens in the United States, and now various animals including chickens, cows, pigs, and fish. It is characterized by:

1. Confinement of animals
2. Concentration of animals in large numbers (as well as ownership of them and their production in very small numbers)
3. Vertical and horizontal integration of industry.

Nearly all animals in the conventional food system live the great bulk of their lives in *confinement*, in cages, pens, and so forth, typically far too small for them to live in any kind of comfort. In confinement, we no longer see animals as being alive, nor as individuals, but rather as units of production, separate and distinct from nature as well as any food chain other than ours. This can be deemed a cost of the conventional food system.

Animals also live together in large numbers, suggestive of *concentration*. This concentration, like the confinement of animals, is often claimed to be inhumane. Take the concentration of hogs as one example:

> Just five slaughterhouses in the United States account for 55 million, or 50 percent, of the hogs consumed in the United States; the Smithfield hog slaughter and processing plant in Tar Heel, North Carolina, kills over about 10 percent of this total, or 32,000 pigs a day. Perdue processes about 2 billion pounds of broiler poultry per year, or over 20 percent of the US total.
>
> *(Silbergeld, 2016, p. 204)*

The costs of concentration to animals, workers, and the environment are discussed in Chapter 7.

*Vertical integration* was defined earlier as when a company controls more than one stage of production, as in the case of owning, slaughtering, and then selling animals, as well as feed operations and so on. *Horizontal integration* is when a company comes to dominate an area of commerce, as in the case of a company buying up competitors or merging with other companies. Silbergeld (2016, p. 54) identifies four developments supporting integration, specifically "expansion of poultry production and the centralization of chicken slaughter and processing: food sanitation, increasing consumer affluence, innovations in transport, and government policies." With regard to government policies,

> new food sanitation laws and regulations favored the position of larger animal slaughter and processing because, from the perspective of government, these operations could be more readily identified and regulated. From the point of view of the industry, the larger plants were better able to absorb the costs of meeting the new standards.
>
> *(p. 55)*

So, it is government that is responsible for the increased concentration of industry, so its fingerprints are on any acts of crime or deviance furthered by industry concentration.

Silbergeld suggests that confinement and then concentration are often viewed as part of the technological changes within the conventional food system that were deemed necessary and good for the survival of the system. As noted earlier, first came steam power with trains and ships, then automotive transport, and later refrigeration. Then there is

> chemically assisted crop production, biotechnology, and confinement as a means of raising animals. Chemical assistance in the form of fertilizers and pesticides improved arability, which is the ability of land to support growth of crops for consumption and forage. Biotechnology broke the environmental requirements of certain plants . . . and the first defining aspect of industrialization is confinement of animals, which eradicated the need for pasturage and arable land and water to grow crops in the same location where the animals were farmed.

After confinement, concentration was born,

> the production of large numbers of animals within a small area. This concentration was economically critical because it supported the investments in larger-scale growing operations while keeping prices low for consumers. Concentration of food animal production then drove concentration of crop agriculture to produce feeds for confined flocks and herds.
>
> *(Silbergeld, 2016, p. 34)*

Concentration, in particular, has a really bad reputation among animal right activists, but it "is central to making large profits on small margins. Geographic concentration also serves the needs of the central focus of animal slaughter and processing by reducing the costs of transporting animals to and consumer product distribution form the slaughterhouse" (p. 42).

Confinement and concentration are all indicative of industrialization. Silbergeld (2016, pp. 29–30) writes:

> Modern food animal production deserves the nomenclature of industrialization because it involves modes of production characteristic of industry. Work in agriculture now

involved performance of specific and relatively limited tasks in production, the replacement of human labor by mechanical energy, and an assembly line to unify separate operations into the production of consumer goods like food animal products. Finally, modern food animal production is industrial in terms of its economic and structural organization, with one economic unit that directly or indirectly controls the central processes of production, from inputs of raw materials to retail products.

Silbergeld (2016, p. 38) clarifies the meaning of some important terms, writing:

> Confinement and concentration are the key aspects of *intensive* food animal production. Integration is the key aspect of *industrial* food animal production. . . . The words *intensive food animal production* accurately describes the scale of modern food animal production, whereas *industrial food animal production* describes its organization.

Silbergeld compares concentration of animals for industrial food production to "badly run and overcrowded hospitals" (p. 103).

With regard to the issue of integration, the agricultural industry is centralized in its organization of ownership and profit. This centralization allows for complete control of all operations related to a food product: "Integration refers to an organizational structure in which the overall control of production is centered in one economic entity (the integrator), and all activities related to production are carried out either through direct control by or contractual agreements with the integrator" (p. 43).

According to Silbergeld (2016, p. 44): "Integration of the industry was dependent upon concentration, which in turn was dependent on confinement," two terms defined earlier. This means for industry to be in complete control of production of animal food products, animals had to be kept confined and in great, concentrated numbers.

Silbergeld (2016) claims that integration began with Arthur Perdue's purchases of hatcheries, feed mills, and slaughterhouses: "by occupying these nodes in the production system, integration permits industry to control both costs and prices. Integration can also increase profitability for the enterprise by excluding, or in economic terms externalizing, higher risks and cost centers from the enterprise" (p. 44). Consistent with the ideas of Chickenization as well as McDonaldization, "Perdue's chickens conquered the United States with their high degree of standardization, reliable availability, promise of safety, and continuous improvements in production as well as innovation" (Silbergeld, 2016, p. 61). So, there are benefits of both industrialization and Chickenization, both to business and to customers; yet, the costs of industrialization and Chickenization are also enormous, as will be illustrated in Chapter 7.

## McDonaldization

The concept of Chickenization is highly related to that of McDonaldization (Ritzer, 2000), a term meant to characterize any societal institution that shares four characteristics of fast-food restaurants. McDonaldization is "the process by which the principles of the fast-food restaurant are coming to dominate more and more sectors of American society as well as the rest of the world" (Ritzer, 2000, p. 1). McDonaldization is made up of four parts—efficiency, calculability, predictability, and control.

*Efficiency* refers to "the optimum method for getting from one point to another" or for achieving some goal (Ritzer, 2000, p. 12). In the fast-food industry, efficiency is imperative, as the term "fast" implies, as in the case of the assembly line mentioned earlier. Yet, efficiency is also imperative in the production of all food, as suggested in the discussion of the chickenization of animal food production, noted earlier. *Calculability* is "an emphasis on the quantitative aspects of products sold . . . and services offered. . . . In McDonaldized systems, quantity has become equivalent to quality; a lot of something, or the quick delivery of it, means it must be good" (Ritzer, 2000, p. 12). In the fast-food industry, more for your money is better than less for your money. Super-sizing makes sense in America when people are raised to want more and have more for less (Moss, 2021). Clearly, animal food production is all about quantity; to keep up with consumer demand, chicken-, cow-, and hog-slaughtering must process tens of thousands of animals each and every day.

*Predictability* refers to "the assurance that products and services will be the same over time and in all locales" (Ritzer, 2000, p. 13). In the fast-food industry, the goal is to make one's entire dining experience completely consistent with all previous visits; no matter where you go, the product will be exactly the same. The specific and limited tasks in production noted earlier are needed to make products the same everywhere. In animal food production, one overriding goal is producing a predictable product; this is accomplished by the routinization of processes in animal slaughtering. *Control*, the final element of McDonaldization, means that many aspects of production and consumption are governed by strict rules and an emphasis on a single way of doing things. In the fast-food industry, control is often achieved through the use of nonhuman technology (Ritzer, 2000, p. 236), as in the case of the replacement of human labor by mechanical energy noted earlier. The same is true in large-scale animal production.

McDonaldization, like Chickenization, is seen across all of the conventional food systems. The system strives for efficiency of food production, calculability of all products, and predictable food products across time and space, and every aspect of the process is controlled, including often labor through the use of technology. At no point in either of these processes is the healthiness of food products ever considered, much less their lack of healthiness; their concern is instead (short-term) profit.

While McDonald's gets credit for McDonaldization, the Burger King Whopper was the first national fast-food hamburger, and standardization and predictability in fast food came before McDonald's, for example, in the case of White Castle. A White Castle brochure from 1932 outlined in great detail how each of its restaurants had the "same kind of counter" while you drink liquids "made in accordance with a certain formula" and served in identical cups being used by thousands elsewhere simultaneously and ate food "prepared in exactly the same way over a gas flame of the same intensity" (Chandler, 2019, p. 13). The same standards of cleanliness and service would be found at every location, as well.

## Back to Industrialization

Whether you call it Chickenized or McDonaldized, the main point is that the contemporary food system is industrialized in nature. And this process has become global as industrialized food

> is rapidly displacing local agricultural practices in many countries, resulting from a speeded-up pace in events otherwise not different from the changes that unfolded over

half a century in the United States from 1930 to 1980: economic consolidation of small-holder and small-company operations into larger economic unites; the adoption of the integrated model, including many of the stages in production; transfer of technology from US production (including intensive production and marketing, confinements, specially formulated feeds with additions of antimicrobial drugs); as well as supplanting of traditional breeds (especially in poultry) with breeds developed in the US specifically for the conditions of intensive production and rapid growth.

*(Silbergeld, 2016, p. 69)*

Silbergeld claims that the industrialization of food around the globe is not being driven by companies in the United States "but rather by the activities of national companies within each country." Yet, she notes that some US companies—she names ConAgra Foods, Cargill, Syngenta, Monsanto, and Cobb-Vantress—have contributed to Chickenization (and McDonaldization) around the world. Cargill, for example, operates around the world, including in China and Brazil.

According to Silbergeld (2016, p. xiii), the industrialization of food has "largely supplanted traditional animal husbandry in countries such as Brazil, China, and Indonesia, and they are making inroads in Africa and the rest of Asia and the Americas." Traditional means of agricultural farming are disappearing, with huge implications for cultures and rural areas. In the words of Silbergeld (2016, p. 152): "It is indisputable that the traditional social structure of rural communities has been altered over the past seventy years, mostly for worse." One way this has occurred is by placing more space and distance between producers of food products and consumers. Another impact has been the replacement of family farms with contract farming.

Silbergeld (2016, p. 46) explains how little control farmers have as a result of contract farming:

Integrators maintain . . . control in food production through binding contracts with the farmers who "grow out" the animals from chicks or shoats to market weight. Through these contracts, the integrators also control all the conditions under which animals are grown, including animal feeds and veterinary oversight. Under the contract system, the integrator is able to assign costs and risks to contracting farmers or growers, which usually include the costs of constructing, modernizing, and operating the confinement buildings in which the chickens or pigs are raised and supplying water and energy for the houses, including ventilation. The farmer supplies the relatively minimal labor needed to manage the growing process, carry out repairs as needed, and intermittent "cleanout" of the buildings.

Benefits to the farmer reportedly include reduced costs and guaranteed price at the sale that is set by the integrator. But farmers, called "growers" in the industry, lose their autonomy and control over their own work and are so controlled by the system:

The integrator defines the conditions of growing, feeding, and managing the chicken house; for contractual agreements, any deviation from the contract is grounds for the integrator to refuse to buy the birds at the end of the growing cycle. If a purchase is refused, the grower is left with the expenses of feed purchases, energy, water, and maintenance as well as labor.

(p. 48)

Can farmers thrive in these conditions? Walker (2019, pp. 136–137) seems to suggest not:

The contracts reward performance—even though most of the performance variables are outside the farmer's control. The health of the coming pigs . . . animal genetics, feed quality, inherent disease susceptibility, extreme weather, and climate are just some examples. Contracts may also require farmers to install and pay for company-mandated technology upgrades, disallow lawsuits by using forced-arbitration clauses, or omit transparency safeguards pertaining to standards and weights. Compensation is typically tied to a "tournament system"—a scheme whereby farmers compete against each other over a fixed amount of compensation.

Silbergeld (2016, p. 59) notes that the US government did not always just go along with the consolidation of the industry at the expense of farmers:

The US government, in the early days before its unwavering support for integration (typified by its provision of loans and subsidies to integrators rather than to farmers), intervened three times to object to the restrictive aspects of integration, but lost each case in court.

Now, however, the government not only supports consolidation of the industry but typically encourages and even rewards it.

Silbergeld (2016, p. 58) compares the contract system to

the model of sharecropping developed in the South after the end of slave labor in agriculture; it is economically similar to crop liens in the days of cotton and tobacco, the main products in the states where intensive poultry production began. . . . During Reconstruction, crop liens replaced slavery by debt peonage, this time an economic indenture but just as stringent, and they contain much the same language as a broiler contract.

In these contracts,

instead of renting land, the farmer essentially "rents" the birds from the integrator during the six to seven weeks it takes to grow a chick into a market-weight broiler chicken. Farmers must raise and then sell the product back to the integrator in accordance with all the terms of the contract, including using the integrator's feeds and operating and even modernizing the broiler house according to the integrator's requirements.

Incredibly, the farmer owns the wastes generated by poultry or swine as they grow, yet, when uses for those wastes are found, the integrator or "producer" of the meat products owns those as well! Silbergeld (2016) thus notes that producers are able to eliminate the costs of wastes from their considerations and calculations, whereas farmers or growers cannot. Again, this exemplifies the issue of control of labor inherent in the conventional food system, typical of a McDonaldized system.

Profits for contract farmers are too low to help build and sustain rural communities. Workers in the 50 poultry processing plants also make low wages and "are overwhelmingly drawn from poor minority populations, not only from the local population but also actively recruited by industry from other sectors." Silbergeld notes that this includes "recent emigrants to the United States, including refugees, undocumented immigrants and the poor" (p. 156). Silbergeld (2016, pp. 153–154) argues: "This new arrangement has been a driver in the loss of economic focus for rural communities, and in many countries, rural villages and towns are disappearing." This is an additional cost of industrialized food production.

Interestingly, there is also a notable relationship between poverty, unemployment, and the presence of industrialized food production. Take the southern United States as one example: "The map of rural poverty in the United States coincides with that of the Broiler Belt in the Southeast" (Silbergeld, 2016, p. 154). Then there is the diminishment of life for people living near "locally unwanted land uses" or LULUs, including terrible odors, dead animal carcasses, and many busy, heavy trucks regularly coming in and out of the area, as well as declining property values. These are additional costs of the conventional food system.

Take Tar Heel, North Carolina, as an example of entrenched rural poverty. It "is the location of the largest hog slaughter and processing plant in the world, through which over 32,000 hogs enter alive and leave as pork every day" (Silbergeld, 2016, p. 159). It is operated by Smithfield Foods and employs about 4,500 workers (Barnes, 2020). According to World Population Review (2022), the average household income in Tar Heel is just short of $59,000, and the poverty rate is 24.6%! The income level is substantially lower than the state average, whereas the poverty rate is significantly higher. This is one real-life example of how the conventional food system creates, maintains, and benefits from poverty.

Silbergeld (2016, p. 30) describes

> the social, agronomic, and economic transformation of agriculture from its traditional grounding in rural societies, pasturage, and smallholder enterprises into the integrated model of confinement, concentration, and integration, with long market chains that extend from breeding animals to selling products to consumers.

In places such as Europe, most chicken and pork "has been produced in the industrial model, and the several Common Agricultural Policies of the European Union and its predecessor the European Economic Community have been built around furthering and supporting this model" (p. 31). So, it is not just in the United States where governments are organized for the survival of the conventional food system.

Silbergeld (2016, pp. 69–70) claims that what drives Chickenization includes increased demand for meat-based protein as a result of increasing affluence and changes in food preference and patterns of consumption as well as increased urbanization. Also, within developing countries, "there is an increasing role of food animal products for export as part of deliberate national economic policies. Affluence and urbanization are historically connected with upscale changes in diet, with increasing consumption of the food of elite populations, including meat." So, to a significant degree, consumers are responsible for the industrialization of food. One could argue that people are the ultimate drivers of the industrial food system, for it is their desire for convenience, price, and availability that really mandates massive production of foods (Silbergeld, 2016). The degree to which consumers are responsible for the conventional food system is discussed later in the book.

## Corporatization of Food

Given the discussion of factors such as McDonaldization and even Chickenization (suggestive of control of chicken food manufacturing by corporations), it should be obvious that our food has become corporatized. It is the corporation, assisted by the government, that is responsible for the industrialization of the conventional food system as well as industrialized food—poor food products widely available in the United States and across the world today. Nesheim et al. (2015) note: "Like other sectors of the capitalist economy, the food system is thus characterized by driving forces of profit, productivity, competition, innovation, product differentiation, and corporate mergers." Obviously, since corporations lie at the heart of the conventional food system, profit is really the most important consideration in all decision-making. Fuhrman (2017) outlines who profits from the conventional food system as it currently stands. This includes food companies and the medical establishment at the least. It is the medical establishment, after all, that deals with all the deleterious health outcomes associated with poor diet and weight gain (e.g., obesity, diabetes, hypertension, heart disease, etc.). Nestle (2013, p. 41) adds to the dieting and weight loss industry, as well. She notes that Americans spend at least $60 billion each year on diet books, diet plans and programs, pills, and even weight loss surgery.

Most of our efforts to lose and keep off weight fail, so this industry profits repeatedly from our poor diets. For example, Fuhrman (2017) reports on a review of 17 studies that followed people from between 3 and 14 years, and it found that 85% of people who were dieting failed to keep weight off. This issue is revisited in Chapter 6.

Moss (2013, pp. 337–338) points out:

> It's simply not in the nature of these companies to care about the consumer in an empathetic way. They are preoccupied with other matters, like crushing their rivals, beating them to the punch . . . food companies are also deeply obligated toward their shareholders.

Moss claims: "Making money is the sole reason they exist—or so says Wall Street, which is there, at every turn, to remind them of this."

Here, it is important to note the nature of corporations themselves. Simon (2006, p. 3) calls the corporation "amoral," writing that "corporate actors are not guided by precepts of right and wrong, but rather by a set of fiduciary principles that have little to do with personal morality and everything to do with growing profits." My own work with corporations has gone further, calling them psychopaths (Pardue et al., 2013a, 2013b) and noting that their main concern is greed rather than morality.

While corporations were intended to be created and maintained for a limited, specific purpose (e.g., building a bridge over a river in one city), they have morphed over time into very powerful permanent entities. They have been transformed

> from a creature of the state or some monopoly interest in society with clearly defined public purpose into an all-purpose legal mechanism for facilitating the carrying-on of

business within a market economy, its character no longer subject to any meaningful review.

*(Rowland, 2005, p. 82)*

Research on corporations shows that their goal is unlimited acquisitiveness; thus, they have no "social good principle." They are, according to many scholars, "singularly self-interested and unable to feel genuine concern for others in any context" (Bakan, 2004, p. 56). This is not to say they do no good; rather they are "adiphoristic—indifferent to right and wrong, good and bad, except insofar as these can be expressed in terms of the corporate equivalents of pleasure and pain—which are profit and loss" (Rowland, 2005, p. 84). The corporation is founded on moral relativism, where circumstances rather than principles determine right and wrong (Rowland, 2005, p. xxi).

The paradox

is that given the good and bad and right and wrong have real existence and are not just semantic distinctions, and that humans are endowed with a moral drive or impulse that not only enables us to distinguish between these oppositions but impels us toward the good, we nevertheless consent to be governed in our daily lives by institutions that reflect the view that morality is relative and that humans are innately self-serving. Despite the certain knowledge of our moral essence, we acquiesce to an ideology of the market and the corporation that denies it.

*(Rowland, 2005, p. 74)*

In fact, corporations are designed to be greedy. They are simply not expected to be responsible for social welfare because it is assumed that if they maintain their focus on profit, social good will flow automatically, via the automated processes of the market (Rowland, 2005, p. 93). Amazingly, when early American corporate executives sought to serve interests other than profit, even courts rejected this idea.

For example, in the case of *Dodge v Ford* (1916), two brothers took Henry Ford to court to challenge Ford's plans to reduce the price of his Model T automobile for the sake of consumers, but at the expense of corporate profit. The judge agreed with the brothers,

reinstated the dividend and rebuked Ford—who had said in open court that "business is a service, not a bonanza" and that corporations should run only "incidentally to make money"—for forgetting that "a business corporation is organized and carried on primarily for the profit of stockholders'" it could not be run "for the merely incidental benefit of shareholders and for the primary purpose of benefiting others."

*(Rowland, 2005, p. 94)*

Well-known economists, including Milton Friedman and Theodore Leavitt, argued that the fundamental duty of the corporation is to earn profits. These are the words of Friedman: "I call [social responsibility] a fundamentally subversive doctrine in a free society, and say that there is one and only one social responsibility of business to use its resources and engage in activities designed to increase its profits" (Friedman, 1970, p. 32). And Leavitt asserted that

welfare and society are not the corporation's business. Its business is making money. . . . Government's job is not business, and business's job is not government. And unless these functions are resolutely separated in all respects, they are eventually combined in every respect. . . . Altruism, self-denial, charity . . . are vital in certain walks of our life. . . . But for the most part those virtues are alien to competitive economics.

*(Leavitt, 1958, p. 138)*

In fact, under the "best interests of the corporation" principle, corporate social responsibility is illegal (Bakan, 2004, p. 36). The duty of corporate managers is simply to make money; any compromise with this duty is dereliction of duty (Bakan, 2004). This is perhaps the best evidence that when corporations make claims about social responsibility, they do not really mean it.

Wade Rowland goes further, saying that even when codes of ethics state "do the right thing," it means something different to corporations. As noted by Rowland, although truth and contrition are

sometimes simulated in the guise of public-relations devices . . . it is the appearance of virtue, rather than virtue itself, that interests corporations . . . Having a good reputation—having the appearance of virtue—is always a benefit and always, by definition, widely known, while being virtuous may not be known to others at all, and it may not be particularly advantageous.

*(Rowland, 2005, p. 100)*

Examples are the ads by "big tobacco" where one major company urges parents to "talk to your kids about not smoking." Are we to believe that cigarette companies really want children to not start smoking? Given the following facts, this is simply impossible to believe: (1) tobacco companies, like all corporations, are in the business of making money; (2) every dollar spent on advertising is aimed at generating brand loyalty as well as new customers; (3) children and young teenagers are highly influenced by tobacco ads and easily recognize the leading brands; (4) virtually everyone starts smoking before the age of 18 years, meaning the vast majority of their cash-paying customers started as juveniles; and (5) one out of every two smokers will die from smoking-related conditions (Glantz et al., 1997; Kluger, 1997; Mollenkamp et al., 1998; Zegart, 2001). If almost no one starts smoking as an adult, who will replace the customers lost to deaths associated with tobacco companies' products? Surely, tobacco companies know that without youth smokers, their companies simply cannot maintain profitability in the United States.

Lost in the advertising claims affiliated with this supposed "anti-smoking" campaign is that tobacco companies got this issue of smoking back on television. It is currently illegal for cigarette companies to advertise their products on television. Yet, this "talk to your kids about not smoking" campaign offered a major tobacco company the opportunity to get cigarettes (and their own logo and company name) back on television. After discussing but a handful of deviant acts by a large tobacco company, Wade Rowland asserts that

the kind of people who run tobacco companies do not appear to resemble any of the people most of us know, unless we happen to work in a clinic for the treatment of mental pathologies. . . . The corporate managers seem to be devoid of the moral sensitivity that is a hallmark of a healthy person.

*(Rowland, 2005, pp. 111–112)*

By the way, large tobacco companies (e.g., RJ Reynolds and Phillip Morris) would go on to buy large food companies (e.g., Kraft and Nabisco), so later when you learn about some questionable behaviors of food companies, try to remember this link. More will be shared on this later in the book, as well.

Simon (2006, p. 4) notes that corporations are legally required to maximize profits for shareholders, and as such, they must continue to grow or they will die and go out of business: "The truth is that under a free enterprise system, corporations cannot place moral or ethical concerns ahead of profit maximization because doing so risks being driven out of business by competing firms that are not similarly plagued by social conscience." Since "food has very little internal growth . . . food companies are always fighting for market share. That always leads to aggressive marketing" (p. 5). This marketing inevitably is aimed at getting us to buy both unhealthy foods loaded with sugar, fat, and salt—foods that transform healthy, natural products into unhealthy, manufactured products—as well as more food than we need, and thus, we can confidently say that food corporations are inherently bad for us in at least that way. With regard to the transformation of food products and marketing efforts to get them into our hands, Simon (2006, p. 6) writes that "food companies refashion natural nourishment into industrial products bearing little resemblance to anything humans were designed to eat, and persuade people to consume as much as possible—public health consequences be damned."

So, as long as food corporations are concerned solely with profit and expanding markets to achieve greater profits, they will continue to sell us ultra-processed food products high in fat, sugar, and salt, to hell with our health and well-being. Like all other corporations, they internalize profits while externalizing costs, meaning it is always someone else who will pay for their reckless decision-making when it comes to food production. One estimate, reported by Simon (2006, p. 7), found that the costs generated by food companies in the form of health-care expenses, unsafe products, and pollution were four times greater than the profits earned by food companies in one year. If true, a costs–benefits analysis of the conventional food system would suggest the system fails to provide more benefits than costs. A rough costs–benefits analysis of the system is offered in Chapter 8.

Since companies do not commit suicide by making changes that would put them out of business, "food makers initiate only marginal, largely cosmetic changes to their traditional business practices—and then greatly exaggerate the benefits of these efforts for maximum PR effect" (Simon, 2006, p. 8). Simon reasons that mega-chain restaurants are "incapable" of offering healthier fare; it is simply not possible for them to do so and to survive. As long as the priorities of shoppers include "price, convenience" (along with cleanliness), it will be impossible for restaurants to offer healthier foods on a large scale. One of the important goals of corporate PR is to create and maintain the impression that government regulation of business (in this case food manufacturing) is unnecessary and that self-regulation by big business is sufficient to take care of any issue that needs to be monitored and resolved:

> In making their case for this approach, food makers insist that they can be trusted to police themselves and behave as "responsible corporate citizens"—even when it comes to controversial issues such as junk food marketing to children. The idea is to keep government regulators at bay at all costs.
>
> *(Simon, 2006, p. 10)*

As an example of how ridiculous this is, consider the pledge in 2005 by the ABA to provide lower calorie and/or nutritious beverages to schools and to limit the availability of soft drinks, something it has lobbied against in state legislatures for years! The pledge had no enforcement mechanism or government oversight and required voluntary participation from schools, and it only applied to vending machines in the first place. According to Simon (2006, p. 15): "The real aim of the ABA's pronouncement was to beat state legislators to the punch and initiate a beverage policy with the convenient advantage of being exceedingly friendly to industry interests." Imagine being in the room when such a policy was proposed and then discussed; it strikes me as crafty and devious behavior, at the very least. And even though the policy never actually materialized, "the ABA announced a multimillion dollar campaign 'to run print and broadcast advertising to educate the public about the new policy,'" suggesting it was all just PR in the first place (p. 16). Simon (2006, p. 91) calls this "nutriwashing."

## Monopolization of Food

Not only has food become corporatized, but it has also become monopolized. A *monopoly* is defined as "exclusive ownership through legal privilege, command of supply, or concerted action," "exclusive possession or control," and "a commodity controlled by one party" (Merriam-Webster, 2022b). A more accurate depiction of big food is that it is controlled by a handful of oligarchs.

Nestle (2013, p. 8) argues:

> The history of agriculture policy in the United States is one of increasing concentration and consolidation, with big driving out small in the name of efficiency. It is also one of cozy relations between corporate agriculture, Congress, and the [US Department of Agriculture].

In the following, I will show the major food corporations and illustrate how many different food companies are owned and operated by just a handful of large companies. The wide variety of company names on the food products we buy suggests that many, many more companies are working to provide our food than actually are. It also creates a false impression of competition in the marketplace.

As one example, there are more than 800 slaughterhouses that are inspected by the federal government, while another 1,800 are operated by states or customs houses with specific rules (Linnekin, 2016). Linnekin (2016, p. 44) notes: "The thirteen largest US cattle slaughterhouses account for 56 percent of all cattle killed in this country. The figures are similar for hogs (twelve plants account for 57 percent of all slaughters) and other livestock." Linnekin suggests that regulatory rules are written for large, monopolized producers that cannot be effectively followed by smaller producers, who thus go out of business, thereby enlarging already big food producers: "This consolidation, in turn, is often used to justify the need for more stringent regulations" (p. 61).

Consolidation generally refers to combining more than one thing into one thing, as in the case of individual companies being combined into one company. Farms are largely consolidated or concentrated in the United States. Walker (2019, p. 218) notes:

All but the top 1 percent of farms are called family farms. One in ten farms account for more than three-quarters of value produced. Nine in ten farms account for less than one-quarter. Household income for the top 10 family farms ranges from $186,000 to $1.7 million. Nine percent of farms receive two-thirds of all government subsidies. Seventy percent of farms receive nothing.

Consistent with consolidation, Constance (2019, p. 88) shows that the number of farms in the United States has significantly decreased over time. The overwhelming majority of farms are considered "family farms," and most of those are small and do not generate substantial revenue. However, "large family farms ($1 million or more in sales) make up 3 percent of farms and 20 percent of farmland [and] they account for 45 percent of sales and 56 percent of farm income."

Consolidation is now the norm in today's food markets: "Today, nearly 90 percent of agriculture sales comes from 12 percent of all farms" (p. 91). Consider pork production as another example: "In one generation, the number of farms producing hogs fell by almost three-quarters—while the media number of hogs per farm climbed from 1,200 to 40,000. Four companies . . . control nearly two-thirds of all hogs processed" (pp. 135–136).

According to Silbergeld (2016, pp. 22–23), 67% of poultry products and 42% of pork products come from factory farms. Factory farms are large farms that have become industrialized and run like assembly lines. An example of a factory farm is the dairy farm. Walker (2019, p. 32) claims that "dairy farms survive by treating cows as machines and by scrutinizing costs." According to FoodPrint (2021), dairy cows are milked for 10–12 months a year: "Average US milk production is about 17,000 pounds per cow annually, though herds with averages of up to 24,000 pounds per cow are not unusual." They spend their lives making milk for humans and are killed years before their normal life span of 15–20 years (Humane League, 2021). Further, cows are often fed grains they cannot digest, leading to gastrointestinal problems. Male dairy calves are not useful for milk and so they are sold for veal or beef production, and nearly "half of female calves are raised as replacement milk cows, as the older ones slow down; the rest are sold for veal or beef" (FoodPrint, 2021). This issue of animal welfare in the conventional food system is discussed in Chapter 7.

According to Walker (2019, p. 130):

> Continuing consolidation has produced a seed industry now dominated by three companies: Bayer-Monsanto, Dow-Dupont, and Syngenta-ChemChina . . . All follow the same strategy: bundle chemicals with patented seeds, pursue new patents, offer rebates to distributors and retailers, vigorously enforce use, and eliminate competition though acquisitions.

In fact, two seed companies—DuPont and Monsanto, control about half of the seed market. Walker discusses the implications, writing: "Seeds are part of nature. Nature determines survival. Whoever controls the seeds holds the power to control life. Such power is now concentrated in the hands of three corporations" (p. 131). And the largest producer is also the larger seller, meaning Smithfield Foods (now owned by the Chinese company Shuanghui International Holdings) exemplifies vertical integration, where one company controls multiple parts of the chair of the production of products. Incredibly, it is the largest four corporations that now own 95% of the hogs, and they provide feed to the farmers to raise them, even though the farmers do not own them: "The farmer puts up the land and finances the construction of the confinement buildings per the company's detailed specifications. The corporation prescribes the exact rearing practices. The farmer provides the labor" (p. 136).

According to Walker, it was the Tyson corporation that pioneered these practices, and 71% of its farmers lived in poverty!

Walker also describes working conditions, including long hours, and a crowded and hot environment filled with toxic wastes and gases. Even with non-performance, farmers are on the hook for land and building mortgages. So, Walker calls this kind of farming the "antithesis of independence and opportunity" (p 137).

Domination of the food industry by handfuls of corporations is common in milk, milling flour, refining sugar, and more. Walker (2019, p. 138) says:

> In seventeen of nineteen categories . . . the top four companies control from 46 to 95 percent of the market. In nine of the seventeen categories, market dominance exceeds 70 percent. At the global level, the top four companies in crop seeds, agricultural chemicals, animal health, animal breeding, and farm machinery control more than half of each market.
>
> *(p. 138)*

All of this was permitted and encouraged by the deregulation of business by Congress (a reminder of the connections between government and the conventional food system), which took over during the 1980s and continued largely unabated since:

> Food is no longer valued for its ability to sustain life, but only for its ability to generate profits. Whether higher returns come from squeezing farmers under contract to grow pigs or poultry, creating a monopoly on seeds that can be doused with chemicals, or selling food laden with cheap calories makes no difference.
>
> *(p. 141)*

According to Howard (2019), most cereal brands are owned by just four companies—General Mills, Kellogg Company, Post Consumer Brands, and Quaker Oats (owned by Pepsi). These four companies control about 85% of US sales. This oligopoly or shared monopoly often results in higher prices for products, even by the admission of at least one of the companies. Howard (2019, p. 199) discusses efforts by the Obama Administration from 2009 and 2010 to investigate alleged abuses of power, including bid rigging, price manipulation, and one-sided contracts. The Department of Justice reportedly concluded that "many potential enforcement options would be unsuccessful because of the way judges now interpret anti-trust laws" (Howard, 2019, pp. 119–120). So, it appears the system is tipped in favor of some interests over others. One allegation of bid rigging was alleged against milk producers who keep prices artificially high due to limited competition.

The Federal Trade Commission (FTC) can and occasionally has opposed mergers when they result in a company controlling too much of a share of the industry. Still, in the meat processing industry, four companies control as much as four out of every five dollars. Increased concentration in the meat industry means "farmers have little choice about whom they sell to and how their animals are raised." Meatpacking plants are largely controlled by Cargill, JBS (Brazil), Smithfield (China), and Tyson. These companies control more than 80% of beef processing and 70% of pork processing. And Dairy Farmers of America is the result of a merger of four companies in 1998. This company was charged federally and fined tens of millions of dollars for price-fixing-type behaviors.

The four largest grain companies are often referred to as ABCD—Archer Daniels Midland, Bunge, Cargill, and (Louis) Dreyfus; these companies control about 85% of the market. Milk and soybean farms also show evidence of consolidation, with fewer farms in control of the products each year.

Only six and soon only four chemical companies will control the chemical industry. Del Prado-Lu (2019, p. 100) writes that "a small oligopoly of chemical companies manages the world's supply of pesticides, fungicides, and insecticides, enforcing power through lobbying of governments and international trade organisations." She claims:

> This comes at the expense of concern for the health and safety of farmers and farming communities. In other words, global agriculture is organised around an economy built from the (over)use of harmful chemicals, which unfairly victimises vulnerable populations of humans, and negatively impacts non-human animals and environments.

We see this concentration in so many other areas of food. For example, "in the hog and beef industries . . . production has shifted to large specialized farms" and in the poultry market, few powerful companies control "feeding, hatching, and processing poultry," meaning poultry growers have significantly less power than before (Nesheim et al., 2015, p. 53). Even the entire seed industry is now controlled by just eight companies. Further, the "four largest firms in each of the other input industries (agricultural chemicals, farm machinery, animal health, and animal genetics) also have more than 50 percent of the global market sales" (Nesheim et al., 2015, p. 54).

The same pattern is found in manufacturing and processing. According to Nesheim et al. (2015, p. 55): "The 12 percent of plants with more than 100 employees ship 77 percent of all of the value of food, and mergers and acquisitions continue to occur often." Further, the "top four beef processing companies increased their share of the slaughter market from 36 to 79 percent between 1980 and 2005, while the four firm concentration ratio in hog and poultry reached 64 and 53 percent by 2005." Poultry processing, soybean processing, and pork packing have also become far more concentrated: "The most dramatic increase in the concentration of food processing sectors since 1990 may be in pork packing, where the four-firm concentration ratio increased from 40 to 66 percent in 2007" (Nesheim et al., 2015, p. 55). Large pork producers are notorious for polluting the environment. Two European firms—EQ group and Hendrix Genetics control the genetics of 99% of turkeys and 94% of egg-laying chickens. And four firms control about two-thirds of pork production. Even the grocery store market is becoming controlled by fewer and fewer companies over time.

Taken as a whole, the typical layout of the agricultural and food industry emulates an oligarchy, with 90% of workers accounting for food production jobs that earn the least and require the most productivity. Only 10% of food employees work inside office jobs with higher paying salaries instead of wages, protection from labor laws, and benefits such as health care and retirement. Again, the system seems designed to maintain poor labor conditions across much of the industry.

On top of this, Howard (2019, p. 117) claims:

> Many . . . negative impacts may result from high levels of market concentration, including reduced innovation, fewer consumer choices, and environmental impacts (such as

pollution and higher energy consumption). Frequently, these consequences dispropor-tionately affect certain populations, such as recent immigrants, ethnic minorities, people of lower socioeconomic status, women, and children.

All of these can be seen as costs of the conventional food system.

According to McMahon and Glatt (2019, p. 29), consolidation and concentration of industry were driven in part by food safety in the European Union. James, Jr. (2019, p. 69) notes: "Virtually every sector of the agri-food system has become more concentrated over time." Citing Hendrickson and James (2005, p. 283), James, Jr. (2019, p. 69) claims that

> industry concentration "limits or inhibits" the choice, options or kinds of decisions that farmers can make as well as "compels or obliges" the choices of farmers 'by forcing them into the kinds of decisions that they otherwise would not have chosen for ethical or other reasons.

An analysis of corporate control of food was published in 2015 (EcoNexus, 2015). The study presented evidence of widespread corporate ownership of food, typically in very few hands. One conclusion speaks volumes, saying

> big corporations buy smaller companies and thus increase their market share and power. Hence, companies can dictate prices, terms and conditions and, increasingly, the political framework. Much of what we consume in the North is being produced more cheaply in the Global South. The profits are made by only a few, predominantly Northern companies. The big losers are the plantation workers and small farmers in the South, as they are the weakest links in the "value chain." In no other section of the population is hunger so widespread. More and more ecosystems are being degraded and destroyed.

So, the implications of corporate control over the conventional food system are enormous. Specifically, the study shows the following realities:

- Animal feed—the top ten corporations control 16% of the market
- Livestock breeding—the top four corporations control 99% of the market
- Seeds—the top ten corporations control 75% of the market
- Fertilizer—the top ten corporations control 75% of the market
- Pesticides—the top 11 corporations control 98% of the market
- Trade—the top four corporations control 75% of the market
- Processing—the top ten corporations control 28% of the market
- Retail—the top ten corporations control 11% of the market

An example of consolidation in food corporations is Cargill. According to EcoNexus (2015, p. 4), "Cargill extends credit to farmers, produces food and feed, trades in energy, stock exchange products, and much more." Howard (2019, p. 126) agrees, writing:

> Cargill . . . has joint ventures with seed/chemical firms, contracts with grain and live-stock farmers, and processes grains and meats. Because it is a privately held corporation,

even less information about its operation is disclosed when compared to publicly traded corporations.

Cargill delivers feed, seeds, and fertilizer to farmers, it contracts with farmers to produce cereal and to fatten cattle and pigs, it buys, transports, and exports grains and soy, then it processes beer, pork, and soy, and finally, it has a contract to supply Koger Supermarkets. Cargill is thus an excellent example of vertical integration.

The study also shows consolidation in the coffee value chain, with five international traders in control, along with only three roasters: "Just three companies roast 40% of the global coffee harvest and five companies trade in 55% of the coffee" (EcoNexus, 2015, p. 5). The study also finds that, in the area of genetics, there is increased control in big business, as shown earlier. Specifically:

> Between 1989 and 2006, the world's number of suppliers of poultry genetics was reduced from eleven to four companies: in the laying hen sector from ten to three companies. Just three companies supply the world market for turkey genetics, and worldwide only two companies breed the ducklings and day-old-chicks that are flown around the world packed in cartons for fattening and egg production factories.
>
> *(p. 7)*

According to the study:

> Almost unnoticed by the public, animal breeding has been converted into a highly concentrated biotech-based industry. The biggest corporations control the genetics of several livestock species. The chemical corporation Monsanto, already the world's largest seed producer, has entered the lucrative business of animal genetics.
>
> *(p. 8)*

Seed companies themselves are highly concentrated, as noted earlier, with companies like Monsanto, DuPont, Sygenta, Dow, and others, owning so many subsidiary corporations with different names. Major companies now also own majorities of patents on plants.

According to Lakhani et al. (2021), it is only a "handful of powerful companies [who] control the majority of market share of almost 80% of dozens of grocery items bought regularly by ordinary Americans." The analysis showed the specific share of items produced by the specific number of corporations that control them. Some of these findings are shown in Table 2.2.

Based on their study, Lakhani et al. (2021) conclude that "a few powerful transnational companies dominate every link of the food supply chain: from seeds and fertilizers to slaughterhouses and supermarkets to cereals and beers." Figure 2.2 shows the monopolization of food, visually.

All of this is possible, they argue, because of lobbying efforts by companies, as well as weak regulatory structures that are unable to stop large mergers and acquisitions. Lakhani et al. (2021) explain the implications of their study: "This matters because the size and influence of these mega-companies enables them to largely dictate what America's 2 million famers grow and how much they are paid, as well as what consumers eat and how much groceries cost." And the revenue that is produced largely goes to the powerful corporate

**TABLE 2.2** Market Share of Specific Items by the Number of Firms

| | |
|---|---|
| Dry dinner mixes with meat | 98% by 4 firms |
| Single-serve yogurt/yogurt drinks | 97% by 4 firms |
| Single-serve prepared pasta dishes | 94% by 4 firms |
| Single-serve prepared sloppy joe sauces | 94% by 4 firms |
| Carbonated soft drinks | 93% by 3 firms |
| Dip | 91% by 4 firms |
| Dry mac and cheese mixes | 87% by 4 firms |
| Microwave popcorn | 87% by 4 firms |
| Canned tuna | 85% by 4 firms |
| Baby formula | 85% by 3 firms |
| Mayonnaise | 83% by 3 firms |
| Baby food | 82% by 3 firms |
| Refrigerated soy milk | 81% by 4 firms |
| Refrigerated almond milk | 81% by 4 firms |
| Chocolate confectionery | 80% by 3 firms |
| Beer | 79% by 4 firms |
| Dry pasta | 79% by 3 firms |
| Bagels | 79% by 4 firms |
| Canned pineapple | 75% by 4 firms |
| Yogurt | 75% by 4 firms |
| Cereals | 73% by 3 firms |
| Hard/soft tortillas/taco kits | 71% by 4 firms |
| Processed/imitation cheese | 71% by 4 firms |
| Soup | 70% by 4 firms |
| Wine | 69% by 4 firms |
| Coffee | 68% by 4 firms |
| Single-serve prepared salads | 68% by 4 firms |
| Canned salmon | 67% by 4 firms |
| Snack bars | 66% by 4 firms |
| Frozen pizza | 66% by 3 firms |
| Sour cream | 64% by 4 firms |
| Doughnuts | 64% by 4 firms |
| Cookies and crackers | 61% by 4 firms |
| Bread | 61% by 4 firms |
| Canned green peas | 60% by 4 firms |
| Canned sweet potatoes | 59% by 2 firms |
| Turkey | 58% by 4 firms |
| Canned tomatoes | 58% by 4 firms |
| Tea bags | 58% by 4 firms |
| Canned green beans | 56% by 4 firms |
| Canned corn | 55% by 4 firms |
| Fresh cut salad | 54% by 4 firms |
| Table sauces | 53% by 4 firms |
| Coconut milk | 52% by 4 firms |
| Rice | 52% by 4 firms |
| Bacon | 52% by 4 firms |
| Canned beans | 51% by 4 firms |
| Bottled water | 50% by 4 firms |

*(Continued)*

**TABLE 2.2**  (Continued)

| | |
|---|---|
| Meat (beef, poultry) | 49% by 4 firms |
| Juice | 47% by 4 firms |
| Egg producers | 41% by 4 firms |
| Processed meats | 39% by 4 firms |
| Cheese | 36% by 4 firms |
| Craft beer | 33% by 4 firms |
| Sweet bakery | 32% by 4 firms |
| Sugar | 29% by 4 firms |
| Egg brands | 23% by 4 firms |
| Whole milk | 23% by 4 firms |
| Frozen fruit | 22% by 4 firms |

*Source:* https://www.theguardian.com/environment/ng-interactive/2021/jul/14/food-monopoly-meals-profits-data-investigation

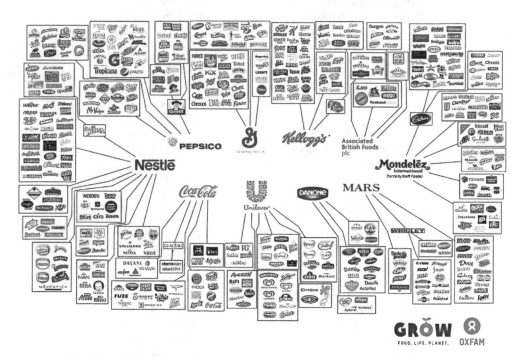

**FIGURE 2.2**   The Monopolization of Food

entities that process and market our foods (about 85 cents out of every dollar, whereas only about 15% goes to farmers). Increased consolidation of the industry has reportedly reduced costs to businesses by increasing debt in farmers. Further, it also means that

> those who harvest, pack and sell us our food have the least power: at least half of the 10 lowest-paid jobs are in the food industry. Farms and meat processing plants are among the most dangerous and exploitative workplaces in the country.

These are costs built into the conventional food system.

For consumers, the main issue is that, when there are fewer corporations in control of our food, there is less competition in the market, and this "means higher prices and fewer choices for consumers—including where they can shop for food" (Lakhani et al., 2021). These authors note, for example: "Until the 1990s, most people shopped in local or regional grocery stores. Now, just four companies—Walmart, Costco, Kroger, and Ahold Delhaize—control 65% of the retail market." So, there are about one-third fewer grocery stores today than there were just 25 years ago.

Overall, about 64% of sales of 61 popular grocery items are controlled by just a handful of corporations: "Four firms or fewer controlled at least 50% of the market for 79% of the groceries. For almost a third of shopping items, the top firms controlled at least 75% of the market share." Lakhani and colleagues (2021) give an example of PepsiCo, which owns five of the most popular potato chip brands (including Lay's, Fritos, and Tostitos), and controls 88% of the market. Another example is Kraft Heinz, a company as the result of a huge merger in 2015, which appears 12 times in the top four firms for groceries. Then there is Anheuser-Busch InBev, which acquired 17 different craft breweries between 2011 and 2020 (including an independent, local brewery where I once worked!).

Another analysis of sales of packaged foods (i.e., baby foods, bakery foods, canned and preserved foods, chilled and processed foods, confectionery foods, dairy, dried and frozen processed foods, dressing and condiments, ice cream, meal replacements, noodles, oils and fats, pastas, ready-to-eat meals, sauces, snack bars, soups, spreads, and snacks) found that the top ten companies accounted for just over 15% of sales worldwide, with each company contributing less than 3.3%. The top ten packaged food companies at the time were Nestle, Kraft Foods, Unilever, PepsiCo, Mars, Danone, Cadbury, Kellogg's, General Mills, and Ferrerro.

The same analysis, this time of soft drink sales (i.e., bottled waters, carbonates, concentrates, functional drinks, packaged fruits and vegetable juices, ready-to-drink teas and coffees, and Asian specialty drinks) found that the top ten soft drink companies accounted for more than 52% of sales worldwide. The study found that Coca-Cola led with about 26% of sales, followed by Pepsi with about 12% of sales (Alexander et al., 2011). The top ten soft drink companies at the time were Coca-Cola, PepsiCo, Nestle, Suntory Holdings Ltd., Dr. Pepper Snapple Group, Danone, Kirin Holdings Co. Ltd., Red Bull GmbH, Tingyl (Cayman Islands) Holdings Corp., and Asahi Breweries Ltd.

The reality of the majority of the conventional system is that it is powered and controlled by large, multinational companies. For example, approximately "80 to 90 percent of U.S. food production is now provided by the 10 to 20 percent of farmers who farm full-time" (Nesheim et al., 2015, p. 53). The result is that farmers "have lost some entrepreneurial autonomy and decision-making power over assets due to unbalanced relationships in bargaining power with agribusiness firms" (Nesheim et al., 2015, p. 176). This is another cost of the conventional food system.

Information collected by Oxfam demonstrates that only a dozen companies control nearly all of the world's food supply. These companies include Mondelez, Kraft, Coca-Cola, Nestlé, PepsiCo, P&G, Johnson & Johnson, Mars, Danone, General Mills, Kellogg's, and Unilever (Ryan, 2017). That only 12 companies dominate the conventional food system is highly suggestive of the incredible power that these actors have over the production, processing, marketing, and sales of food in the United States and across the world. According to McGrath (2017), the top 25 companies in the food industry brought in $741.2 billion in revenue in 2016, resulting in $86 billion in profit. Jusko (2017) reports that 31 food companies are among the top 500 revenue-generating publicly held manufacturing companies in the United States. Some of the top revenue-generating food companies from 2017 are described in Appendix 1. Notice their global reach!

In addition to these individual global corporations, there are additional organizations that act on behalf of them as well as their particular interests (e.g., restaurants, sugar, corn, milk, etc.). These include, but are not limited to, the NRA, the Sugar Association, the American Sugar Alliance, the Corn Refiners Association, the American Dairy Association, the National Dairy Council, the International Dairy Foods Association, the Dairy Alliance, the North American Meat Institute, the American Association of Meat Processors, and the National Meat Association. Each of these organization lobbies on their own behalf, as well as that of the aforementioned companies. Finally, it should be pointed out that these food companies do not include corporations such as Coca-Cola and Pepsi, although they undeniably share culpability for medical problems such as diabetes. According to Mourdoukoutos (2018), Coca-Cola ranks sixth among the nation's most valuable brands (with about $34 billion in revenue in 2017), and Pepsi ranks 29th (with about $64 billion in revenue in 2017).

## Key State Agencies Involved in Food

A very large number of government agencies at many levels of government (e.g., local, state, and federal) are involved in the US food system. The agencies are responsible for the quality and safety of and/or regulation of farming, fishing, ranching, water, land, labor, shipping, handling, processing, storing, trading, and shipping, as well as the overall safety of workers and the food they produce. These agencies include but are not limited to those most known to the American people—the USDA, the Food and Drug Administration (FDA), and the Environmental Protection Agency (EPA)—but also many other agencies at the federal, state, and local levels. Each is discussed in the following.

First, it is important to note that important issues like food safety are not actually well-served by our current system of government regulation. For example, Simon (2006, p. 57) writes: "Various federal agencies set standards for food safety, provide nutrition advice, subsidize agriculture, and otherwise oversee the types of foods we eat and the information we receive about them." Yet, she also asserts that "Uncle Sam is more aligned with Big Food than with the citizens it's supposed to represent" (p. 143). The Institute of Medicine and National Research Council (1998) also claims: "The food safety system in this country is complex and multilevel. It is also essentially uncoordinated. As a consequence, the government's role is also complex, fragmented, and in many ways uncoordinated." It goes on to say: "In fact, surveillance and reporting systems are insufficient in scope, resources, and statutory authority to generate reliable current measures of foodborne illness, much less to establish trends." This is in large part due to the fact that there are "at least a dozen federal agencies implementing more than 35 statutes make up the federal part of the food safety system."

### USDA

The USDA "provide(s) leadership on food, agriculture, natural resources, rural development, nutrition, and related issues based on public policy, the best available science, and effective management" (US Department of Agriculture, 2018). USDA not only regulates food products to assure consumer safety (Food Safety and Inspection Services or FSIS) but also promotes health and well-being of consumers by creating dietary guidelines (Center for Nutrition Policy and Promotion), assists in the regulation of animals and plants (Animal and Plant Health Inspection Service), helps protect forests and grasslands (Forest Service), helps conserve natural resources (Natural Resources Conservation Service), directly helps farmers through loans, conservation efforts, marketing, and risk insurance (Farm Service Agency and Risk Management Agency), assists in the development of rural areas (Rural Development) conducts different types of research (Agricultural Research Service, Economic Research Service, and National Institute of Food and Agriculture), helps market agricultural products in and beyond the United States (Agricultural Marketing Service, Foreign Agricultural Service, Grain Inspection, and Packers and Stockyards Administration), maintains a library about agriculture (National Agricultural Library), and provides helpful data for people involved in food production (National Agricultural Statistical Service). Most importantly, FSIS aims to "ensure that meat and poultry products for human consumption are safe, wholesome, and correctly marked, labeled, and packaged" (Institute of Medicine and National Research Council. Committee to Ensure Safe Food from Production to Consumption, 1998). FSIS shares responsibility with FDA for eggs and processed egg products.

## FDA

The FDA has the simply stated but extremely complex mission of "protecting human and animal health" (US Food and Drug Administration, 2018). The agency has jurisdiction over much of the nation's food products but also deals with drugs, medical devices, radiation-emitting products, vaccines, blood and biologics, animals and veterinary issues, cosmetics, and tobacco! With regard to food, FDA has jurisdiction over both domestic and imported foods except for meat and poultry products. FDA's Center for Food Safety and Applied Nutrition aims to make sure foods are "safe, nutritious, wholesome, and honestly and adequately labeled" (Institute of Medicine and National Research Council. Committee to Ensure Safe Food from Production to Consumption, 1998). FDA also helps prevent food-borne illness and provides information for labeling on most packaged foods. The agency also provides information on food additives, ingredients, allergens, and dietary supplements, as well as on "pathogens, chemicals, pesticides, natural toxins, and metals" in food. The FDA also conducts research on food and food production, issues food recalls due to illness outbreaks, and helps defend the food system from criminal and terrorist attacks. Additionally, the FDA offers "guidance and regulatory information" directly to the industry about many topics related to food, as well as assures compliance with and enforces federal regulations related to food.

## EPA

The mission of the EPA is to help assure Americans have access to "clean air, land and water" (EPA, 2018). The EPA not only creates and enforces regulations meant to assure this goal but also conducts research on environmental problems, provides grants to states, non-profit agencies, educational institutions, and so on, partners with corporations, state and local governments, and non-profits to conserve natural resources, and provides educational materials about environmental protections. The role of the EPA in the production of food primarily centers around protecting natural resources to make food production as effective and efficient as possible. Further, EPA

> licenses all pesticide products distributed in the United States and establishes tolerances for pesticide residues in or on food commodities and animal feed. EPA is responsible for the safe use of pesticides, as well as food plant detergents and sanitizers, to protect people who work with and around them and to protect the general public from exposure through air, water, and home and garden applications, as well as food uses. EPA is also responsible for protecting against other environmental chemical and microbial contaminants in air and water that might threaten the safety of the food supply.
>
> *(Institute of Medicine and National Research Council.*
> *Committee to Ensure Safe Food from Production to Consumption, 1998)*

### CDC

In addition to these agencies, other agencies such as the US Centers for Disease Control and Prevention (CDC) assist with assuring safe foods. The CDC tracks foodborne germs and illnesses, provides information on food poisoning—which it claims impacts one in six Americans every year—and works with federal, state, and local health officials to investigate sources of food poisoning (US Centers for Disease Control and Prevention, 2018c). CDC also collects data on pathogens from states.

### Other Agencies

In addition, the National Marine Fisheries Service (NMFS) conducts voluntary seafood inspections of processing plants, fishing vessels, and seafood products. NMFS coordinates its efforts with the FDA's Office of Seafood Safety. Further, the Grain Inspection, Packers, and Stockyards Administration's Federal Grain Inspection Service assures federal quality and safety standards for the nation's grain products. And then, there are numerous state- and local-level agencies, such as health departments, that engage in activities such as inspecting restaurants for safety: More than 3,000 agencies have some food safety responsibilities. According to the Institute of Medicine and National Research Council. Committee to Ensure Safe Food from Production to Consumption (1998): "States are responsible for the inspection of meat and poultry sold in the state where they are produced, but FSIS monitors the process."

### Government–Corporate Relations

It should be obvious from the brief descriptions of the missions of these agencies that they are closely aligned with the corporations that produce the food that is the main source of food for Americans and citizens around the world. So too are regulatory agencies in other countries. That these state agencies help directly market products for companies and assist with assuring their successes in the marketplace domestically and abroad may surprise some. What is well-known is that regulation of food companies (and other corporations) is often negligent due to the "revolving door" between corporations and regulatory agencies (discussed in Chapter 1)—meaning officials move back and forth between the two—so that companies are being regulated by people with more information about but also stronger loyalties to the companies (Lima, 2018). This makes effective regulation for consumer health and safety less likely. So too does deregulation generally, as in the case of deregulation of fast-food products and its effects on body mass index (De Vogli et al., 2014).

In part because of the cozy relationship between government agencies and corporate entities, food companies are freer to market their products through corporate-friendly studies (Marion, 2018). They are also free to encourage us to eat more and more and more, toward the goal of greater profits in spite of the negative health effects on consumers (Nestle, 2013). And the system of the US government generally allows companies who produce and market unhealthy products such as sodas to guarantee billions of dollars in sales through advertising to vulnerable populations, lobbying government officials and political parties, financial donations to health organizations and researchers to muddy conclusions of scholarship on food and health, and producing advertisements claiming corporate responsibility (Marion, 2017). The close relationships between government agencies and the corporations that produce, market, and sell food exemplify state-corporate criminality given the detrimental health outcomes associated with the conventional food system.

Moss (2013, p. 211) points out that "when it comes to nutrition, the role the government plays is less a matter of regulation than it is promotion of some of the industry practices deemed most threatening to the health of consumers." Consider the case of meat and dairy. Moss (2013, p. 213) notes that "the biggest deliverers of saturated fat—the type of fat doctors worry about—are cheese and red meat, and it is in producing and selling these two products that the food industry has shown its greatest ability to influence public policy." He notes that the Department of Agriculture "doesn't regulate fat as much as it grants the industry's every wish. Indeed, when it comes to the greatest sources of fat—meat and cheese—the Department of Agriculture has joined industry as a full partner in the most urgent mission of all: cajoling the people to eat more."

How can this happen? Part of it is owing to the conflicting missions of the Department; on one side are the more than 300 million Americans whose health the Department is charged with protecting, and on the other are the hundreds of companies that form the $1 trillion food manufacturing industry who produce the foods who harm us. Of how little importance is nutrition to the agency? Far less than 1% of its budget goes to its Center for Nutrition Policy and Promotion (Moss, 2013). Moss (2013) compares the $6.5 million this Center gets each year with the $2 billion it uses for its marketing programs called "checkoffs" to "sell American on more beef."

Walker notes the conflicting duties of the USDA, writing

> there are programs within the department working at cross purposes. One program administers subsidies that favor diets steeped in calories from processed grains, sugar, and fat—while another provides dietary guidelines that promote a healthy diet from nutritious food. To straddle the divide, works like *nutrition* are featured instead of *calories*. In recent farm bills, provisions that provide easier access to diets rich in subsidized calories are titled "Nutrition Assistance Programs."
>
> *(p. 93)*

But, then, the USDA is a political operation, with the fourth highest number of political appointed spots in the US government (Walker, 2019).

## Conclusion

In this chapter, you read a brief history of food and came to see just how food processing began and then started to understand some of its significant limitations. You also saw just how complex the conventional food system is. Most importantly, you learned that contemporary food is industrialized, corporatized, monopolized, and globalized. And you also saw how food is connected to terms like Chickenization, and McDonaldization. Finally, the chapter examined key state agencies involved in various aspects of food within the United States.

It is this conventional food system that is responsible for providing food not only to Americans but also to citizens around much of the world. It is a remarkable feat to feed billions of people all over the globe. For that, key actors in the conventional food system deserve much credit. Yet, as you will see in future chapters, there are enormous costs associated with both the foods we eat and the system itself. Stated simply, this food system is not sustainable, meaning change must come to assure healthier food options, better working conditions, more humane treatment of animals, and so forth. The conventional food system is thus broken and cannot stand as it is.

## Appendix 1: Top Companies of the Conventional Food System

The leading revenue-generating food company, Archer-Daniels-Midland Co. (ADM), has been formally in business since 1923. According to its website (Archer-Daniels-Midland, 2018), the company is "one of the world's largest agricultural processors and food ingredient providers, with approximately 31,000 employees serving customers in more than 170 countries." Its business includes roughly "500 crop procurement locations, 270 ingredient manufacturing facilities, 44 innovation centers and the world's premier crop transportation network." The company, like all major players in the conventional food system, is global in presence and reach.

The company manufactures acidulants (citric acid products), alcohols, beans and pulses, dried fruits, dry bakery mixes and fillings/icings, emulsifiers and stabilizers (thickeners), flavors and extracts/distillates, flours and ancient grains, nutrition and health supplements, nuts and seeds, oils and fats, pastas, proteins, starches, sweeteners and sweetening solutions, and non-GMO and organic products. Thus, many of the company's products actually comprise the ingredients of a wide variety of food products manufactured by other companies.

Similarly, the second-ranked revenue-generating food company on the list, Bunge Ltd., began as a trading company in 1818 and expanded into a grain trading company in 1884—expanded to South America in 1908 and North America in 1919—and now operates in more than 40 countries (Bunge, 2018). The company makes products in the categories of oilseeds (including feed for animals, cooking oils, margarines, and shortenings), grains (including feed for animals as well as grains for use in beer, cereal, snacks, and baked goods), and sugarcane (including sugar and ethanol). The company is involved in producing and processing crops (e.g., corn, wheat, and milled rice) and foodstuffs (e.g., edible oils like soybean, rapeseed, and sunflower seed), transportation and logistics, as well as marketing and distribution. The company also manufactures fertilizers used on crops by farmers.

The third-ranked revenue-generating food company, Tyson, is also located in many countries across the globe. It began as a chicken delivery company in Arkansas in 1931. Its website claims that the company now produces 20% of all the chicken, beef, and pork in the United States. It also provides proteins to many national restaurant chains. Its major brands include Tyson, as well as Aidells, Ball Park, Golden Island, Hillshire Farm, Jimmy Dean, Sara Lee, State Fair, Wright Brand, and many other companies (Tyson, 2018).

The fourth-ranked revenue-generating food company is Kraft-Heinz Co. According to its website, the company is located in at least 28 countries on nearly every continent in the world. It is known for many of the world's best-known brands, including Heinz, Kraft, as well as Caprisun, Classico, Grey Poupon, Kool-Aid, Jello, Lunchables, Maxwell House, Philadelphia, Planters, OreIda, Oscar Meyer, Velveeta, and many other companies (Kraft-Heinz Co., 2018). Number five among the top-ranked revenue-generating food companies is Mondelez International Inc. According to the company's website, the company operates in 165 countries on at least six continents. Among its best-known brands are BelVita, Bubbaloo, Cadbury, Chips Ahoy, Dentyne, Halls, Honey Maid, Nabisco, Nilla, Nutter Butter, Oreos, Premium, Ritz, Sour Patch Kids, Tang, Tobleron, Trident, Triscuit, Wheat Thins, and many more (Mondelez International Inc., 2018).

The sixth-ranked revenue-producing food company is General Mills Inc. It provides a wide range of food products in more than 100 counties located on six continents. Officially created in 1928, the company is known for its products including General Mills but also Betty Crocker, Cheerios, Gold Medal, Nature Valley, Pillsbury, Wheaties, Yoplait, as well as organics and natural brands. The company also makes numerous toys and games (General Mills Inc., 2018).

The seventh-ranked revenue-producing food company is Kellogg Co., which has been in business for more than 100 years. Known widely for its products such as All-Bran, Cheez-It, Chips Deluze, Coco pops, Corn Flakes, Corn Pops, Eggo, Famous Amos, Fiber Plus, Frosted Flakes, Frosted Mini-wheats, Fruit Loops, Garden Burger, Keebler, Morning Star, Nutrigrain, Poptarts, Pringles, Rice Krispies, Special K, and Townhouse, the company makes far more than cereal (Kellogg Co., 2018).

The eighth-ranked revenue-producing food company is Conagra Brands Inc. The company was formed as Conagra in 1971, after previously being known as Nebraska Consolidated Mills since 1919; that company reportedly started as four flour mills in the state of Nebraska. Now housed in Chicago but in existence in at least 40 places, the company is known for such brands as Duke's, Frontera, Healthy Choice, Hebrew National, Hunt's, Marie Calender's, Orville Redenbacher's, Pam, P.F. Chang's, Peter Pan, Redi Wip, Rotel, Sandwich Bros, and more (Conagra Brands Inc., 2018).

The ninth top revenue-producing food company is Hormel Foods Corp., founded in 1891 in Minnesota and now operating in 75 countries. Existing as a family of companies, Hormel is known for brands including Hormel chili, as well as Applegate, Black Label, Buffalo, Chi-Chi's, Curemaster Reserve, Dinty Moore, Don Miguel, Embasa, Evolve, Fontanini, Jennie-O, Justin's, Little Sizzlers, Lloyd's, Mary Kitchen, Muscle Milk, Natural Choice, Old Smokehouse, Sandwich Makers, Skippy, Spam, Valley Fresh, and Wholly Guacamole, including many other brands (Hormel Foods Corp., 2018).

Finally, the tenth revenue-producing food company is the Campbell Soup Co. Known mostly for its soups, the company began in 1869 in New Jersey but is now found on nearly every continent. Among its widely popular brands are not only Campbell soups but also Arnott's, Bolthouse Farms, Chunky, Goldfish, Milano, Pace, Pacific, Pepperidge Farm, Plum Organics, Prego, Spaghettios, Stockpot, Swanson, TimTam, and V8 (Campbell Soup Co., 2018).

# 3
# WHAT AMERICANS EAT

## Introduction

Eating is pleasurable and it is something we do every day, mostly without really thinking about it. In this chapter, the major topics of consideration include what people are advised to eat and what people actually tend to eat. While eating habits vary across the globe, the fact is that most of us are purchasing products from the same conventional food system; a focus on the United States shows that what we tend to eat does not match what we are supposed to eat if we want to be healthy.

Early food guidance in the United States is examined, followed by an analysis of the various food pyramids, food plates, and similar tools used by governments meant to illustrate healthy eating. Problems with food guidance are identified, including that major food companies (referred to collectively as "industry") have exerted enormous influence over food guidance in the United States. As a result, governments have historically offered dietary evidence that runs counter to both good science and good health. A couple of quick examples are the government's advice to eat lots of grains (but not specifically whole grains) and the government's advice to avoid fats (not just bad fats).

Additional focus in the chapter is placed on the issue of physical activity, which, the chapter shows, is also part of a healthy life. Yet, I show in the chapter that major food companies have latched onto this idea and promoted the concept of "energy balance" to push exercise as the solution to deleterious health outcomes such as obesity and diabetes. I characterize these efforts as meant to divert attention away from the harmful nature of many of the food products corporations produce, even though energy balance is also promoted by the government in its dietary guidance to Americans.

The chapter ends with an analysis of why people eat what they eat. The issue of whether people have free will when it comes to their eating—whether people are actually in control of what they eat—is examined. The evidence, perhaps surprisingly, questions the idea of free will. Why we eat what we eat is far more complex than a simple choice.

DOI: 10.4324/9781003296454-3

## What Is a Healthy Diet?

### Early Guidance

Davis and Saltos (2022, p. 33) show that the first dietary recommendations were published by the USDA in 1894, when "specific vitamins and minerals had not even been discovered." This was titled, *Farmers' Bulletin*, and it was "based on content of protein, carbohydrate, fats, and 'mineral matter'" (p. 34). But, the first USDA food guide was published in 1916, and it was titled, *Food for Young Children*. Foods were placed into five categories—"milk and meat, cereals, vegetables and fruits, fats and fatty foods, and sugars and sugary foods" (p. 35). One year later, in 1917, *How to Select Foods* was published, focused on these five food groups. Another guide was released in 1921, showing the amounts of foods one should purchase each week among these five food groups.

In the 1930s, during the Great Depression, the USDA "developed food plans at four cost levels to help people shop for food. The plans were outlined in terms of 12 major food groups to buy and use in a week to meet nutritional needs" (Davis & Saltos, 2022, p. 35).

In 1941, the first-ever Recommended Daily Allowances (RDAs) were released by the Food and Nutrition Board of the National Academy of Sciences. These "listed specific recommended intakes for calories and nine essential nutrients—protein, iron, calcium, vitamins A and D, thiamin, riboflavin, niacin, and ascorbic acid (vitamin C)" (Davis & Saltos, 2022, p. 35). Then, in 1943, the USDA released its Basic Seven food guide as a leaflet titled, "National Wartime Nutrition Guide." It "specified a foundation diet that would provide a major share of RDA's for nutrients, but only a portion of caloric needs. It was assumed that people would include more foods that the guide recommended to satisfy their full calorie and nutrient needs" (p. 36). Beginning in 1943, these seven foods included (1) green and yellow vegetables; (2) cabbage or salad greens grapefruit, oranges, and tomatoes; (3) fruits, vegetables, and potatoes; (4) milk-based products; (5) meat and eggs; (6) bread, flour, and cereal; and (7) butter or margarine. The specific number of servings was included, but no serving sizes were specified (Minger, 2013).

In 1956, a new food guide, known as the Basic Four foundation diet, was released. It "recommended a minimum number of foods from each of four food groups—milk, meat, fruits and vegetables, and gain products" (Davis & Saltos, 2022, p. 36). This guide was used for the next couple of decades.

Then, during the 1970s, the Senate Select Committee on Nutrition and Human Needs held:

- Healthy diets could play an important role in promoting health, increasing productivity, and reducing health-care costs.
- The American diet has changed within the last 50 years, and people need guidance to improve their health through better nutrition.
- The government has a role to provide nutrition guidance to Americans and encourage the advancement of nutrition research and industry food reformulation.

*(Dietary Guidelines for Americans, 2022)*

### Dietary Guidelines

In 1977, the government released the *Dietary Goals for the United States*, which held:

- To avoid overweight, consume only as much energy as is expended; if overweight, decrease energy intake and increase energy expenditure.

- Increase the consumption of complex carbohydrates and "naturally occurring" sugars from about 28% of intake to about 48% of energy intake.
- Reduce the consumption of refined and processed sugars by about 45% to account for about 10% of total energy intake.
- Reduce overall fat consumption from approximately 40% to about 30% of energy intake.
- Reduce saturated fat consumption to account for about 10% of total energy intake, and balance that with polyunsaturated and monounsaturated fats, which should account for about 10% of energy intake each.
- Reduce cholesterol consumption to about 300 mg a day.
- Limit the intake of sodium by reducing the intake of salt to about 5 g a day.

*(Dietary Guidelines for Americans, 2022)*

You'll note not only detailed recommendations with regard to specific substances but also an allegiance to "energy balance" in the first bullet point, the idea that one's weight is directly attributable to the amount of energy one consumes and expends. This topic is revisited later in the chapter; you will see that food companies exploited this idea to divert attention away from the harmful nature of many of their products.

According to Davis and Saltos (2022, p. 36), in the 1970s: "The focus shifted from obtaining adequate nutrients to avoiding excessive intakes of food components linked to chronic diseases. The 1977 guidelines specified quantitative goals for intakes of protein, carbohydrate, fatty acids, cholesterol, sugars, and sodium." In its 1979 publication, *Food*, the USDA "began addressing the role of fats, sugars, and sodium in risks for chronic diseases." A new food guide, the *Hassle-Free Guide to a Better Diet*, modified the Basic Four to include a fifth group with fats, sweets, and alcohol, arguing for moderation in these substances.

In 1980, the USDA and Health and Human Services (HHS) collectively released *Nutrition and Your Health: Dietary Guidelines for Americans*. This report laid out seven principles for a healthful diet. According to Dietary Guidelines for Americans (2022):

> This edition was based, in part, on the 1979 Surgeon General's Report on Health Promotion and Disease Prevention and the findings from a task force convened by the American Society for Clinical Nutrition, which reviewed the evidence relating six dietary factors to the Nation's health. The focus of the 1980 Dietary Guidelines was to offer ideas for incorporating a variety of foods in the diet to provide essential nutrients while maintaining recommended body weight.

The guidelines also recommended limiting sugar, fat, saturated fat, cholesterol, sodium, as well as alcohol, promoting moderation in consumption.

According to Davis and Saltos (2022, p. 37): "The guidelines called for a variety of foods to provide essential nutrients while maintaining recommended body weight and moderating dietary constituents—fat, saturated fat cholesterol, and sodium—that might be risk factors in certain chronic diseases." The guidelines

> suggested numbers of servings from each of the five major food groups—the bread, cereal, rice, and pasta group; the vegetable group; the fruit group, the milk, yogurt, and cheese group; and the meat, poultry, fish, dry beans, eggs, and nuts group—and recommended sparing use of a sixth food group—fats, oils, and sweets.

In 1984, USDA presented *A Pattern for Daily Food Choices* in the form of a food wheel graphic as well as in tabular form.

After this, due in part to controversy about whether the industry was heavily influencing dietary advice to Americans from the government,

> Congress directed the USDA and HHS to convene a Federal advisory committee to seek outside scientific expert advice prior to the Departments developing the next edition of the Dietary Guidelines. Thus, a Dietary Guidelines Advisory Committee [DGAC] was established, composed of scientific experts entirely outside the Federal sector, and the advisory committee's Scientific Report helped to inform the development of the 1985 Nutrition and Your Health: Dietary Guidelines for Americans.
>
> *(Dietary Guidelines for Americans, 2022)*

In 1985, these guidelines were released, which

> promoted a dietary pattern that emphasized consumption of vegetables, fruits, and whole-grain products—foods rich in complex carbohydrates and fiber—and of fish, poultry without skin, lean meats, and low-fat dairy products selected to reduce consumption of total fat, saturated fat, and cholesterol.
>
> *(Davis & Saltos, 2022, p. 41)*

Then, in 1989,

> USDA and HHS established a second scientific advisory committee to review the 1985 Dietary Guidelines and make recommendations for the next revision. The guidance of earlier Dietary Guidelines was reaffirmed. The 1990 Nutrition and Your Health: Dietary Guidelines for Americans promoted enjoyable and healthful eating through variety and moderation, rather than dietary restriction.

For the first time, the guidelines suggested numerical goals for total fat—30% or less of calories—and for saturated fat—less than 10% of calories. These goals were for diets over several days, not for one meal or one food. (Davis & Saltos, 2022, p. 42).

Figure 3.1 shows a summary of the history reviewed earlier. The figure illustrates the number of food groups suggested across time, examples of different foods within each group, and numbers of suggested servings/serving sizes/frequency of servings.

### Food Pyramid

According to Davis and Saltos (2022), work began in 1988 to create a visual for Americans that would graphically represent nutritional guidelines, and the *Food Guide Pyramid* was born in 1992. The original pyramid in the United States, created in 1992, recommended:

- Six to 11 servings of grains
- Three to five servings of vegetables
- Two to four servings of fruit
- Two to three servings of meat and beans

| Food guide | Number of food groups | Protein-rich foods Milk/Meat | Breads |
|---|---|---|---|
| 1916 Caroline Hunt buying guides | 5 | Meats/other protein-rich food 10% cal milk; 10% cal other *1 cup milk plus 2–3 svg other* (based on 3-oz. serving) | Cereals and other starchy foods 20% cal 9 svg (based on 1 oz. or 3/4 cup dry cereal svg) |
| 1930's H.K. Stiebeling buying guide | 12 | Milk-2 *cups* / Lean meat/poultry/fish—9–10/week / Dry mature beans, peas, nuts—1/Week / Eggs—1 | Flours, cereals— *As desired* |
| 1940's Basic Seven foundation diet | 7 | Milk and milk products *2 cups or more* / Meat, poultry, fish, eggs, dried beans, peas, nuts—1–2 | Bread, flour, and cereals—*Every day* |
| 1956–70's Basic Four foundation diet | 4 | Milk group—2 *cups or more* / Meat group—2 *or more* (2-3 oz. svg) | Bread, cereal— 4 *or more* (1 oz. dry, 1 slice, 1/2-3/4 cup cooked) |
| 1979 Hassle-Free foundation diet | 5 | Milk-cheese group— 2 (1 cup, 1 1/2 oz. cheese) / Meat, poultry, fish, and beans group—2 (2–3 oz. svg) | Bread-cereal group—4 (1 oz. dry, 1 slice, 1/2 to 3/4 cup cooked) / Milk-cheese group— 2 (1 cup, 1 1/2 oz. cheese) |
| 1984 Food Guide Pyramid total diet | 6 | Milk, yogurt, cheese- 2–3 (1 cup, 1 1/2 oz. cheese) / Meat, poultry, fish, eggs, dry beans, nuts—2–3 (5–7 oz. total/day) | Breads, cereals, rice, pasta— 6-11 svg • Whole grain • Enriched (1 slice, 1/2 cup cooked) |

| Food Guide | Vegetables/Fruits | Other (incl. fats) |
|---|---|---|
| 1916 | Vegetables and fruits 30% cal 5 svg (based on an average 8 oz. svg.) | Fatty foods (20% cal)—9; Sugars (10% cal)—10 (based on 1 tbsp. svg) |
| 1930's | Leafy green/yellow—11–12/week / Potatoes, sweet potatoes—1 / Other veg grill—3 / Tomatoes and citrus—1 | Butter—*na* / Other fats—*na* / Sugars—*na* |
| 1940's | Leafy green/yellow—*1 or more* / Potatoes, other fruit/veg—2 or more / Citrus, tomato, cabbage, salad greens—1 *or more* | Butter, fortified margarine— *Some daily* |
| 1956–70's | Vegetable-fruit group— 4 *or more* (incl. dark green/yellow veg frequently and citrus daily; 1/2 cup or average-size piece) | |
| 1979 | Vegetable-fruit group—4 (incl. vit. C source daily and dark green/yellow veg. frequently 1/2 cup or typical portion) | Fats, sweets, alcohol— *Use dependent on calorie needs* |
| 1984 | Vegetable—3–5 • Dark green/deep yellow • Starchy/legumes • Other (1 cup raw, 1/2 cup cooked) Fruit-2-4 • Citrus • Other (1/2 cup or average) | Fats, oils, sweets— Total fat not to exceed 30% cal *Sweets vary according to calorie needs* |

**FIGURE 3.1**  Food Guide Advice From 1916 to 1984

- Two to three servings of dairy
- Few sweets and oils

*(Minger, 2013)*

Figure 3.2 shows the original food pyramid. As you can see, the grains make up the base, but there is no differentiation between whole grains and processed grains. You also notice that meat and beans get their own category, but there is no differentiation between types of meat. Dairy has its own category, as well. Finally, oils are grouped together with sweets—located at the top of the pyramid, meant to use them sparingly—even though some oils are healthy sources of unsaturated fats that are found to be related to lower levels of cholesterol, reduced risks of heart disease and weight gain, and better brain functioning and maintenance of blood sugar. Fats and oils are also integrated throughout each of the food group levels.

The Food Pyramid was revised in 2005—called MyPyramid—to specifically visually illustrate the importance of exercise in conjunction with healthy eating. Figure 3.3 illustrates the 2005 version of the pyramid. First, note the stick figure running upstairs on the new pyramid and the phrase, "Steps to a healthier you," and second, the fact that all numbers of servings were removed. The first shows the devotion to *energy balance*. Simon (2006, p. 147) claims that "MyPyramid's emphasis on activity plays right into the food

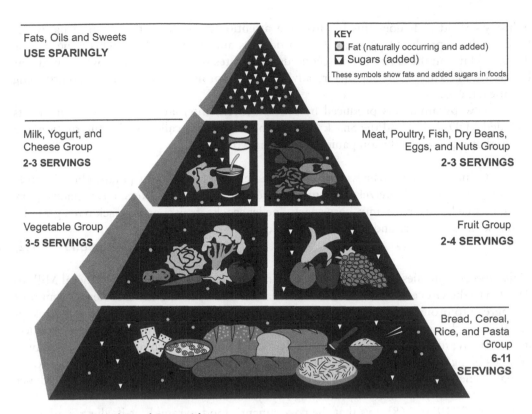

FIGURE 3.2    Original Food Pyramid

FIGURE 3.3    MyPyramid

industry's hands." It does this by diverting attention away from the harmful nature of food products produced by major food producers and manufacturers. This issue will be discussed later in the chapter. The second demonstrates the fact that the new pyramid was less useful in terms of offering dietary advice in terms of how much of each food grouping to eat each day.

The new pyramid was produced by Porter Novelli International, whose prior clients included McDonald's and the Snack Food Association (now called SNAC International, a group that represents 400 companies around the world):

> SNAC International business members include manufacturers of potato chips, tortilla chips, cereal snacks, pretzels, popcorn, cheese snacks, snack crackers, meat snacks, pork rinds, snack nuts, party mix, corn snacks, pellet snacks, fruit snacks, snack bars, granola, snack cakes, cookies and various other snacks.
>
> *(Potato Pro, 2022)*

Major food companies from cereal manufacturer (i.e., candy for breakfast) General Mills to PepsiCo to the Grocery Manufacturer's Association made public announcements of support and agreed to promote the new Food Pyramid to their customers, often on product packaging. Perhaps this is because MyPyramid does not recommend *not* eating foods they produce. It doesn't say, for example, *not* to eat ultra-processed foods, which make up the great bulk of foods being produced in the first place! Whether a company whose clients produce foods that are generally not healthy should be involved in any way in federal nutritional guidance seems completely lost on those promoting MyPyramid.

Walker (2019, p. 168) says that the government's food pyramid, or its dietary advice—focused on moderation and variety—is "outgunned by a well-funded modern food system bent on producing and selling more." Maybe this is one major reason for our place in the world:

> When compared with sixteen peer countries, the United States is near the top in terms of wealth but near the bottom in terms of health. Per person, we spend more on health care and less on food. We also consume more calories per person than any other country. And we are dead last in chronic maladies like obesity, diabetes, and heart disease, the leading cause of death.
>
> *(p. 169)*

In essence, many of the foods we eat that are high in fat, sugar, and salt have no nutritional value whatsoever. Clearly, the food pyramid is not working!

The first food pyramid, if you will, was actually created in 1974 in Sweden: "a three-tiered triangle with cheap commodities like bread, potatoes, pasta, margarine, and milk at the base; supplementary fruits and vegetables in the middle; and pricier meat products as the apex" (Minger, 2013, p. 17). The base of the pyramid was made up of essential foods, and the center and top were meant to be supplemental or complementary foods that provided vitamins and minerals not found in base-level foods. It was introduced with the message of "A good, healthy diet at a reasonable price." Other countries soon followed suit, including Denmark, Japan, Sri Lanka, and West Germany. The United States would catch up to this image a couple of decades later, as shown earlier.

## 1995 Guidelines

The 1990 National Nutrition Monitoring and Related Research Act then mandated that dietary guidelines be issued, and 1995 was the first year of those guidelines. The 1995 guidelines were not unlike earlier versions, and they recommended eating a variety of foods, balancing food eaten with physical activity, choosing a diet with many grains, vegetables, and fruits, and eating low amounts of fats, saturated fats, cholesterol, sugars, salts, and alcohol. Here, we start to see helpful nutritional advice that runs counter to the wishes of the industry, which is devoted to the sale of products high in fat, sugar, and salt. The similarities and differences in dietary guidance from 1980 to 1995 are shown in Figure 3.4. You will notice a high degree of similarity in dietary advice over the years, but slight differences in wording.

| 1980 | 1985 | 1990 | 1995 |
|---|---|---|---|
| Eat a variety of foods. | Eat a variety of foods. | Eat a variety of foods. | Eat a variety of of foods. |
| Maintain ideal weight | Maintain desirable weight | Maintain healthy weight | Balance the food you eat with physical activity—maintain or improve your weight. |
| Avoid too much fat, saturated fat, and cholesterol. | Avoid too much fat, saturated fat, and cholesterol. | Choose a diet low in fat, saturated fat, and cholesterol | Choose a diet with plenty of grain products, vegetables, and fruits.* |
| Eat foods with adequate starch and fiber. | Eat foods with adequate starch and fiber. | Choose a diet with plenty of vegetables, fruits, and grain products | Choose a diet low in fat, saturated fat, and cholesterol.* |
| Avoid too much sugar. | Avoid too much sugar. | Use sugars only in moderation. | Choose a diet moderate in sugars |
| Avoid too much sodium. | Avoid too much sodium. | Use salt and sodium only in moderation | Choose a diet moderate in salt and sodium. |
| If you drink alcohol, do so in moderation. | If you drink alcoholic beverages, do so in moderation. | If you drink alcoholic beverages, do so in moderation. | If you drink alcoholic beverages, do so in moderation. |

**FIGURE 3.4** Similarities and Differences in Dietary Guidelines From 1980 to 1995

*Source:* Davis, C., & Saltos, E. (2022). Chapter 2. Dietary recommendations and how they have changed over time. Downloaded from: https://www.ers.usda.gov/webdocs/publications/42215/5831_aib750b_1_.pdf

## 2000 Guidelines

The 2000 dietary guidelines directed Americans to "aim for fitness, build a healthy base, and choose sensibly." Figure 3.5 shows this visually, from the guidelines. Under aiming for fitness, the guidelines say to strive for a healthy weight and be physically active each day. Under building a healthy base, the guidelines suggest being guided by the Food Pyramid advice on nutrition, choosing whole grains where possible, and eating a variety of fruits and vegetables. Under choosing sensibly, the guidelines report eating foods low in saturated fats and moderate in total fats, moderating one's intake of sugars and alcohol, and reducing one's intake of salt. Again, this dietary advice runs counter to the goals of the food industry.

## AIM FOR FITNESS...

▲ Aim for a healthy weight.

▲ Be physically active each day.

## BUILD A HEALTHY BASE...

▇ Let the Pyramid guide your food choices.

▇ Choose a variety of grains daily, especially whole grains.

▇ Choose a variety of fruits and vegetables daily.

▇ Keep food safe to eat.

## CHOOSE SENSIBLY...

● Choose a diet that is low in saturated fat and cholesterol and moderate in total fat.

● Choose beverages and foods to moderate your intake of sugars.

● Choose and prepare foods with less salt.

● If you drink alcoholic beverages, do so in moderation.

**FIGURE 3.5** Advice From the 2010 Dietary Guidelines

The guidelines report the suggested number of servings for different groups by age and gender and specify what counts as a serving. The report also gives example foods for some food groups, with examples of whole grains provided and specific food sources of vitamins, as well as illustrates what are healthy oils and dairy products versus unhealthy oils and dairy products. The report shows the amount of saturated fat in some foods and even notes a concern about the amount of added sugars being consumed in soft drinks in the United States. The 2000 guidelines even show how to read a food label.

## 2005 Guidelines

The 2005 dietary guidelines also suggest eating fewer calories, being more active, and making wiser food choices. It urges Americans to choose "nutrient-dense" foods—rich in fiber and high in whole grains—and provides specified amounts of some foods as examples. The guidelines advise consumers to limit saturated and trans fats, cholesterol, added sugars, salts, and alcohol, again, contrary to the wishes of major food corporations. Specifically, the guidelines advise getting less than 10% of calories from saturated fats and no more than 20–25% of calories from fats. The 2005 guidelines urge a "balanced eating pattern," and they promote energy balance in the form of "balance calories from food with calories expended." There is dietary advice for specific populations, including people trying to lose weight, for example.

## 2010 Guidelines

The 2010 guidelines are more of the same, with the advice to consume fewer calories, make informed food choices, and be physically active. They again advise Americans to consume nutrient-dense foods, which it says include vegetables, fruits, whole grains, non- or low-fat milk and dairy foods, seafood, lean meats, eggs, beans and peas, and seeds and nuts. The report also says that "Americans currently consume too much sodium and too many calories from solid fats, added sugars, and refined grains." The 2010 guidelines list specific foods to reduce and others to increase and again make recommendations for specific populations. The report also shows very detailed sources of saturated fats, solid fats, and added sugars in the American diet, and the largest portion of added sugars are shown to be from sodas and energy and sports drinks. This seems like a salvo against big food, so it would be hard to argue that the industry is exerting too much influence over dietary guidelines at this point. Finally, the report shows stunning data with regard to the "heavy toll of diet-related chronic conditions" including cardiovascular disease, hypertension, diabetes, cancer, and even osteoporosis.

## MyPlate

In 2011, the MyPlate concept was introduced to the American people. The image, shown in Figure 3.6, illustrates a plate divided into four roughly equal sections, including proteins, grains, vegetables, and fruits, plus a circle for a glass of dairy (e.g., milk). Though the general division of foods seems relatively healthy—for any meal that is half fruits and vegetables will likely be healthy—there is still no guidance in the figure about which proteins (e.g., red or white meat) to choose or which grains (e.g., whole or refined) to choose. Yet, on the website associated with the program, www.myplate.gov/, one can click on each piece of the plate to read about what is and should be included within each category. There, good advice with regard to these issues is offered.

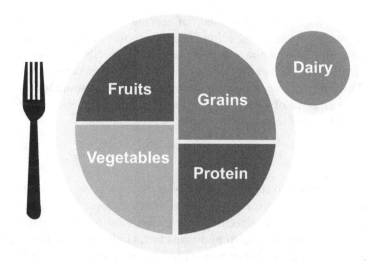

**FIGURE 3.6** MyPlate

Overseen and controlled by the USDA, confusion and failure associated with the food pyramid or My Plate program should not be surprising given that nutrition really is not their job; instead, USDA is responsible for the business of agriculture (Minger, 2013), as noted in Chapter 2. Minger (2013, p. 25) writes that

> Asking the Department of Agriculture to promote healthy eating as like asking Jack Daniels to promote responsible drinking: the advice could only come packaged with a wink, a nudge, and a complementary shot glass. As the appointed guardian for all things agriculture, the USDA wasn't in a position to discourage food sales; yet its anomalous duty to improve America's eating habits called for this very feat.

I agree that MyPyramid and MyPlate are not particularly useful guides for Americans; yet, federal dietary guidelines contain nearly all of the nutritional advice people need to eat healthy foods, should they bother to read them.

Silbergeld (2016, p. 108) characterizes the MyPlate program as "a deliberate maneuver to avoid criticism from both industry and nutritionists" because it "completely obscures the goal of providing a pictographic statement of nutritional advice that includes anything other either quantitative or qualitative." Given this, it is not surprising that it

> has a lot of endorsements from industry, including the Alliance for Potato Research and Education, the beef industry, the frozen and canned food industries, Boston Chicken, the Beef Cattle Institute, Frito-Lay, the National Dairy Council, General Mills, the Food Marketing Institute, and Sodexo—keeping company with the American Medical Association, the Center for Science in the Public Interest, and the American Cancer Society.

Why these professional and medical organizations would support a dietary guidance so thoroughly absent of any detail is surprising.

## 2015–2020 Guidelines

The most recent guidelines, *Dietary Guidelines for Americans, 2015–2020*, held that since "dietary pattern may be more predictive of overall health status and disease risk than individual foods or nutrients . . . dietary patterns, and their food and nutrient components, are at the core of the Dietary Guidelines for Americans, 2020–2025" (Dietary Guidelines for Americans, 2022). A review of the enormous document that makes up these guidelines reveals some pretty simple principles to follow when it comes to diet and nutrition. It offers the following five guidelines to follow, although some are redundant:

1. Follow a healthy eating pattern across the lifespan. All food and beverage choices matter. Choose a healthy eating pattern at an appropriate calorie level to help achieve and maintain a healthy body weight, support nutrient adequacy, and reduce the risk of chronic disease.
2. Focus on variety, nutrient density, and amount. To meet nutrient needs within calorie limits, choose a variety of nutrient-dense foods across and within all food groups in recommended amounts.
3. Limit calories from added sugars and saturated fats and reduce sodium intake. Consume an eating pattern low in added sugars, saturated fats, and sodium. Cut back on foods and beverages higher in these components to amounts that fit within healthy eating patterns.
4. Shift to healthier food and beverage choices. Choose nutrient-dense foods and beverages across and within all food groups in place of less healthy choices. Consider cultural and personal preferences to make these shifts easier to accomplish and maintain.
5. Support healthy eating patterns for all. Everyone has a role in helping to create and support healthy eating patterns in multiple settings nationwide, from home to school to work to communities.
The redundancy comes in terms like nutrient-dense, defined in the guidelines as:
A characteristic of foods and beverages that provide vitamins, minerals, and other substances that contribute to adequate nutrient intakes or may have positive health effects, with little or no solid fats and added sugars, refined starches, and sodium. Ideally, these foods and beverages also are in forms that retain naturally occurring components, such as dietary fiber. All vegetables, fruits, whole grains, seafood, eggs, beans and peas, unsalted nuts and seeds, fat-free and low-fat dairy products, and lean meats and poultry—when prepared with little or no added solid fats, sugars, refined starches, and sodium—are nutrient-dense foods. These foods contribute to meeting food group recommendations within calorie and sodium limits. The term "nutrient dense" indicates the nutrients and other beneficial substances in a food have not been "diluted" by the addition of calories from added solid fats, sugars, or refined starches, or by the solid fats naturally present in the food.

The key recommendations include: "Consume a healthy eating pattern that accounts for all foods and beverages within an appropriate calorie level." And the Guidelines say:
A healthy eating pattern includes:

- A variety of vegetables from all of the subgroups—dark green, red and orange, legumes (beans and peas), starchy, and other

- Fruits, especially whole fruits
- Grains, at least half of which are whole grains
- Fat-free or low-fat dairy, including milk, yogurt, cheese, and/or fortified soy beverages
- A variety of protein foods, including seafood, lean meats and poultry, eggs, legumes (beans and peas), and nuts, seeds, and soy products
- Oils.

Further: "A healthy eating pattern limits:

- Saturated fats and trans fats, added sugars, and sodium."

Figure 3.7 illustrates recommended daily intake of different types of foods. As you can see in the figure, there are specific recommendations for different kinds of vegetables, plus fruits, whole and refined grains, dairy, oils, and different kinds of proteins.

The Guidelines go on to say:

Key Recommendations that are quantitative are provided for several components of the diet that should be limited. These components are of particular public health concern in the United States, and the specified limits can help individuals achieve healthy eating patterns within calorie limits:

- Consume less than 10% of calories per day from added sugars
- Consume less than 10% of calories per day from saturated fats
- Consume less than 2,300 mg per day of sodium
- If alcohol is consumed, it should be consumed in moderation—up to one drink per day for women and up to two drinks per day for men—and only by adults of legal drinking age.

The most recent dietary guidelines again confront major food corporations in the business of selling ultra-processed foods high in fat, sugar, and salt.

The Guidelines are specific with regard to intake of saturated fats, trans fats, dietary cholesterol, and sodium, among other elements such as caffeine and alcohol. They say "intake of saturated fats should be limited to less than 10 percent of calories per day by replacing them with unsaturated fats and while keeping total dietary fats within the age-appropriate AMDR" (Acceptable Macronutrient Distribution Range). The report notes:

Strong and consistent evidence shows that replacing saturated fats with unsaturated fats, especially polyunsaturated fats, is associated with reduced blood levels of total cholesterol and of low-density lipoprotein-cholesterol (LDL-cholesterol). Additionally, strong and consistent evidence shows that replacing saturated fats with polyunsaturated fats is associated with a reduced risk of CVD [cardiovascular disease] events (heart attacks) and CVD-related deaths. Some evidence has shown that replacing saturated fats with plant sources of monounsaturated fats, such as olive oil and nuts, may be associated with a reduced risk of CVD. However, the evidence base for monounsaturated fats is not as strong as the evidence base for replacement with polyunsaturated fats.

| Food Group[a] | Amount[b] in the 2,000-Calorie-Level Pattern |
|---|---|
| **Vegetables** | 2½ c-eq/day |
| Dark Green | 1½ c-eq/wk |
| Red and Orange | 5½ c-eq/wk |
| Legumes (Beans and Peas) | 1½ c-eq/wk |
| Starchy | 5 c-eq/wk |
| Other | 4 c-eq/wk |
| **Fruits** | 2 c-eq/day |
| **Grains** | 6 oz-eq/day |
| Whole Grains | ≥ 3 oz-eq/day |
| Refined Grains | ≤ 3 oz-eq/day |
| **Dairy** | 3 c-eq/day |
| **Protein Foods** | 5½ oz-eq/day |
| Seafood | 8 oz-eq/wk |
| Meats, Poultry, Eggs | 26 oz-eq/wk |
| Nuts, Seeds, Soy Products | 5 oz-eq/wk |
| **Oils** | 27 g/day |
| **Limit on Calories for Other Uses (% of Calories)** | 270 kcal/day (14%) |

[a]Definitions for each food group and subgroup are provided throughout the chapter and are compiled in Appendix 3.

[b]Food group amounts shown in cup-(c) or ounce-(oz) equivalents (eq). Oils are shown in grams (g). Quantity equivalents for each food group are defined in Appendix 3. Amounts will vary for those who need less than 2,000 or more than 2,000 calories per day. See Appendix 3 for all 12 calorie levels of the pattern.

[c]Assumes food choices to meet food group recommendations are in nutrient-dense forms. Calories from added sugars, added refined starches, solid fats, alcohol, and/or to eat more than the recommended amount of nutrient-dense foods are accounted for under this category.

NOTE: The total eating pattern should not exceed *Dietary Guidelines* limits for intake of calories from added sugars and saturated fats and alcohol and should be within the Acceptable Macronutrient Distribution Ranges for calories from protein, carbohydrate, and total fats. Most calorie patterns do not have enough calories available after meeting food group needs to consume 10% of calories from added sugars and 10% of calories from saturated fats and still stay within calorie limits. Values are rounded.

**FIGURE 3.7**  Dietary Advice from the 2015–2020 Guidelines

With regard to trans fats, the Guidelines say: "Individuals should limit intake of trans fats to as low as possible by limiting foods that contain synthetic sources of trans fats, such as partially hydrogenated oils in margarines, and by limiting other solid fats."

When it comes to cholesterol, the Guidelines indicate "individuals should eat as little dietary cholesterol as possible while consuming a healthy eating pattern. In general, foods that are higher in dietary cholesterol, such as fatty meats and high-fat dairy products, are also higher in saturated fats." The guidelines also note that

> Strong evidence from mostly prospective cohort studies but also randomized controlled trials has shown that eating patterns that include lower intake of dietary cholesterol are associated with reduced risk of CVD, and moderate evidence indicates that these eating patterns are associated with reduced risk of obesity.

Finally, with regard to sodium, the Guidelines say the

average sodium intake, which is currently 3,440 mg per day . . . is too high and should be reduced. Healthy eating patterns limit sodium to less than 2,300 mg per day for adults and children ages 14 years and older and to the age- and sex-appropriate Tolerable Upper Intake Levels (UL) of sodium for children younger than 14 years.

It should be noted that the most recent guidelines feature a shift away from simply identifying dietary components and treating them in isolation. Instead, the focus is on dietary patterns, as noted earlier. According to Dietary Guidelines for Americans (2022):

> The current science base shows that components of a dietary pattern can have interactive, synergistic, and potentially cumulative relationships, such that the dietary pattern may be more predictive of overall health status and disease risk than individual foods or nutrients. Thus, dietary patterns, and their food and nutrient components, are at the core of the *Dietary Guidelines for Americans, 2020–2025*.

Further, the new guidelines also take "a lifespan approach focusing on what to eat and drink at different life stages, and confirms the core elements of a healthy eating pattern."

And then finally, the Guidelines also recommend enough exercise:

> In tandem with the recommendations above, Americans of all ages—children, adolescents, adults, and older adults—should meet the Physical Activity Guidelines for Americans to help promote health and reduce the risk of chronic disease. Americans should aim to achieve and maintain a healthy body weight. The relationship between diet and physical activity contributes to calorie balance and managing body weight. As such, the Dietary Guidelines include a Key Recommendation to: Meet the Physical Activity Guidelines for Americans.

Again, this wording is consistent with an allegiance to energy balance, discussed later in the chapter.

Figure 3.8 illustrates how food guidelines have changed over time. The figure demonstrates the approach used to evaluate science, how long advisory reports were, how long guideline reports were, the focus of the guidance, and the number of guidelines produced. Note how the length of the advisory committee scientific reports grew significantly over

**Evolution of the Dietary Guidelines for Americans Process and Products: 1980 to Present**

The following chart is a brief snapshot of how the development and products of the *Dietary Guidelines* review process have evolved from 1980 to present. Over time, the *Dietary Guidelines for Americans* (*Dietary Guidelines*) has grown from a consumer-oriented brochure to a comprehensive resource for policy makers and health professionals with complementary consumer materials developed by the Departments of Agriculture (USDA) and Health and Human Services (HHS).

| Edition | Approach for Reviewing the Evidence | Advisory Committee Scientific Report | Dietary Guidelines Publication | Focus of Guidance | Number of Guidelines and/or Key Recommendations |
|---|---|---|---|---|---|
| 1980 | Review of current science by select scientists from USDA & HHS, along with collective expertise of scientific community | N/A | *Dietary Guidelines* consumer brochure (19 pages) | Healthy Americans (age not specified) | 7 Guidelines |
| 1985 | Creation of Dietary Guidelines Advisory Committee outside the Federal sector; Advisory Committee's search and review of the scientific literature | Scientific Report (19 pages) | *Dietary Guidelines* consumer brochure (23 pages) | Healthy Americans (age not specified) | 7 Guidelines |
| 1990 | Advisory Committee's search and review of the scientific literature | Scientific Report (48 pages) | *Dietary Guidelines* consumer brochure (27 pages) | Healthy Americans, ages 2 years and older | 7 Guidelines |
| 1995 | Advisory Committee's search and review of the scientific literature | Scientific Report (52 pages) | *Dietary Guidelines* consumer brochure (43 pages) | Healthy Americans, ages 2 years and older, to help promote health and prevent disease | 7 Guidelines |
| 2000 | Advisory Committee's search and review of the scientific literature and data analyses | Scientific Report (87 pages) | *Dietary Guidelines* written for consumers, policy officials, and health professionals (40 pages) | Healthy Americans, ages 2 years and older, to help promote health and decrease risk of certain diseases | 10 Guidelines (clustered into 3 themes) |
| 2005 | Advisory Committee's search and review of the scientific literature, data analyses, food pattern modeling analyses, and other scientific reports. | Scientific Report (364 pages) and online appendices | *Dietary Guidelines* written for health professionals and policy makers (84-pages) | Americans, ages 2 years and older, to help promote health and decrease risk of chronic diseases | 41 Key Recommendations (23 for general population, 18 for specific population groups) |
| 2010 | Advisory Committee's systematic review of scientific literature using USDA's Nutrition Evidence Systematic Review (NESR) (formerly the Nutrition Evidence Library), data analyses, food pattern modeling analyses, and other scientific reports. | Scientific Report (453 pages), online appendices and full systematic reviews (available at NESR.usda.gov) | *Dietary Guidelines* written for health professionals and policy makers (108-pages) | Americans ages 2 years and older, including those at risk of chronic diseases, to help promote health and decrease risk of chronic diseases | 29 Key Recommendations (23 for general population, 6 for specific population groups) |
| 2015 | Advisory Committee's systematic review of scientific literature using USDA's NESR, data analyses, food pattern modeling analyses, and existing systematic reviews, meta-analyses, and evidence-based reports | Scientific Report (567 pages), online appendices and full systematic reviews (available at NESR.usda.gov) | *Dietary Guidelines* written for health professionals and policy makers (144-pages) | Americans ages 2 years and older, including those at risk of chronic diseases, to help promote health and decrease risk of chronic diseases | 5 overarching Guidelines with 13 supporting Key Recommendations |
| 2020 | Advisory Committee's systematic review of scientific literature using USDA's NESR, data analyses, and food pattern modeling analyses | Scientific Report (835 pages–first print), online appendices and full systematic reviews (available at NESR.usda.gov) | *Dietary Guidelines* written for health professionals and policymakers | All Americans from birth through older adulthood including those at risk of chronic diseases, to promote health and reduce risk of chronic disease. | 4 overarching Guidelines with supporting Key Recommendations embedded within the Guidelines |

**FIGURE 3.8**  How the Food Guidelines Have Changed Over Time

time, as did the lengths of the actual dietary guideline publications. It should not be surprising that an overwhelming majority of Americans never read these guidelines, given how long they are. This greatly diminishes their potential utility.

## Is This the Advice of Experts?

We might like to think of these guidelines as the opinion of the top dietary experts in the country. And, in a way, they are. Yet, as you will see, they were historically the result of intense efforts by food companies to influence the science.

For example, according to Nestle (2018, p. 115): "Because [the guidelines] influence the rules governing nutrition education, school meals, food assistance, and food labels, every word on them is the target of intense food industry lobbying." As one example, Nestle (2002) shows how opposition from groups such as the National Cattleman's Association (representing beef interests) and the National Milk Producers Federation (representing dairy interests) led to the withdrawal of the 1990 food pyramid. Nestle (2002, p. 81) also shows how the sugar industry used its influence to change the terminology in federal dietary guidance from "*go easy* on beverages and foods high in *added* sugars" and "choose beverages and foods that *limit* your intake of sugars" to "choose beverages and foods to *moderate* your intake of sugars."

The Dietary Guidelines Advisory Committee (DGAC) is appointed to review the published research on food to make recommendations. Then, the agencies write the guidelines; this secondary process began with the George W. Bush Administration, taking the guidelines "out of the hands of scientists and . . . into the hands of agency political appointees." This reality raises at least the possibility of political interference or impact on dietary guidance.

Nestle (2018) shows clear examples of potential conflicts of interest in members of the guideline committees in her book. She shows, in pretty good detail, how Coca-Cola organized against the proposed language to the federal dietary guidelines: "eliminating sugar-sweetened beverages from schools, taxing them, and restricting advertising of foods and beverages with 'high' sodium or sugars for all populations" (p. 4). And she claims that Coca-Cola's government relations team worked hard to make sure such language never appeared in the guidelines, and it did not, including any references to taxes.

Nestle (2013) notes that recommendations from the government about what to eat, and how much, were seen as a "red flag" to the meat and dairy industries since written and even visual depictions of recommendations (such as in the famous "Food Pyramid") recommended smaller and less servings. Minger (2013) illustrates the steps taken by the meat and dairy industry to try to stop the Food Pyramid from discouraging Americans from consuming their products; their efforts successfully stopped the release of the guidelines in 1991. Even the salt industry, such as the Salt Institute, asserted that "there is definitely no need for a dietary goal that calls for the reduction of salt consumption" (Minger, 2013, p. 44). Incredibly, it wrote that "degenerative diseases inevitably accompany old age: and "healthcare expenditures increase if the life span is prolonged." That similar arguments were once made by tobacco companies about the benefits of their customers dying earlier speaks volumes to the morality of big salt when it seems to take a stand that letting people die earlier might be a good thing.

Minger (2013, p. 23) outlines how an original set of dietary advice guidelines developed by experts in nutrition "came back a mangled, lopsided perversion of its former self." She gives the examples of how the proposed guidelines suggested fewer servings of grains, and only whole grains instead of

> ultra-processed wheat and corn products, plus an extra serving of dairy from what was proposed, a slashing of the number of proposed servings of fruits and vegetables which were to form the heart of the guidelines, and, finally, rather than encouraging Americans to cut back widely on sugars the language choosing a diet "moderate in sugar" appeared, with no explanation for what that means.

Nestle (2018) addresses the question of whether experts with ties to corporations who might profit from dietary guidance be allowed to serve on advisory committees. Evidence from the UK Health Forum shows inappropriate influence in deliberations in the countries of Chile, Fiji, Guatemala, Mexico, and more. Such a condition could occur with the National Academy of Sciences' Food and Nutrition Board, which sets forth RDAs; some have alleged they are "in the pocket of the food industry" (Nestle, 2018, p. 107). She provides evidence that the critics may be right about the Board, at least in the past. She writes that the first edition of the dietary guidelines of 1980, "Advised reductions in intake of fat, saturated fat, and cholesterol (meaning, in effect, meat, dairy, and eggs) to reduce the risk of heart disease." Nestle notes: "The Board opposed the guidelines so vehemently that it issued a counter-report, *Toward Healthful Diets*, arguing that fat restrictions were unnecessary for healthy people" (pp. 108–109). Nestle claims that health advocates charged that six members of the Board had ties to industries impacted by the guidelines, so the counter-report was inappropriate: "The furor over the report so embarrasses the academy that it eliminated the industry panel, removed board members with strong ties to food companies, and appointed new members with fewer industry ties" (p. 109).

The guidelines for 2015–2020 are, again, based on the DGAC review, noted earlier. The 2015–2020 *Dietary Guidelines for America* notes this about the DGAC:

> The federal advisory committee, which was composed of prestigious researchers in the fields of nutrition, health, and medicine, conducted a multifaceted, robust process to analyze the available body of scientific evidence. Their work culminated in a scientific report which provided advice and recommendations to the Federal Government on the current state of scientific evidence on nutrition and health.

Federal dietary guidelines are then based on this review.

Dietary Guidelines for Americans (2022) explains a three-stage process used by the DGAC in going about its business. First, it uses the Nutrition Evidence Systematic Review (NESR),

> a state-of-the-art approach to search, evaluate, and synthesize the body of food and nutrition-related science. This rigorous, protocol-driven approach is designed to minimize bias, increase transparency, and ensure relevant, timely, and high-quality systematic reviews to inform Federal nutrition-related policies, programs, and recommendations.

Second, there is food pattern modeling, which includes "modeling of multiple types of diets informed by the science.

The 2020 Advisory Committee continued the use of food pattern modeling, carrying forward these types of eating patterns and exploring eating patterns for toddlers for the first time." Finally, the DGAC uses data analysis

> to help us understand the current dietary intakes and health status of Americans. These data help to ensure that the *Dietary Guidelines* are practical, relevant, and achievable. Since 1995, the Dietary Guidelines Advisory Committees have used data analysis to support recommendations for changes to the *Dietary Guidelines*.

According to Dietary Guidelines for Americans: "Together, these three complementary approaches provide a robust evidence base for the development of dietary guidance." Each of these steps seems to reduce the impact of industry on dietary guidelines. This would explain the guidelines' direct language of reducing fat, saturated fat, salt, and sugar, each of which runs counter to the raison d'être of food companies in the business of profiting off of these substances.

## Problems With the Guidelines

Still, problems with the dietary guidelines remain. According to Herz (2018, p. 45), at least some of what we know from the dietary guidelines is wrong. First released by the Department of HHS in 1980, they said that "Americans should cut down on their intake of saturated fat—such as beef, butter, and eggs—as well as dietary cholesterol. In an effort to lower the risk of cardiovascular disease and curb obesity." She claims that the message to Americans was oversimplified and seemed to suggest avoiding all fats and "now seems to be have been largely wrong." For example, most people do not have to worry about cholesterol because the amount in food does not impact the amount in our blood.

According to Nestle (2018, p. 119), the 2015 DGAC "omitted advice to limit cholesterol to three hundred milligrams per day as a means of reducing heart-disease risk." It said it "will not bring forward this recommendation because available evidence shows no appreciable relationship between consumption of dietary cholesterol and serum cholesterol. . . . Cholesterol is not a nutrient of concern for overconsumption." This, obviously, is a benefit to the egg industry, since eggs are a major source of dietary cholesterol for consumers.

Nestle (2018, p. 119) writes that the DGAC report cited two review articles funded by independent researchers: "One found insufficient evidence to decide whether lowering dietary cholesterol reduces blood cholesterol. The second . . . found no association of eggs with an increased risk of type 2 diabetes and of heart disease in patients with diabetes." She asserts:

> "Insufficient evidence" is not surprising. Studies on eggs and cholesterol are particularly difficult to interpret because saturated fat raises blood cholesterol levels more than does dietary cholesterol, blood cholesterol levels in the general population are already so high that adding an egg or two makes little difference, and so many people take statins that the effects of dietary cholesterol are blunted.

So, dropping the cholesterol advice from dietary guidelines raised the appearance at least of the possibility that DGAC members' ties to the egg industry might have influenced the decision. The Physicians Committee for Responsible Medicine sued, arguing that the guidelines violated the Federal Advisory Committee Act that prohibits influence by special interests on committee members: "They charged that the egg industry, through the Egg Nutrition Center (the education and research arm of the USDA-sponsored American Egg Board) deliberately organized research to cast doubt on a linkage between eggs and high blood cholesterol levels" (Nestle, 2018, p. 120). Nestle's preliminary review in her book suggests a very real possibility that this is true.

Incredibly, though the 2015 guidelines do not advise one to limit consumption of dietary cholesterol to 300 mg a day, they do say "individuals should eat as little dietary cholesterol as possible while consuming a healthy eating pattern. . . . Eating patterns that include lower intake of dietary cholesterol are associated with reduced risk of CVD [cardiovascular disease]." If that is not confusing to you, nothing else should be. On the one hand, the evidence is inconclusive about dietary cholesterol, and on the other hand, you should eat as little as possible. Such advice makes nutritional guidance on cholesterol pretty worthless.

Herz (2018, p. 46) also claims that the "data seem to have been inaccurate about saturated fats." She cites a 2006 Women's Health Initiative study of tens of thousands of women that found eating a low-fat diet did not reduce their risk of heart disease, stroke, or colorectal cancer. Further, not eating enough fat seems to disrupt "the natural flora and fauna in our bodies and cause inflammation, and may lead to physiological mis-signaling that causes us to eat too much."

Meanwhile, it is clear that unsaturated fatty acids including those found in nuts and seeds, some plant oils, as well as fatty fish are good for us. Diets that emphasize such foods are "linked to a lower risk of heart attack, dementia, stroke, diabetes, obesity, and death from cardiovascular disease" (Herz, 2018, p. 47). It is "unnatural fats" such as trans fats that are problematic: "Trans fats are created when cis fatty acid molecules—the versions found in nature—are reconfigured by the manufacturing process of partial hydrogenation" (p. 47). Many ultra-processed foods contain these fats, and the problem is "trans fats are unhealthy and directly associated with a host of ills that hasten death" (p. 47).

Herz (2018, p. 42) explains:

The human body can synthesize most of the fats it needs from carbohydrates, but two essential fatty acids, linolenic and linoleic acid, cannot be manufactured by the body. They must be directly consumed to produce omega-3 and omega-6. These two fatty acids are critical for the normal functioning of all out physiological systems, from physical growth to live and immune function to the condition of the skin.

Meanwhile, saturated fatty acids (SFAs) are unhealthy. Nestle (2018, p. 64) argues that the highest sources of SFAs are found in meat and dairy foods. Studies show that consumption of SFAs is associated with higher blood cholesterol levels. Of course, meat industry-funded studies find that eating red meat has no effect on blood cholesterol levels and that reducing SFAs is associated with reductions in protein intake. Meanwhile, the American Heart Association (AHA) says there are undeniable health benefits with SFAs replaced by unsaturated fats.

The most significant problem I see with federal guidelines is that they do not discuss processed and ultra-processed foods. Further, they do not advise Americans to avoid such foods, especially the latter, which have no nutritional value at all and are linked to clear deleterious health outcomes such as obesity, diabetes, illness, and even death. Perhaps this is a bridge too far for the government, who still has a vested interest in protecting the integrity of the USDA in its efforts to sell more food products to the American people, regardless of what type of food it is or what is in it.

Simon (2006, p. 144) uses the Dietary Guidelines for Americans to show "how much more closely aligned the federal government is with industry than with the general public." Nowhere in the guidelines are Americans told to avoid ultra-processed foods because this would be a threat to the nearly trillion dollar business behind them. As major sources of trans fats, you would think the government would want to tell us to avoid them, but "Trans fat was left vague because otherwise they would have to say where trans fats *are*—in processed foods" (p. 145). As long as the government is concerned with the welfare of the big business, say, more than that of the health of Americans,

it cannot say: don't eat too many of the major sources of saturated fats: meat, cheese, milk, and eggs. Nor could Uncle Sam tell us to avoid the main sources of trans fats: fried and baked goods such as chips, cakes, and cookies. That would ruffle too many industry feathers.

*(p. 145)*

"Regulatory capture" refers to the situation "where regulatory agencies are working for special interests rather than for the public interest." Simon (2006, p. 57) provides the following examples: "the US sugar industry opposed any mention of limiting sugar intake in the dietary guidelines and lobbyists worked hard to insert the words 'choose a diet moderate in sugars; rather than reducing sugar intake'"; "the meat, sugar, soda, and processed food industries have successfully avoided the message of 'eat less meat, soda, and processed foods.' replacing it with the more euphemistic 'reduce saturated fat, added sugar, and sodium.'" Kimura (2019) provides the example of the dairy industry. It worked with scientists for decades to convince us that milk was a nutritionally perfect food and thus should be recommended as part of our daily diet. To the degree regulations serve corporate interests rather than public interests, regulation will not work to assure healthy outcomes for consumers.

## Comparing a Healthy Diet to What Americans Eat

Americans in general do not eat well. Specifically, the typical American diet is too low in minerals, vitamins, whole grains, fruits, and vegetables, as well as too high in fat, saturated fat, cholesterol, processed grains, sugar, and salt. This statement is based on an assessment of what Americans actually eat, compared with the measure of the Daily Value (DV), which is a combination of the Daily Reference Value (DRV) and Reference Daily Intake (RDI) (U.S. Food and Drug Administration, 2018).

Table 3.1 shows the DV for adults and children over the age of 4 years, based on a caloric intake of 2,000 calories. Actual caloric needs for individuals vary by size (i.e., weight) as well as activity level, but this table gives an idea of what the average person needs.

When the DV is compared to what Americans actually eat, the disparities are startling. According to the Office of Disease Prevention and Health Promotion (ODPHP)—created in 1976 as part of the US Department of HHS—Americans eat too little of what they are supposed to eat and too much of what they are not supposed to eat. Specifically:

- About three-fourths of the population has an eating pattern that is low in vegetables, fruits, dairy, and oils.
- More than half of the population is meeting or exceeding total grain and total protein foods recommendations, but not meeting the recommendations for the subgroups within each of these food groups.
- Most Americans exceed the recommendations for added sugars, saturated fats, and sodium.
- The eating patterns of many are too high in calories.
- The American diet falls short in potassium, fiber, choline, magnesium, calcium, and vitamins A, C, D, and E.

*(Office of Disease Prevention and Health Promotion, 2017)*

Figure 3.9 shows the percentage of the US population age 1 year and older who are at, below, or above each dietary goal or limit.

Data from ODPHP show that Americans eat too few vegetables (including dark green vegetables, red and orange vegetables, legumes, starchy vegetables, and other vegetables), too few fruits (but too much sugar from fruit juices), not enough whole grains, not enough dairy, and not enough "nutrient-dense foods"—foods that do not contain "added calories

**TABLE 3.1** Recommended Daily Values (DV) for Adults and Children 4 Years and Older

| Food Component | DV |
| --- | --- |
| Total fat | 65 g |
| Saturated fat | 20 g |
| Cholesterol | 300 mg |
| Sodium | 2,400 mg |
| Potassium | 3,500 mg |
| Total carbohydrate | 300 g |
| Dietary fiber | 25 g |
| Protein | 50 g |
| Vitamin A | 5,000 International Units (IU) |
| Vitamin C | 60 mg |
| Calcium | 1,000 mg |
| Iron | 18 mg |
| Vitamin D | 400 IU |
| Vitamin E | 30 IU |
| Vitamin K | 80 µg |
| Thiamin | 1.5 mg |
| Riboflavin | 1.7 mg |
| Niacin | 20 mg |
| Vitamin B6 | 2 mg |
| Folate | 400 µg |
| Vitamin B12 | 6 µg |
| Biotin | 300 µg |
| Pantothenic acid | 10 mg |
| Phosphorus | 1,000 mg |
| Iodine | 150 µg |
| Magnesium | 400 mg |
| Zinc | 15 mg |
| Selenium | 70 µg |
| Copper | 2 mg |
| Manganese | 2 mg |
| Chromium | 120 µg |
| Molybdenum | 75 µg |
| Chloride | 3,400 mg |

from components such as added sugars, added refined starches, solid fats, or a combination." Americans also eat too many refined grains. Figure 3.10 illustrates how far Americans fall short of eating enough whole grains and that we tend to eat far too many refined grains.

Americans also eat a lot of meat. This is problematic, in part, because of the types of meat we typically eat. For example, Minger (2013, p. 182) points out: "Muscle meat is notoriously high in the amino acid *methionine* . . . whereas the oft-neglected parts of an animal—most notably skin, bones, tendons, and connective tissue—are rich in the amino acid *glycine*." She notes that

methionine isn't an inherent dietary villain, it can stir up trouble if it's not balanced out by a proper array of other nutrients. For one, a high methionine intake increases your

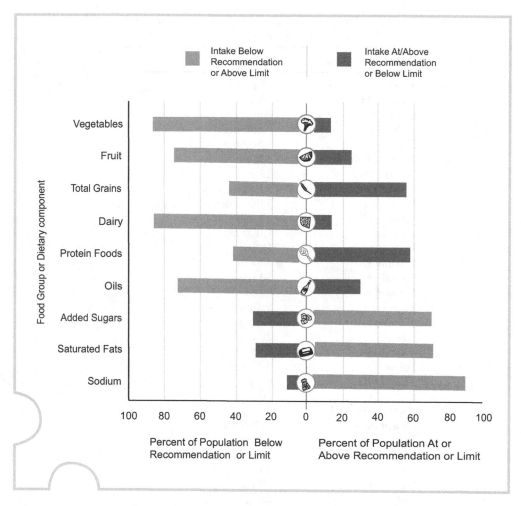

**FIGURE 3.9** Dietary Intakes Compared to Recommendations. Percent of the US Population Ages 1 Year and Older Who Are Below, at, or Above Each Dietary Goal or Limit

*Source:* Office of Disease Prevention and Health Promotion (2015). (https://health.gov/dietaryguidelines/2015/guidelines/chapter-2/current-eating-patterns-in-the-united-states/)

need for vitamin B-12, vitamin B-6, folate, choline, and betaine, which help neutralize *homocysteine*, one of the methionine's most noxious byproducts.

Further, muscle meat "is a sad lightweight as far as vitamins and minerals go. While it may be rich in protein and iron, muscle tissue misses out on a host of other goodies like copper, vitamin A, potassium, magnesium, pantothenic acid, and more" (Minger, 2013, p. 183). It would appear, therefore, that we are not eating the right part of animals in the first place.

In the second place, consuming too much meat may lead to health problems for consumers (but many studies reviewed by Minger call this conclusion into question). Cooking meat at high temperatures can also produce bad outcomes. Minger (2013, p. 184) notes that

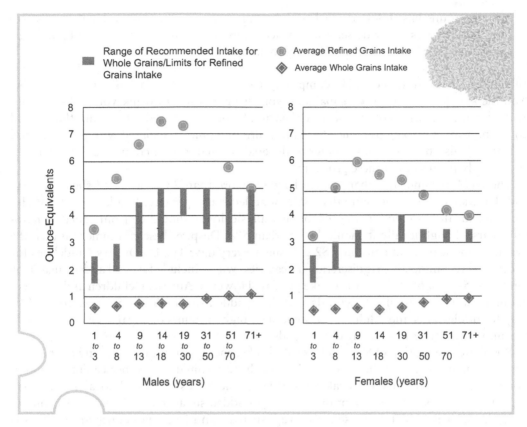

**FIGURE 3.10**    Average Whole and Refined Grain Intake in Ounce-Equivalents Per Day by Age-Sex Groups, Compared to Ranges of Recommended Daily Intake for Whole Grains and Limits for Refined Grains

*Source:* Office of Disease Prevention and Health Promotion (2015). https://health.gov/dietaryguidelines/2015/guidelines/chapter-2/a-closer-look-at-current-intakes-and-recommended-shifts/

"every time you cook meat at very high temperatures, especially exposed to direct flame, you invite carcinogenic properties into your meal." She notes that high-temperature cooking generates heterocyclic amines (Has), polycyclic aromatic hydrocarbons (PAHs), and "Despite making food pretty delicious, Has and PAHs can result in DNA mutations after being 'bioactivated' by certain enzymes in your body—creating a setting ripe for cancer formation."

With regard to sugars, ODPHP (2017) writes: "Added sugars account on average for almost 270 calories, or more than 13 percent of calories per day in the U.S. population." The problem of elevated sugars is particularly bad among children, adolescents, and young adults. According to ODPHP (2017):

The major source of added sugars in typical U.S. diets is beverages, which include soft drinks, fruit drinks, sweetened coffee and tear, energy drinks, alcoholic beverages, and flavored waters. The other major source of added sugars is snacks and sweets, which includes grains-based desserts such as cakes, pies, cookies, brownies, doughnuts, sweet rolls, and pastries; dairy desserts such as ice cream, other frozen desserts, and puddings; candies; sugars; jams; syrups; and sweet toppings.

All of these are foods that could be completely avoided, given their lack of nutritional value. And the simplest way to lose or maintain your weight is not to drink your calories. But, at the least, Americans could moderate their intake of these "food" items and that would go a long way in helping to reduce harmful health outcomes associated with poor diet and nutrition. This obviously runs counter to the goals of major food companies, which profit enormously from sales of these products.

The AHA recommends that women have no more than 25 g of added sugar per day and that men consume no more than 38 g of added sugar per day. In addition, the WHO recommends that no more than 10% of an adult's calories should be from added or natural sugars, though ideally it should be less than 5%. Despite these recommendations, the average American adult consumes 82 g of sugar every day. The US Dietary Guidelines for Americans recommends that discretionary calories, which include fats and added sugars, be limited to 5–15% of total caloric intake per day. However, American children and teens get about 16% of their total calories a day just from added sugars (Sugar Science, 2022). It is easy for adolescents and adults to consume such high amounts of sugar when you look at the amount of sugar in so many of our foods.

Regular consumption of SSBs such as soft drinks, energy drinks, iced tea, and fruit drinks has been increasing over the past few decades. In the United States, per capita consumption of SSBs rose from 64.4 kilocalories per day in the 1970s to 141.7 kilocalories per day in 2006, and SSBs are now the primary source of added sugar in the US diet. SSBs contain sweeteners, such as high fructose corn syrup, sucrose, and fruit juice concentrates. This is problematic because it is believed that these added sugars lead to weight gain. High consumption of SSBs is also associated with the development of type 2 diabetes. Obviously, corporations produce, market, and sell these unhealthy products. And government agencies allow it, in spite of their addictive and harmful nature. There is thus clear culpability in these actions; the issue of culpability is revisited in Chapter 9.

The same is true for other sugary products. For example, cereals, especially children's cereals, also have large sugar contents. Children's cereal contains about 40% more sugar per serving than adult cereals. The serving size listed is unrealistically small for most people; FDA scientists have calculated that the average person consumes 30% more than the serving size listed on the box, so sugar intake ends up being higher than what the labels indicate (Environmental Working Group, 2014).

Many of the foods in America have high sugar contents, which can partially explain unhealthy eating habits, but healthier foods are also more expensive, which can also make it harder for people to eat healthy foods. A meta-analysis of 27 studies determined that in the United States, a diet of healthy foods costs an average of $1.49/day more than unhealthy foods (Rao et al., 2013). This is partly because of the foods most subsidized by taxpayers—including meat, cheese, corn, and grains that are heavily processed and refined.

Figure 3.11 shows the average intake of added sugars in the typical American diet, in contrast with the recommended maximum limit. And Figure 3.12 illustrates that nearly half of all the added sugars we eat come from beverages.

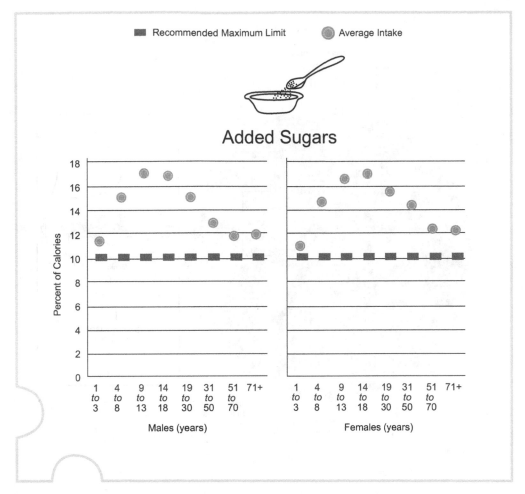

**FIGURE 3.11** Average Intake of Added Sugars Compared to the Recommended Maximum Limits

*Source:* Office of Disease Prevention and Health Promotion (2015). https://health.gov/dietaryguidelines/2015/guidelines/chapter-2/a-closer-look-at-current-intakes-and-recommended-shifts/

Not only do Americans consume too much sugar, but they also consume too much sodium. This is depicted in Figure 3.13.

According to Pew Research Center (2016), Americans in 2010 consumed far more than in 1970. Pew reports "we eat a lot more than we used to: The average American consumed 2,481 calories a day in 2010, about 23% more than in 1970." This includes "a lot more corn-derived sweeteners" in 2010 than in 1970, such as high fructose corn syrup. It also includes more chicken, as chicken consumption has "more than doubled since 1970" while consumption of "beef has fallen by more than a third." We also ate 29% more grains in 2010 than in 1970, but these were "mostly in the form of breads, pastries and other baked goods." Data reported by the Pew Research Center (2016) illustrate that, from 1970 to 2010, Americans consumed far more grain products, fats and oils, and sugars/sweeteners, as well as slightly more legumes, nuts, soy, and fruits/fruit juices. This is depicted in Figure 3.14.

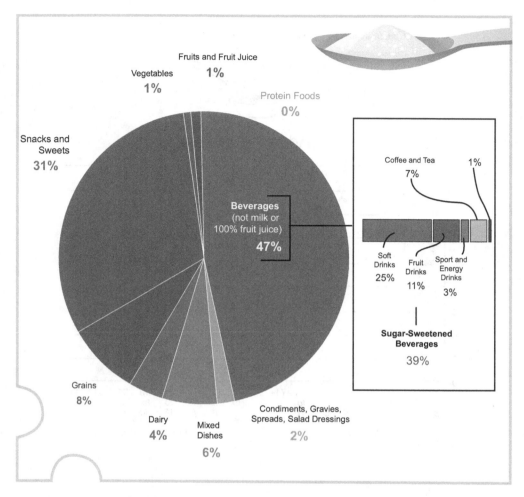

**FIGURE 3.12**  Sources of Added Sugars in the Typical American Diet

*Source:* Office of Disease Prevention and Health Promotion (2015). https://health.gov/dietaryguidelines/2015/
guidelines/chapter-2/a-closer-look-at-current-intakes-and-recommended-shifts/

Figure 3.15 shows increases and decreases in various foods consumed by Americans from 1970 to 2010. Perhaps it is not surprising, given our dietary patterns, that 54% of Americans say they believe their eating patterns were less healthy in 2010 than in 1970. Further, 58% said that "most days they probably should be eating healthier" (Pew Research Center, 2016).

A majority of calories consumed in 2010 came from just two food groups: flours and grains (582 calories, or 23.4%) and fats and oils (575 calories, or 23.2%). These foods can be healthy or unhealthy, depending on the source of the flours and grains and the fats and oils. Unfortunately, Americans tend to choose unhealthy sources, such as refined grains rather than whole grains, and saturated fats rather than unsaturated fats. Consumption of saturated fats is illustrated in Figure 3.16.

According to ODPHP (2017): "Current average intake of saturated fats are 11 percent of calories. Only 29 percent of individuals in the United States consume amounts of saturated

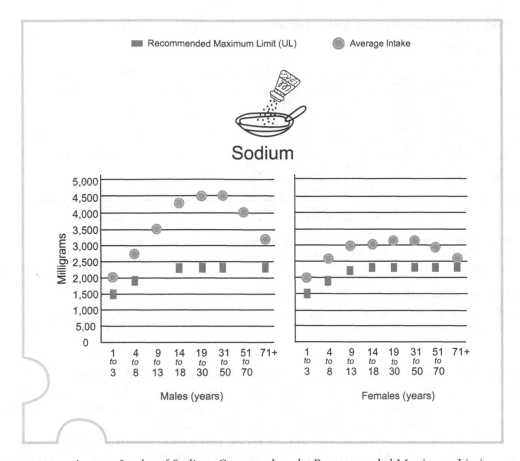

**FIGURE 3.13**    Average Intake of Sodium Compared to the Recommended Maximum Limits

*Source:* Office of Disease Prevention and Health Promotion (2015). https://health.gov/dietaryguidelines/2015/guidelines/chapter-2/a-closer-look-at-current-intakes-and-recommended-shifts/

fats consistent with the limit of less than 10 percent of calories." In addition to not eating too much saturated fat, Americans also eat too much solid fats—"the fats in meats, poultry, dairy products, hydrogenated vegetable oils, and some topical oils" (ODPHP, 2017). The average consumption of solid fats accounts for 325 calories per day, or greater than 16% of the average caloric intake for Americans, even while providing few to no actual nutrients.

Data presented by Fuhrman (2017, p. 178) show the following changes in per-person food consumption from 1900 to 2000:

Sugar—5–170 pounds per year
Soft drinks—0–53 gallons per year
Oils—2–30 pounds per year
Cheese—2–30 pounds per year
Meat—131 to 11 pounds per year
Homegrown produce—131 to 11 pounds per year
Calories—2,100–2,757 per day

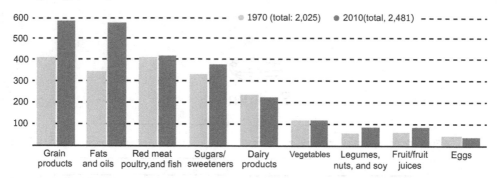

**FIGURE 3.14**  American Dietary Choices From 1970 to 2010

*Source:* Pew Research Center (2016). www.pewresearch.org/fact-tank/2016/12/13/whats-on-your-table-how-americas-diet-has-changed-over-the-decades/

So, you can clearly tell that Americans are generally eating and drinking more over time, and they generally are choosing very unhealthy foods and drinks. Much of this change occurred over the past four decades, as noted earlier, as the food environment changed, where unhealthy food is now ubiquitous and largely unavoidable.

Why is this occurring? The simple answer is that major food companies are producing more and more junk food. Interestingly, even the CED (CED, 2017, p. 6), an organization devoted to sustainable capitalism, admits that only about 40% "of new foods and beverages are formulated with positive nutrition or health attributes." This presumably means that 60% of new foods and beverages are not formulated with positive nutrition or health attributes. CED (2017, p. 35) adds that "over a quarter of new food product introductions were in candy, gum, and snacks . . . and over a fifth were in beverages." Research thus suggests that much of what is celebrated as the benefits of the conventional food system (e.g., new food products) actually turns out to impose significant costs on society (e.g., obesity and diabetes). It is food companies who are culpable for these realities; the issue of culpability is analyzed in Chapter 9.

Research also demonstrates that Americans buy the junk food being produced by large, profitable companies. For example, CED (2017, p. 21) discusses a recent study that used retail scanner data to determine what consumers purchase. It found that more than 75% of the calories that would be consumed at home "came from highly processed (61 percent) and moderately processed (16 percent) foods and drinks." Highly processed were composed of "multi-ingredient industrially formulated mixtures" and included "soda, sausages, ready-to-eat dishes, ice cream, and candy." Moderately processed foods are "directly recognizable as original plant or animal source food, and they include items such as sweetened fruit juices, cheese, and potato chips" (CED, 2017, p. 9). These foods lead directly to weight-related health conditions such as obesity and diabetes, discussed in Chapter 6.

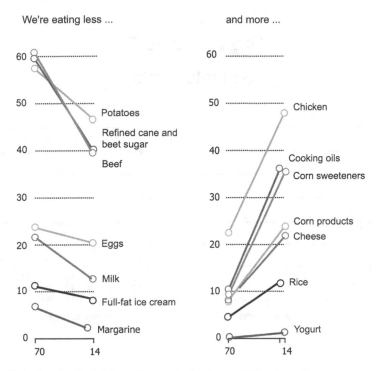

# How the American diet has changed since 1970

*Average annual per capita  availability,   in pounds*

We're eating less ...                    and more ...

Potatoes
Refined cane and beet sugar
Beef

Eggs
Milk
Full-fat ice cream
Margarine

Chicken
Cooking oils
Corn sweeteners
Corn products
Cheese
Rice
Yogurt

Note: Figures adjusted for spoilage  and  other losses.  Milk and yogurt are measured in gallons. Most recent  available year for  "cooking  oils,"  "rice,"  and "margarine"  is 2010.  "Potatoes"  includes  fresh, frozen, dehydrated,  canned, shoestring,  and chips.
Source: USDA Economic Research Service; Pew Research Center analysis

PEW  RESEARCH  CENTER

**FIGURE 3.15**   Changes in American Diets From 1970 to 2010

*Source:* Pew Research Center (2016). www.pewresearch.org/fact-tank/2016/12/13/whats-on-your-table-how-americas-diet-has-changed-over-the-decades/

According to Winson and Choi (2019, p. 40):

Two trends characterize dietary change in our world today: (1) the degrading of the quality of whole foods through the process of simplification, speed-up, and adulteration over the past one hundred years or so; and (2) diffusion of a host of nutrient-poor-food-like products, or what we call pseudofoods, into all manner of food environments.

*(p. 40)*

Indeed, our standard food fare has been described as "a jamboree of processed snacks, high fructose corn syrup, multivarious forms of corn and wheat, lab-concocted flavorings and chemicals, and other sketchy items that only vaguely resemble food" (Minger, 2013,

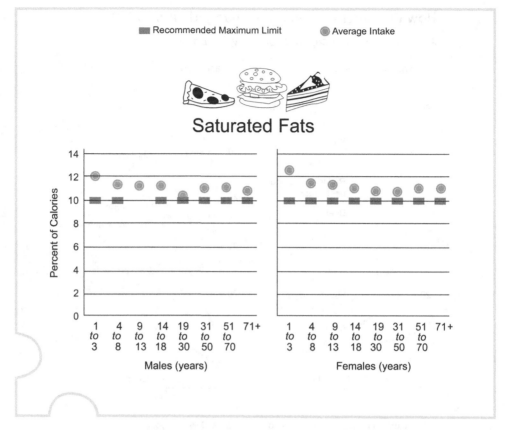

**FIGURE 3.16**   Average Intake of Saturated Fats as a Percent of Calories Per Day Compared With Maximum Recommended Limits

*Source:* Office of Disease Prevention and Health Promotion (2015). https://health.gov/dietaryguidelines/2015/guidelines/chapter-2/a-closer-look-at-current-intakes-and-recommended-shifts/

p. 145). Minger (2013, p. 175) claims that Americans are, in essence, "federally coaxed toward diets higher in PUFAs (polyunsaturated fatty acids) than the human body has ever experienced before in history." This is problematic due to the serious deleterious health outcomes associated with consumption, yet, in 2009, the AHA called PUFAs healthy, safe, and beneficial. According to Minger (2013, p. 178), the lead author of the AHA paper "received significant funding from the bioengineering giant Monsanto, in addition to serving as a consultant for them." Additionally, "two additional authors of the AHA paper—plus all three of its reviewers—had received grants, been given other financial support, or served on advisory boards for places with a vested interest in PUFAs being healthy." This is a reminder of the issue of potential conflicts of interest in dietary advice.

### Physical Activity Matters, Too

With all this focus on diet and nutrition, it also cannot be ignored that sedentary living poses a significant threat to one's well-being. This was made clear by the review of the dietary guidelines over the years, shown earlier. Minger (2013, p. 137) explains that not moving enough (i.e., not exercising enough) "induces rapid changes in how the body handles the food we eat, compromising our insulin sensitivity and glucose tolerance—both factors in conditions ranging from heart disease to diabetes, and influencing our ability to metabolize saturated fat and sugar."

Not only are Americans not eating particularly well, but they are also not physically active enough. According to ODPHP (2017): "Only 20 percent of adults meet the Physical Activity Guidelines for aerobic and muscle-strengthening activity." ODPHP (2017) reports that males are more likely than females to do regular physical activity (24% of males vs. 17% of females meet recommendations). Still, a large majority of Americans do not engage in enough exercise. Almost "30 percent of adults report engaging in no leisure time physical activity" at all. And "individuals with lower income and those with lower educational attainment have lower rates of physical activity and are more likely to not engage in leisure time physical activity."

The amount of exercise engaged in by Americans has declined over the most recent decades. ODPHP (2017) asserts that this "can be attributed to less active occupations; reduced physical activity for commuting to work, school, or for errands; and increased sedentary behavior often associated with television viewing and other forms of screen time." Again, these are realities heavily influenced by corporations and the state.

## Energy Balance

All of this raises the issue of energy balance. Instead of systemic change to our food environment, which Simon notes, "represents a genuine threat to the corporate bottom line," companies promote education about healthy eating and exercise, such as is the case in the "Energy Balance" campaigns of major corporations. *Energy balance* is a concept promoted by multinational food corporations with the specific intent of diverting attention from any moral and legal culpability they hold for the global epidemics of obesity and diabetes that stem largely from the heavily processed foods they produce, advertise, and sell. The Global Energy Balance Network (GEBN), funded by companies such as Coca-Cola, placed blame for deleterious medical outcomes associated with eating on individuals and asserted it is their own responsibility to burn more calories than they consume.

The concept of energy balance has been endorsed and promoted by large, multinational food corporations such as Coca-Cola, PepsiCo, General Mills, and more, in an effort to divert attention away from the harmful nature of their products. Both SSBs and "pseudofoods" such as ultra-processed food products that are high in salt, sugar, and fat are produced, marketed, and sold to at least partially uninformed consumers in spite of the fact that they have no or nearly no nutritional value (Ross, 2013).

A focus on energy balance helps divert attention from the unhealthy foods produced, marketed, and sold to us by large multinational food companies. Kimura (2019, p. 63) gives us an example with the soda industry which "funded research to respected scientists whose research tended to emphasize the importance of exercise as opposed to eating less and drinking less of their products." When it comes to Coke, Ross (2013, p. 111) notes that the company promotes "the efficient use of energy" and a program it calls "active healthy living" as a means to shift focus from the unhealthy nature of its products. Lest you think Coca-Cola is a moral company with the welfare of its consumers at the top of its priorities, recall that the company is a profit-seeking, amoral company (like virtually any other large business) (Robinson & Murphy, 2008), and it uses its "Coca-Cola Retailing Research Council" to "plumb() the social science of shopping to identify the ways in which both teens and adults can be made more vulnerable to persuasion" so that they will purchase their least healthy products—high calorie, high sugar sodas (Ross, 2013, p. 112).

According to Nestle (2018, p. 91), Coke got heavily involved in health research starting in 2004 when it established the "Beverage Institute for Health and Wellness" to raise awareness of "active, healthy lifestyles" and of "beverages as effective delivery systems for hydration." An analysis of 389 articles published in 169 journals that had financial ties to Coca-Cola found that the studies "typically concluded that physical activity is more effective than diet in weight control, sugars and soft drinks are harmless, evidence to the contrary is wrong, and industry-funded research is superior to that funded by other sources" (Nestle, 2018, p. 91).

Coca-Cola also helped fund the "GEBN," whose primary message is a "lack of physical activity is responsible for obesity—not diet, and certainly not soft drinks" (Nestle, 2018, p. 92). An examination by Nestle (2018, p. 98) into GEBN found that Coca-Cola was "actively involved in every aspect of the organization, from conception to recruitment of members to dissemination of research results." She writes that "Coca-Cola executives worked closely with GEBN scientists to influence the direction of the research, hide its funding sources, and promote the energy-balance strategy professionals and the media" (p. 98). Not surprisingly, its studies tended to conclude that the most important correlates of obesity in children were things such as low levels of physical activity, not getting enough sleep, and watching too much television, rather than excessive consumption of sugary sodas.

Meanwhile, researchers affiliated with Coca-Cola's GEBN wrote a study questioning data from the "National Health and Nutrition Examination Survey" that shows weight gain is associated with higher levels of consumption of sugary drinks. According to Nestle (2018, p. 179), they claimed the data were "physiologically implausible and should be ignored." Incredibly, such studies would be summarized in the mainstream news media without any reference to the fact that they were funded by Coca-Cola, often because such funding information was kept secret from the news.

There is some truth to the energy balance model. Unsurprisingly, the energy balance model has been legitimized by much of medical science, and even the WHO claims that the cause of obesity "is an energy imbalance between calories consumed and calories expended" (Taubes, 2021). So, it is true that exercise or physical movement matters when it comes to weight management and/or weight loss, as well as some illnesses such as obesity, diabetes, and hypertension. The issue, however, is that food intake is also undeniably important, an issue intentionally ignored by large food companies.

Studies also generally show that exercise reduces the risks of many illnesses, including those related to weight (Anderson & Durstine, 2019). For example, Warburton et al. (2006) state: "We confirm that there is irrefutable evidence of the effectiveness of regular physical activity in the primary and secondary prevention of several chronic diseases (e.g., cardiovascular disease, diabetes, cancer, hypertension, obesity, depression, and osteoporosis) and premature death." Indeed, studies find evidence that exercise may help prevent as many as 35 chronic conditions; these include accelerated biological aging/premature death, low cardiorespiratory fitness (VO2max), sarcopenia, metabolic syndrome, obesity, insulin resistance, prediabetes, type 2 diabetes, non-alcoholic fatty liver disease, coronary heart disease, peripheral artery disease, hypertension, stroke, congestive heart failure, endothelial dysfunction, arterial dyslipidemia, hemostasis, deep vein thrombosis, cognitive dysfunction, depression and anxiety, osteoporosis, osteoarthritis, balance, bone fracture/falls, rheumatoid arthritis, colon cancer, breast cancer, endometrial cancer, gestational diabetes, preeclampsia, polycystic ovary syndrome, erectile dysfunction, pain, diverticulitis, constipation, and gallbladder diseases (Booth et al., 2012).

Yet, serious scholars have come to doubt the theory of energy balance and even criticize it due to major problems and "serious inconsistencies" (Arencibia-Abite, 2020). Sharma (2014), for example, calls it the "Folk Theory of Obesity" and writes: "The notion is fundamentally flawed, for one simple reason: it assumes that weight is the 'dependent' variable in the [energy-in, energy out] equation." Instead, Sharma demonstrates, "it is as much (if not more) body weight itself that determines energy intake and output as vice versa." Specifically, heavier people tend to eat more "because they have a stronger drive to eat and/or need more calories to function." Taubes (2021) agrees, showing that being overweight often produces overeating rather than the other way around. So, one problem with energy balance is that body weight might be the independent variable.

A second problem with the energy balance model is described by Taubes (2021); he calls the energy balance model "fatally, tragically flawed" and states instead that the real problem is a physiological one in the body, or "a hormonal or constitutional disorder, a dysregulation of fat storage and metabolism, a disorder of fuel-partitioning." Sharma (2014) agrees, showing that various physiological processes occurring in the body help determine weight gain, "including leptin resistance, impaired secretion of incretins . . . insulin resistance, alterations in the hypothalamic-pituitary-adrenal (HPA axis), and sympathetic activity." Thus, problems of weight seem to stem at least in part from biological problems within individuals.

In essence, some people, by nature of their genetic makeup, are predisposed to gain and keep on weight. Kang (2012) shares recent studies have identified several obesity genes. These genes may explain why some individuals have an unhealthy set point, the average weight a person maintains over time. Individuals with a genetic susceptibility to obesity may be predisposed to abnormalities in neural function. In essence, obesity genes influence appetite to increase energy intake or affect metabolism to decrease energy expenditure.

One example is a gene that leads to leptin production, a hormone that signals the body to stop eating. While genetics has not significantly changed since the 1980s—when obesity began rising in the United States and around the world—what did change was the food environment. Specifically, a huge variety of highly processed, carb-based snack foods floods the market (and flood it every year with thousands of new products), providing the types of calories required to gain and keep on weight. So, whereas Taubes describes obesity as a physiological disorder rather than a behavioral disorder—a "disorder of fat accumulation"—the problem behaviors (or culpability) lie with the producers of the foods more so than those who eat them.

A third major problem with the energy balance model is that it treats all calories the same when evidence suggests they are not. For example, Moulesong (2021) shows that eating too many carbohydrates can lead to insulin resistance, resulting in excess fat storage in the cells, and ultimately, weight gain. He writes:

> Insulin tells your body to convert excess incoming carbohydrates to fat and store them. Normally, your body would then use that body fat to supply your energy needs throughout the day. But when you become insulin resistance, your body is producing even more insulin than normal.

When there is too much insulin in the blood, "it doesn't allow your body to access its stored fat. You can't burn body fat while too much insulin is present" so "you get stuck storing fat while having no access to it." This suggests that it is not just how much you eat that matters (i.e., energy) but also what you eat that matters. Keep in mind that it is large multi-national food companies that produce, market, and sell a wide variety of high-carb foods, raising the issue of culpability for issues of weight gain and the diet-related health conditions of obesity and diabetes.

Taubes (2021) agrees, showing one of the main problems in weight gain is carbohydrates, in both quantity and quality, that "establish a hormonal milieu that fosters the accumulation of excess fat." He writes:

> Diets that can successfully resolve obesity are not those that induce us to eat less, per energy-balance thinking, but those that reduce circulating levels of insulin, accomplished most effectively by replacing dietary carbohydrates—sugars, starchy vegetables and grains, and the like—with fat.

Arencibia-Albite and Manninen (2021) also agree with the idea that carbs are more problematic when it comes to weight gain. They point out that "macronutrient mass intake is significantly greater under the high carbohydrate (HC) diet than in the high fat (HF) diet" because HF diets generally lead to less mass intake (Arencibia-Abite, 2020). Arencibia-Albite and Manninen (2021) actually posit an alternative to the energy balance model, one they refer to as the "mass balance paradigm." It holds that "body weight fluctuations are dependent on the difference between daily mass intake, in food and beverages, and daily mass excretion (e.g., elimination of macronutrient oxidation products) and not on energy imbalance." The excretion refers to things such as $CO_2$, water, minerals, urea, $SO_3$, fecal matter, and other waste products (Arencibia-Abite, 2020). The validity of this model is beyond the scope of this book, but the point is those authors do not also believe the energy balance model is correct.

So, it is clear that food or caloric intake matters to some degree when it comes to weight management and/or weight loss, as well as the prevention of some illnesses such as obesity, diabetes, and hypertension. Silbergeld (2016, p. 209) agrees, noting: "Consuming more calories, along with lifestyle changes of reduced activity" are likely culprits when it comes to rising rates of obesity. This means that claims by food companies about the importance of movement or physical exercise in the energy balance equation are only partly true.

It is also important to note here that eating less is more effective than moving more at preventing weight gain and reducing illness (Nestle, 2013). Yet, eating less is obviously bad for food corporations, whose business is literally to sell as much food to produce as much profit as possible (Simon, 2006). This may be why they try to focus exclusively on the importance of physical movement rather than the consumption of the foods they produce, market, and sell.

A fourth problem with energy balance is that it posits a rational process whereby people can simply calculate calories consumed versus calories expended, when, in fact, there is little rationale about eating in the first place. While cognitive decision-making stems from the prefrontal cortex at the front of the brain, the majority of decisions in daily life are made beneath the level of consciousness through deeper brain structures. One example is the amygdala, which, Rowland (2017) states, is the structure that directs decisions subconsciously based on past experiences and primitive emotions. This relates to energy balance as we are somewhat powerless to what we consume and how much energy we expend, especially given that nearly all eating behavior is considered automatic and driven by primitive brain structures.

All on to that the fact that eating certain cheap, hyperpalatable foods has been known to cause a spike of dopamine in the brain as they act as a reward. "Dopamine secreting neurons in the brain could conceptually alter physical activity levels either or both by (a) directly stimulating efferent motor pathways and (b) creating motivation for exercise by reinforcing physical behavior through reward (pleasure)" (Rowland, 2017, p. 44). This raises the question of food addiction driving our food behaviors (Robinson, 2022), an issue examined in Chapter 5.

A fifth and final problem with energy balance that tends to be ignored by its proponents is that environmental factors are highly involved in daily energy intake. Our society promotes increased food intake through ready access to food presented in supermarkets, fast-food restaurants, and all-night convenience stores (Kang, 2012, p. 348). The amount and types of unhealthy food available nearly everywhere—even in places where you would not historically expect to find it, such as at the cash registers of even hardware stores—lead to the consumption of enormous amounts of calories and then enormous levels of illness and death. Stated simply, we live in and/or are surrounded by food swamps filled with "foods" that have little to no nutritional value but that are filled with high levels of salt, sugar, and fat (Ross, 2013). Paarlberg (2021, p. 20) notes that

> the excess calories we consume are . . . damaging to our health. Average per capita food calorie consumption in the United States . . . increased 25 percent between 1970 and 2002. Physical activity levels declined at the same time, but this was not the heart of the problem.

Marion Nestle agrees, writing: "Large portions are a sufficient explanation for why people are gaining weight. It's not because of lack of exercise; it's because we're eating more" (Goldberg, 2018, p. 3).

The food companies, however, see the problem differently, basically concluding that "it's time Americans got off their lazy duffs" (Simon, 2006, p. 29). As noted by Simon (2006, p. 29): "Food companies, trade associations, and industry front groups love to portray lack of exercise as the 'true cause' of (and hence the solution to) the obesity epidemic." This effort both deflects blame away from the producers of problematic foods, high in fat, sugar, salt, and calories and changes the subject from food to exercise. An example of an industry group that has embraced the concept of energy balance is the ABA. ABA blames us for being couch potatoes and has even stated "It's about the Couch, Not the Can." This was a statement by ABA in response to efforts to ban sodas in public schools!

According to Simon (2006, p. 31), the term energy balance became "virtually ubiquitous" by 2004 within the food industry. She notes the following examples of how companies have publicly stated support for this notion:

- "We believe this [obesity] is all about energy balance" (Shelly Rosen, McDonald's).
- "Like most experts in the health field, I believe that the ultimate solution to the obesity problem is energy balance" (Susan Finn, American Council for Fitness and Nutrition). Note that the American Council for Fitness and Nutrition is actually a non-profit organization made up of 80 "food and beverage companies, trade associations and nutrition

advocates to work toward comprehensive and achievable solutions to the nation's obesity epidemic," according to its website (SourceWatch, 2022).

- "We believe, as do many nutrition experts, that solving the obesity problem is about maintaining a healthy lifestyle and achieving the proper energy balance."

*(Allison Kretser, Grocery Manufacturers Association)*

Other companies, such as PepsiCo, have funded branded playgrounds for kids, demonstrating the importance of exercise while simultaneously exposing children and their parents to company logos on playground equipment. And McDonald's sent its famous clown ambassador, Ronald McDonald, to schools, in order to promote exercise (Mayer, 2005), as well as started an advertising campaign called "Active Achievers" to promote eating right and staying active (energy balance). McDonald's also began a school-based program delivered to 31,000 schools and seven million children to "motivate children to be more active in unique and fun ways during grade school physical education classes" (Simon, 2006, p. 35). Earlier the company started a "Global Advisory Council on Balanced Lifestyles," announced a "Balanced Lifestyles Platform" (with fitness guru Bob Greene), launched a "Balanced Active Lifestyles" public awareness campaign, and even gave Ronald McDonald a makeover to make him look more fit, active, and athletic.

Not to be outdone, and illustrating the close connections between government and corporate food producers, the US Department of HHS started a "Healthy Lifestyles and Disease Prevention Initiative" aimed at promoting modest daily exercise. Further, the Food and Drug Administration's "Obesity Working Group" recommended to food manufacturers that they put advice on their packaging to say "To manage your weight, balance the calories you eat with your physical activity" (Simon, 2006, p. 151).

In short, energy balance is a tool used by major food corporations to divert attention away from their own culpability when it comes to outcomes such as obesity, diabetes, hypertension, as well as illness and death. So, Simon's (2006, p. 325) definition of food balance seems appropriate:

The oversimplified term that food executives use to explain obesity in a way that sounds objective and scientific, but which conveniently obscures overconsumption of their healthy products. It also has the added benefit of emphasizing weight loss and physical activity, keeping the focus on individual behavioral change rather than corporate food company change.

Why is energy balance so important to food companies? According to Simon:

Food companies know that as long as they keep the nation focused on "education" and "individual choice," the status quo is virtually guaranteed. That is, no changes to the food environment are on the horizon, since virtually no one is demanding it (and those who are, are largely powerless to achieve it).

Simon (2006, p. 21) suggests that food companies are in a quandary:

On the one hand, if they acknowledged rocketing rates of diabetes, heart disease, and other diet-related maladies, they embroil themselves in a public relations nightmare;

talking about their own culpability in creating these problems. On the other hand, plausible denial is extremely risky, given the overwhelming scientific evidence.

Thus, the intermediary approach is to place the blame on individual consumers and insist that only they have the responsibility for the choices they make and the food products they buy and ingest. So, "it's up to each individual to make 'better' choices at supermarkets and restaurants . . . it's every person's duty to make positive lifestyle changes, such as exercising more" (pp. 21–22). And when people have trouble making the necessary lifestyle changes—which, of course, nearly everyone does—the proposed solution of food companies is better education, which comes to them from the companies in the form of corporate PR. This education, according to Simon, "perpetuates overconsumption of their unhealthy products, and leaves big corporations firmly in charge of the information we receive about nutrition and good health" (p. 41).

Even Tommy Thompson, then US Secretary of HHS, said publically at a large, corporate-friendly event in 2004: "We have to do it ourselves" . . . "go out and spread the gospel of personal responsibility." So, at the highest level of the American government, top officials had also bought into the rhetoric of personal responsibility when it comes to food choices. Whether we actually have free will when it comes to what we eat was discussed later in the chapter. Simon (2006) asserts that claims about personal responsibility are really claims in support of no regulation by the government. Imagine if this approach were also taken when it comes to the drug war, thereby eliminating the responsibility of the dealers and only holding the users responsible.

According to Cohen (2014, p. 198), "in the case of obesity, restauranteurs and purveyors of food need to be held responsible for what they serve. . . . If the food industry wasn't selling so much food that makes us sick, we wouldn't be sick." Moss (2013) writes of a secret meeting of corporate food CEOs where he describes "insider admissions of guilt" with regard to the deleterious health outcomes associated with poor diet and nutrition.

Simon (2006, pp. 28–29) summarizes the ways in which major food manufacturers are culpable for bad outcomes associated with their food products. They:

- Make all of the least healthy foods cheap, readily available, and convenient to prepare and eat.
- Concoct mysterious chemical ingredients in laboratories to ensure that the food is exceptionally tasty in ways that are irreproducible by mortal home chefs.
- Market the hell out of these foods, in every form of media available.
- Use highly sophisticated advertising techniques, including subliminal messages, in an effort to make these products even more alluring—so irresistible, in fact, that people would want to consume them frequently and in large quantities, thus guaranteeing addictive tendencies.
- Hijack the scientific process, suppress the truth about good nutrition, and deprive government programs of adequate funding to promote healthy eating.

These efforts are discussed further in Chapter 9 in the context of the culpability of food companies for harms associated with what we eat.

### Why We Eat What We Eat

Though we all like to believe differently, eating well and exercising enough are not as simple as mere choice or willpower. According to CDC, healthy foods are not often (and sometimes rarely) available in schools, childcare centers, and even neighborhoods. The same can be said for opportunities for necessary physical activity or exercise. Other factors in the community that impact nutrition and exercise "include the affordability of healthy food options, peer and social supports, marketing and promotion, and policies that determine how a community is designed" (ODPHP, 2017). Finally, of course, there is the culpability of companies comprising the conventional food system who produce and advertise unhealthy products to consumers.

The problem of poor eating is likely structural in nature. Consider this example:

> Lack of income at the neighborhood level discourages higher-quality food outlets from locating in [poor] neighborhoods; lack of these outlets, in turn, increases the price and availability of better products; and lack of transportation and lack of safety reduce the ability and willingness of residents to travel to better stores in other neighborhoods. Transportation is also a barrier.
>
> *(Silbergeld, 2016, p. 228)*

A higher presence of fast-food companies in poor areas exacerbates the issue. Moss (2021, p. xxi) writes that "it's no secret that fast-food companies and the makers of processed food have had a disproportionate effect on the eating habits of poor communities of color, though the health consequences of bad diets have struck the wealthy, too."

Herz (2018) claims that fast food is disproportionately appealing to people with lower incomes and wealth. Herz (2018, p. 281) explains that "a hallmark of these chains is that they are cheap and disproportionately frequented by people who are financially disadvantaged." This is problematic for the simple reason that eating at fast-food restaurants is associated with higher rates of obesity and other health conditions. Further, Herz claims that the poor cannot afford to add fresh fruits and vegetables to their diets. She claims that "a healthy diet rich in fresh fruits and vegetables costs about $550 more per year than a standard diet high in processed foods and refined carbohydrates" (p. 282). An important point to keep in mind is that the food system is designed to be this way, and it is rooted in capitalism and a strong profit motive over and above all other considerations. Who eats fast food and how much they eat is irrelevant, as long as someone does (and enough people do).

Still, many people "argue that obesity reflects a failure of personal responsibility. They see obesity as a lack of self-control and called for changes in 'lifestyle,' often translated as acceptance of calorie-restricted diets" (Herz, 2018, p. 282). Yes, the dominant view of why people eat too much and/or are obese or overweight says that it is their fault, that they are personally responsible for the free choices they make about what to put in their mouths (and how much to exercise, or not). This view is, at best, presumptuous, and at worst, just wrong.

According to Schrempf-Stirling and Phillips (2019, p. 112), "obesity is the result of many actions by a variety of actors such as governments, businesses, and consumers." And they claim that there are at least three limits on freedom of choice when it comes to what people eat. These include one's level of knowledge (or awareness) of food, the number and quality of alternative foods, "and physical or psychological compulsion" (p. 113). The latter issue of food addiction is discussed in Chapter 5. Schrempf-Stirling and Phillips (2019) also note that consumer choices are impacted by taste, affordability, and convenience, factors that are largely beyond the control of individuals (p. 115).

Walker (2019, pp. 56–57) explains that humans are even hardwired for obesity:

> At the core of America's obesity epidemic is individual DNA hardwired with the imperative to overcome food scarcity. In anticipation of having to live through lean times, whenever possible, we are genetically programmed to pack away excess calories in the form of fat. No matter what others may tell us about health and well-being, our DNA regards this buffer of excess weight as a matter of life and death.

But, the major problem with the claim that conditions such as obesity are the result of individual choice is that it ignores that there is a whole science of food selection that strongly suggests even the foods we choose to eat are influenced by so many factors beyond our control. Consider, first, that our food preferences appear to begin before birth. Herz (2018, p. 82), for example, shows that "becoming familiar with food flavors begins before birth" because fetuses detect aromatic compounds in amniotic fluid from the foods that mothers consume. These aromas help us account in part for cultural differences in preferences for foods, as fetuses are exposed to foods in the womb that become preferences for food after they are born. Such aromas also impact aroma preferences in babies exposed to breast milk, directly impacted by the foods mothers consume. And babies who are breastfed and exposed to a wide variety of foods become less picky eaters than babies who are formula fed.

After birth, of course, the flavor of food matters. Several important points need to be made here. First, it is taste receptors in our gut that tell our brain whether to keep eating or stop, and they also influence the types of foods we like in the first place (Herz, 2018). Second, aroma plays a huge role in food flavor. Studies show aroma matters a great deal in determining foods we like. Herz (2018, pp. 68–69) notes:

> Food aromas gear our body up for eating before our tongues make food contact. First, food aromas are literally mouthwatering. . . . Second, food aromas ignite a cascade of physiological responses, called the cephalic phase response, that prepare the gastrointestinal tract to optimally digest foods, further whetting our appetite before the chow hits our mouth. Finally, aromas, more than any other food cue, seduce us into eating.

Interestingly, obese people may be more able to imagine the aroma of fatty foods, which is associated with greater cravings for those foods. Yet, obese individuals seem to be less able to detect sweet and salt tastes, meaning they may be more likely to add additional sugar and salt to foods to make up for the deficiencies (Herz, 2018). This can increase the likelihood of deleterious health outcomes such as weight gain and hypertension, respectively.

In essence, humans learn to taste with our noses. Herz (2018, p. 79) shows that "aroma . . . is able to trigger a neural response in the taste cortex once it has been associated to a taste through our past experiences." This poses a challenge for food manufacturers who want to offer products with less sugar and salt but still make the products seem sweet and salty to the smell.

Particular aromas are also associated with particular emotions. For example, the scent of vanilla is associated with sweet memories (Herz, 2018). And caramel odors, being associated with sweet taste, produce a "conditioned analgesia" like sweet foods that causes the release of endorphins and can ease pain (p. 81). Importantly, it is past associations with odors that give them their meaning to us.

Science shows that smell is directly wired into parts of our brain that are associated with emotion, memory, and motivation. Herz (2018, p. 74) explains: "When we inhale through our nostrils scent information goes only into the smell center of the brain—the orbitofrontal cortex—which is also where pleasure, emotions, and memories and motivation are processed and why scents are so provocative and evocative." Smell is also linked to the amygdala, hippocampus, and hypothalamus, where motion, motivation, and memory are processed.

Our emotions and tastes for food are intertwined; in fact, there is a whole science of this called *neurogastronomy*, and evidence shows food companies are using this science to market foods to us that link our senses and desires in ways that make us purchase more of their products (Herz, 2018). Keep in mind as you read this book—food companies work tirelessly, behind the scenes, to get us to buy products including a great number of them that they know to be unhealthy. Past experiences with food also clearly play a role in our present relationships with it. Herz (2018, p. 285) says "food is memory and food is emotion, and it is the aroma and flavor of food that most potently ignite . . . wellsprings of and pleasure." Take this example from Herz (2018, p. 116). She writes that "thinking about chocolate or seeing chocolate . . . triggers memories and emotions about chocolate that make you think about chocolate all the more."

Yet, "elaborated intrusions" or cravings caused by cues about food can be interrupted by engaging in complex behaviors or activities, as well as mental strategies like meditation and practicing detachment from food. Mindful eating is also associated with slower and more controlled eating. Herz (2018, p. 261) explains: "Attention plays a major role in our food experiences. Paying more attention makes food flavors richer, and the more intense the flavor, the more satiating the food." Distracted eating is associated with greater consumption, and this is one reason the snack market is so bad for us; we tend to eat them as we do other things, from watching TV to playing with our foods, even to driving. People not only eat more but are also not as good at estimating how much they have eaten when they are distracted, especially when watching TV, when one is not even paying attention to when one is full. So-called smartphones are not very smart when it comes to eating. Studies show that the more time adolescents spend on them, the more likely they are to consume SSBs, the less likely they are to exercise, the worse their sleep, and the more likely they are to be obese (Herz, 2018).

Memory plays a large role in the food choices we make. Moss (2021, p. 57) explains how food manufacturers use this reality to their favor: "The ability of food and of food manufacturers to influence our behavior is, fundamentally, a matter of the information that we absorb, retain, and recall. We remember what we eat and eat what we remember." He continues: "Food resonates so large in our memory because food looms so large in our lives. The act of eating touches everything we experience, everywhere we go, everyone we know, and everything we feel" (p. 59). According to Moss, "when we eat something delicious, we're experiencing not just what we're eating at that moment but also the memory of all the prior experiences" (p. 75).

Incredibly, according to Moss, most of our strongest food memories are of junk foods: "The memory of these processed delights can stay so strong, for so long, that they can hold us back even when we've decided to improve our eating habits" (p. 60). Part of this owes to the fact that memories from adolescence are the most pronounced in people, the easiest to recall. So, who does not remember lots of experiences with different junk foods in adolescence? The "reminiscence bump refers to the ability to recall these memories (Moss, 2021, p. 64). Take our experiences with soda as one example. Moss discusses a Coca-Cola in-house marketer who reported that "people's memories of the soda were as valuable as the recipe for Coke itself" (pp. 66–67). As a result, Coke's logo alone can light up parts of the brain that are associated with desire.

Amazingly, ads do not have to be honest or accurate to impact our memories and thus behaviors. One experiment used a print ad for the fast-food company, Wendy's, and referenced childhood memories of playing on company playgrounds, something the company did not even have (that was McDonald's). Many participants in the study did not catch the false claim, they just had fond memories of eating fast food.

Major food companies use techniques, such as the electroencephalogram, which analyzes the brain's electrical patterns, "to pinpoint the moment when our emotions make us vulnerable to persuasion and branding." They use techniques such as functional magnetic resonance imagery (MRI) "to unlock the secrets of cravings and compulsive behavior, but for the advantage of advertisers, not addiction research" (Moss, 2021, p. 72). I cannot think of any clearer evidence of culpability in food companies than its use of brain mapping technologies to find ways to manipulate us with food.

Next, the color of food matters significantly when it comes to how food tastes (Herz, 2018). Part of this, likely a big part, comes from our past experiences with food. Herz (2018, p. 132) explains that "color increases our liking for various foods, but only when we can see that it correlates with our expectations." She gives the example of people thinking brown M&Ms taste more like chocolate, even though all M&Ms taste the same. Some colors are associated with different taste expectations, like red with sweet and blue with salty. At the same time, red is found to inhibit consumption, logically because we associate it with stopping. Likewise, vision itself is very important to eating. Studies show, for example, that merely seeing high-calorie foods makes bad-tasting foods taste better; this is because it activates the nucleus accumbens in the brain, a hub for the feel-good neurotransmitter dopamine! Since we receive numerous cues from our eyes when we are preparing to and actually eating, not being able to fully see (e.g., in dimly lit restaurants) is associated with diminished flavors. Dim lighting also calms our mood and causes us to stay longer and eat more (Herz, 2018). There is a reason fast-food restaurants are brightly lit. They are trying to get customers in and out as quickly as possible.

Other senses, including hearing, are also involved in our tastes. Certain sounds are associated with sweetness, while others are associated with saltiness: "Our brain uses a cue from one sense. Such as hearing or vision, to inform another sense, such as taste" (Herz, 2018, p. 162). Synesthesia is when one sense crosses over into another. How crispy chips are perceived to be is associated with the sound one hears when opening the packaging. And how crisp and tasty food is rated depends on the degree of noise present. This may be because one of the nerves involved in taste crosses over the ear before going to the brain.

Music also impacts our food choices. Herz (2018) reports the finding, for example, that people buy more French wine with French music playing in the background. Background music also impacts our eating; lighter music is associated with more eating. Ambiance matters, even in fast-food restaurants; they do not want their stores to be too comfortable because people will stay longer and eat more slowly.

The number of foods present impacts how much we eat; the more foods that are present, the more we will eat. This might explain why people tend to eat more from buffets. Herz (2018, p. 228) says it this way: "when a plethora of different foods is at our fingertips we tend to eat well beyond physical or nutritional needs and often don't stop until it's painful to sit up straight."

One reason variety of food is important is a physiological one. Herz (2018, p. 222) writes that

> food pleasure is due to our internal physiology initiating a desire for certain nutrients to fulfill a physical need, and when the nutrient need is met our physiology manifests this fulfillment as a drop in pleasure felt from continuing to eat that food.

Additionally, some sensory properties of food make it turn from appetizing to unappealing. There may also be an evolutionary reason:

> Sensory specific satiety encourages us to switch from a food that has been eaten past the bliss point to a different food, and is believed to be a biologically advantageous response since it produces a desire to seek out different kinds of food, increasing the variety of nutrients we consume.
>
> *(p. 225)*

However, much of the variety we see in foods in our stores is an illusion. Moss (2021) shows that food manufacturers make really small changes in products to be able to offer a wider variety of options to consumers, even when there is very little difference between them. A benefit to food companies that is not advantageous to consumers is, with greater variety, people stay hungry longer. That is, we are not satiated as quickly when we have a variety of foods in front of us. Stated simply, "variety [is] high among the factors that cause[] us to lose control" (p. 116). It "disrupts our ability to put the brake on eating" (p. 117). Perhaps this is why we tend to eat more in buffets.

Even the size of containers that food is served in is found to be associated with how much we eat; larger servings are associated with greater amounts consumed. Larger servings look better to people than smaller sizes, which is why advertising tends to feature larger portions. We also eat faster and more when we are served more food, and since eating interferes with our ability to know when we are full, we will eat more just by continuing to eat (Herz, 2018). It is important to note again that serving sizes have increased over time, and so, the average regular-size plate of food today is about the size of a family serving platter from decades past! And Herz (2018, p. 145) notes that "cookies are now 700 percent too large, pasta dishes 500 percent too generous, muffins 333 percent too big, and steak servings excessive by 224 percent." This is one reason Americans' caloric intake has increased by 500 calories per day since the 1970s. Another possible reason relates to the fact that we are particularly bad at estimating how many calories we eat, especially when we have eaten more. Then, there is biology, as

> high-calorie treats activate the reward pathways that incentivize us to want these foods regardless of how much we say we aren't interested in them. That is, boosted neurological

reward that seeing high-calorie foods induces may inadvertently lead us to overeat them in spite of ourselves.

*(Herz, 2018, p. 149)*

The issue of food addiction is discussed in Chapter 5.

Food also tastes better if its appearance is neat and arranged in appealing ways (Herz, 2018). Then, there are cues in the environment that trigger us to eat. According to Moss (2021, p. 79), a cue such as a billboard is "one of the biggest factors that maintains overeating in America, and how well people pick up these cues totally predicts weight gain. It's the biggest predictor of weight gain that we've stumbled over in a couple of decades of research."

Take all this evidence together and it really questions the existence of free will or control over the foods we ingest. Silbergeld (2016, p. 221) claims that

what people want to eat is not simply the exercise of individual choice. Over the twentieth century, the influence of manufactured desire has played a major role in what people choose to consume, not only in food but also for the entire world of consumption.

That is, companies intentionally aim to create desires for their products, most of which are unhealthy. They are morally and could be legally culpable for the outcomes that result, such as buying and eating unhealthy foods.

Scholars such as Hatanaka (2019, p. 8) assert that "our consumption choices are embedded in a set of institutional arrangements that often channel our food choices in specific directions." She writes: "While many people tend to assume consumer autonomy, in actuality, *many choices are already made for them before they grab something off a supermarket shelf*" (emphasis in original, p. 9). First, we get most, if not all, of our groceries from national or global supermarket chains—basically oligopolies. Second, we can only choose between certain brands, many of which are owned by the same corporations despite having different company names on product labels. Stores are "highly strategic as to what products and how much of the products should be placed in which stores to maximize profits" (p. 9). Third, we are under the influence of heavy marketing, as well as shelf and product placements. So, Hatanaka concludes that our food preferences are socially constructed. Winson and Choi (2019, p. 39) add that "the multiple food environments where we obtain what we eat every day are themselves carefully organized, shaped, and controlled by powerful actors that rarely have your nutritional needs as a priority."

These powerful forces produce and then promote industrial diets, a

mass diet characterized by the consumption of a wide variety of highly processed, nutritionally compromised edible products. . . . They are typically made with highly processed or engineered ingredients like refined white flour, high-fructose corn syrup, or other cheap sweeteners made from corn and are often loaded with salt.

*(Winson and Choi, 2019, p. 40)*

Industrialized foods were not normal until they were normal, meaning that mass advertising worked to make these foods desirable "by delivering on promises of convenience and time saving, by claims of purity, and by suggestions that consumption of certain products would confer social status" (p. 46).

Moss (2021, p. 26) points out that "much processed food is designed to be consumed mindlessly, allowing us to watch TV or play video games or drive a car at the same time." He asserts we are conditioned to behave this way. It is not free will, he asserts, that drives our eating. We choose what we eat, but it is not really up to us. In fact, in the case of some foods, such as sugar and fat, they are very difficult to exert control over, because they exert control over regions of the brain that see free will or self-control disappear as a result. Moss even claims that processed food basically causes us to lose control over our eating by hijacking our biology in ways to keep us seeking it out and consuming it. Again, this issue is revisited in Chapter 5.

Yes, it is possible to eat with purpose and deliberation, "giving some thought to the preparation and consumption of food," and, when we do this, we engage the hippocampus in the brain. "It helps keep the go brain from getting us into trouble." Instead, "when we do things by rote, or by habit, as in eating a candy bar while staring at a computer, this mode of eating shows up in a part of the brain called the *striatum*." In this region of the brain, "we are reacting to inducements, like the sight or smell or memory of candy, without applying the kind of oversight that can put the brakes on a bad decision" (Moss, 2021, p. 63).

So, it appears that outcomes such as obesity are actually produced by many factors, both within and outside of individuals. Within individuals, one example is the human brain. A study of children found brain differences between normal-weight children and obese children when it came to the parts of their brains that were activated at the smell of onion and chocolate. Specifically, the cingulate gyrus (associated with impulsivity) lit up in obese kids, but in the other kids, "the areas of the brain associated with regulating pleasure and decision-making became more active" (Herz, 2018, p. 70). This may help explain why some kids become and remain obese. Herz explains: "This indicates that for children who are at risk for overeating, food odors are potent triggers of the brain chemistry that increases impulsivity," and impulsivity makes it more likely that adults will also overeat.

The source of appetite and satiety lies in the brain, and we have chemicals that signal us to go (i.e., eat) and stop (i.e., stop eating). It turns out that the go signals are more primitive and stronger than the stop signals, so it is literally a battle not to eat too much for many people. Another internal factor important for eating is the hormone ghrelin, secreted by the gut when the stomach is empty. When it rises, the signal to the body is that it is time to eat. It also slows down metabolism, resulting in less calories burned. "After a hearty meal, ghrelin levels drop, reducing appetite and signaling the brain to quit noshing. This drop in ghrelin also revs up the metabolism so that we can burn the calories we've just ingested" (Herz, 2018, p. 70).

As an example of a factor outside the individual, the fact that "government policies . . . have made less nutritious food cheaper than high-quality fruits and vegetables." This is reportedly a result of food subsidies, although this is debated by many food scholars.

The US Farm Bill is a law renewed by Congress every five years. It puts forth subsidies that, according to the Government Accountability Office (GAO), are meant "to help farmers manage the risk inherent in farming" (Linnekin, 2016, p. 66). According to Linnekin (2016), farmers received about $300 billion in farm subsidies between 1995 and 2012. Though these subsidies have come under intense scrutiny and have faced serious criticisms in the media and by the public, Congress did enact legislation to move away from direct payments to farmers toward helping them purchase crop insurance, although the 2014 Farm Bill made recipients of farm subsidies secret.

Between 1995 and 2020, farmers received about $424 billion in subsidies, and nearly half of that went to just corn, wheat, and soybeans. Corn subsidies make up about 27% of the total, but Lakhani et al. (2021) clarify: "Very little corn grown in the US is eaten these days. Instead, more than 99% goes into animal feed, additives like corn syrup used in sugary junk food and, increasingly, ethanol, which produces toxic air pollutants when burned with gasoline." Lakhani et al. (2021, p. 91) state: "Over a decade, corn received 30 percent of all farm subsidies; five crops (corn, wheat, soybeans, rice, and cotton) received two-thirds. Subsidizing major crops has also benefited concentrated animal feeding operations that rely on milling cheap grains into animal food."

According to Alkon (2019, p. 356):

Federal agricultural subsidies were designed to bolster the production of commodity crops such as corn, rice, and soy during the Great Depression and continue to ensure a cheap supply of the ingredients necessary for processed foods. Fruits and vegetables, ironically referred to in agricultural parlance as *specialty crops*, receive minimal subsidies, and there is no federal subsidy for organic production that could help to offset the higher labor cost associated with these foods.

Higher prices of produce are cited as the main reason why poor people of color often avoid such foods (Alkon, 2019).

Yet, Packer and Guthman (2019, p. 247) review the evidence with regard to whether government subsidies of corn and soy lead to greater consumption of unhealthy foods and thus higher rates of obesity. They write:

In general, the reason that processed food costs less than fresh fruits and vegetables has little to do with subsidies and much more to do with production procedures. Simply put, many processing ingredients, such as soy, corn, and wheat, are far less costly to produce on a mass scale than fresh fruits and vegetables.

They conclude that removing US subsidies on corn and soy "would cause only minimal decreases in caloric consumption, while removal of all US agricultural policies, including both direct subsidies like commodity payments and indirect subsidies like import tariffs, would cause an increase in total caloric intake" (pp. 247–248).

One criticism of farm subsidies is that, since they encourage more crop production, they ultimately lead to more environmental damage. Another is that they may also contribute to poor health outcomes such as obesity since amounts of money in farm subsidies "have been directed to the production of high fructose corn syrup and other corn-based sugars, which are used to sweeten packaged foods like cookies, crackers, cereal, and soda" (Lakhani et al., 2021, p. 74).

Subsidies "encourage many farmers to plant [only] a few types of crops—particularly monocultures of corn and soy—and to plant more of those crops than consumers would otherwise purchase" (Linnekin, 2016, p. 66). Indeed, the main crops subsidized by taxpayers include corn, wheat, soybeans, rice, and sorghum (Linnekin, 2016). Thus, two-thirds of all the calories that we currently eat come from corn, soybeans, wheat, and rice.

Yet, these "crops are grown mainly for animal feed," so the link between subsidies and poor health outcomes becomes weaker when you consider this reality. Further, at least 40%

of the corn grown in the United States is for ethanol rather than food (Nestle, 2013). This, Nestle argues, is "the single most important reason for the recent sharp rise in worldwide food prices and worldwide hunger" (p. 28). Still, fruits and vegetables remain seen as "specialty crops" not worthy of subsidies.

Fuhrman (2017, p. 174) claims that food subsidies are a major reason "the cheapest, unhealthiest, and most fattening foods are the most available foods in poor neighborhoods," because subsidies have allegedly made unhealthy food cheaper, while simultaneously, the costs of healthier foods continue to rise. Fuhrman (2017, p. 156) notes: "Numerous studies have found that the price of food goes down as the added sugar and oil content goes up. Consequently, inner cities in the United States have the lowest cost, lowest nutrient, most dangerous food supply in the world."

Nestle (2013, p. 5) notes: "Federal dietary guidelines may encourage consumption of fruits and vegetables, but federal subsidies go almost exclusively to the growers of food commodities such as corn and soybeans." This may be one reason why the Consumer Price Index shows that the price of fruits and vegetables since the 1980s has risen by 40%, whereas the price of desserts, snack foods, and sodas fell 20–30% (Nestle, 2013, p. 65).

Still, a study by the USDA's Economic Research Service found that one could satisfy the basic recommendations from the 2010 Dietary Guidelines for Americans for about $2.00 to $2.50 per day. And Cohen (2014, p. 74) suggests much of what we think about subsidies may be wrong. She writes: "Although a common belief is that agricultural subsidies are responsible for the low price of ingredients used in grain-based snack foods, processed foods, and drinks with added sugars, like high-fructose corn syrup, there is little hard evidence to support that." She continues: "In fact, the reason that crop subsidies were introduced in the first place was to keep cost of grains high, rather than make them cheap" because, with a very successful crop yield, prices would go down, making profitable farming more difficult: "Government subsidies persuaded many farmers to grow less so the market would not be flooded with grain, keeping the price at a level where farmers wouldn't go bankrupt."

Yet, Nestle (2013, p. 12) claims that Congress "began rewarding farmers for growing as much food as they could fit onto their land," and she suggests this is why so many calories per capita are produced each year. Cohen does not blame subsidies for rises in obesity (whereas Nestle seems to) but rather a very successful snack food market. Snacks are obviously a segment of the industry that major food corporations believe to be a place to invest in for the future, as in the case of Kellogg's, which recently announced it is increasing investment in the snack food industry over cereals because it is a more profitable enterprise.

Finally, subsidies tend to go to very large and profitable farmers who do not need them in order to be successful at farming. Nestle (2013, p. 5) claims that federal subsidies go to "larger and richer farms" as a result of greater consolidation of agricultural production. She even claims that payments often go to "the owners of the largest farms, many of them wealthy landowners who live in cities, rent out the land, never set foot on the farms, and simply collect the checks" (p. 6). Fuhrman (2017, p. 174) notes: "These policies primarily benefit huge agricultural corporations, fast-food giants, and processed food manufacturers, not the family farm trying to grow fresh fruits and vegetables."

The term "obesogenic environment" blames weight gain on the fact that people "are surrounded by cheap, fast, nutritionally inferior food and a dearth of opportunities for physical activity" (Packer & Guthman, 2019, p. 243). Yet, it is important to note that a comprehensive review of 93 studies found that only 29 studies supported the obesogenic environment hypothesis, leading to the conclusion that "a large body of research has failed robustly to identify direct causal pathways between the physical environment and weight status" (Packer & Guthman, 2019, p. 244).

Still, there is evidence that

> many of the chemical substances that researchers have identified as possible obesogens are prevalent in the food system, including certain pesticides, bisphenol A, and phthalates (chemicals present in plastic and other types of packaging materials). Food ingredients and processing agents maybe be part of the mix as well.
>
> *(Packer & Guthman, 2019, p. 249)*

Then there are food additives that mimic human hormones, which may also play a role. Packer and Guthman (2019, p. 249) conclude: "It appears that obesogens are present all along food supply chains, from farm production, to transportation and storage, to food processing and retail." So, it appears there is plenty of blame to go around for the epidemic of obesity, and some of it lies at the feet of the conventional food system after all.

A final piece of evidence speaks directly to the issue of whether we have free will or control over what we eat. And that is the failure of our dieting efforts. According to Walker (2019, p. 185): "The average American dieter attempts to lose weight four times a year. Most who try will eventually put the weight back on, plus some." Moss (2021, p. 173) agrees, writing:

> For the vast majority of people, dieting just doesn't work. It fails because of our physiology: the body plays a game of sabotage by lowering its metabolism or otherwise undercutting our efforts. It fails because life intervenes, with layoffs or new babies or sick parents. It fails because no amount of willpower can be sustained forever. And it fails because when that willpower *is* working for the dieter, the price that's being paid is really high. Successful dieting ruins your relationship to eating.

A 2005 review published in the *Annals of Internal Medicine* of Weight Watchers and four other major dieting programs found that weight loss tended to be incredibly modest, only about five pounds. Further, much of the weight loss was fleeting, as participants gained some of the weight back over time. And this is after people tried the programs multiple times. Over time, Weight Watchers has really just become a "new lifestyle" company, promoting not just healthy eating but also exercise and mindfulness. It even changed its name just to WW.

According to Moss (2021), there is evidence that weight gain is associated with an increased desire for food, so future overeating is more difficult to avoid as people get pleasure from it. Further, the nature of fat itself makes it hard for us to lose weight. Moss (2021, p. 98) writes that "if you *are* dieting, then fat is your worst enemy. It steals your determination to lose weight without you ever knowing it." Fat tells your body to burn less calories, as predicted by the set point theory. It holds that "the body finds a comfortable weight from which it refuses to budge for very long." Moss describes the amazing thing that happens when people fall off the dieting wagon, so to speak: "your shriveled fat cells replenish themselves with a rush of triglycerides from the excess food" and "the fat cells send chemical signals to the nearby veins, causing them to sprout out toward the fat. This increases the blood supply, which helps produce new fat cells." Even with liposuction to remove fat from where it is visible on your body, "new fat will accumulate around the heart and other inner organs, which is associated with an increased risk of heart disease." It is as if our own bodies work against us. So much for free will and self-control! Consider bariatric surgery patients, for example, who have the sizes of their stomachs drastically reduced through surgery. According to Moss (2021), their appetites return with a force, and they stop losing weight.

Herz (2018) claims that about 95% of people trying to diet fail. Yet, the nature of some foods can help people lose weight. For example, strong food aromas cause people to take smaller bites, and complex aromas can help us feel full. Further, the scent of olive oil can help people control their weight, as it is associated with fatty foods and thus feelings of fullness. The aroma of chocolate and that of vanilla are associated with greater eating among people trying to diet, whereas the smell of oranges is associated with less eating (Herz, 2018). Non-food aromas can also reduce overeating, as they interfere with food cravings. Aromas can also make healthier foods more desirable to consume. Desire to eat can also be reduced by being engaged with different forms of technology, as in using one's brain in engrossing ways (Herz, 2018).

### A Focus on School Lunches

There are nearly 100,000 schools in the United States. Table 3.2 shows the number of schools in the United States.

According to the School Nutrition Association (2022), before the pandemic hit, about 29.6 million students each day received free meals. This included:

- 20.1 million free lunches
- 1.7 million reduced price (student pays $0.40)
- 7.7 million full price

And about 4.9 billion lunches are served annually! The reported cost of these meals was $14.2 billion.

**TABLE 3.2**  Number of Schools in the United States

| | | |
|---|---:|---:|
| Public schools | | 98,469 |
| Prekindergarten, elementary, and middle | 70,039 | |
| Secondary and high | 23,529 | |
| Other | | 4,901 |
| Private schools | 30,492 | |
| Prekindergarten, elementary, and middle | 18,870 | |
| Secondary and high | 3,626 | |
| Other | | 7,996 |

*Source:* National Center for Education Statistics (2022). Educational institutions. Downloaded from: https://nces.ed.gov/fastfacts/display.asp?id=84

In terms of breakfast, 14.77 million students each day were served, including:

- 11.80 million free breakfasts
- 0.74 million reduced price (student pays $0.30)
- 2.23 million full price

A total of 2.45 billion breakfasts are served annually! The reported cost of these meals was $4.6 billion.

One might think politicians, parents, teachers, and other societal leaders would aim to assure that the nation's school lunches are healthy, especially given how important they are to kids and their health. Nestle (2013, p. 83), for example, writes:

> School meals provide a large portion of what many kids eat during the day. Whether schools want to or now, they set an example of what kids are supposed to eat. If schools serve poor quality food, and permit and encourage sales of sodas and snacks, they convey the idea that such foods are what kids are expected to eat every day.

Amazingly, there are significant problems with school lunches. One major problem is that schools serve a wide variety of unhealthy food and drink products. Take sodas as just one example. According to Nestle (2002, p. 216): "The well-financed promotion in schools of soft drinks and other foods of poor nutritional quality directly undermines federal efforts to improve the dietary intake of children and to reduce rates of childhood obesity." How we tolerate food of low nutritional quality being offered to our children, especially sodas (which are basically just carbonated, flavored sugar water) is beyond me.

Coca and Barbosa (2019, p. 348) claim: "The school-aged demographic of Western countries have largely adopted food habits based primarily on processed goods of low nutritional value." Interestingly, countries such as Canada and Brazil have taken steps to try to get healthier foods in their schools by attempting to get fresh foods directly from local farms to students.

Yet, in the United States, with tales such as objections by pizza companies to the proposal by the USDA that tomato paste meet the same volume standards as fruits and vegetables (one-half cup to be considered a serving), you get the sense that other priorities often take precedence over health. In this case, pizza makers lobbied Congress and Congress responded with an amendment to a bill that stated "None of the funds made available by this Act may be used to . . . require crediting of tomato paste and puree based on volume" (Nestle, 2013, p. ix).

Further, with companies being on record as admitting that schools are where they get to target children, you would think we would want to do something about it. Simon reports a quote from the CEO of Coca-Cola: "The school system is where you build brand loyalty." And she claims that the battle for public health in schools is one against "allowing corporations unfettered access to impressionable children and youth" (p. 220). Coca-Cola launched a program titled "Your Power to Choose" claiming that kids should have the freedom to choose to drink sodas in schools or not. But of course, this ignores the fact that schools were trying to choose *not* to feature sodas in their schools.

Incredibly, when Congress directed the USDA to limit access to sodas and junk food in 1977, the soda industry filed a lawsuit and, in 1983, a federal court asserted that the agency overstepped its authority. Now, entire school districts sign "pouring rights" deals with soda companies that require them to only serve sodas made by that brand: "Often these deals are presented as being very lucrative to districts, with schools offered enticing incentives such as sports marquees or cash bonuses to sign" (p. 221).

Fast food has also invaded our nation's schools. Food such as chicken nuggets, pizza, hamburgers, French fries, soda, and sports drinks are common in elementary, middle, and high schools, as well as colleges and universities. Even brand-name products such as Pizza Hut and Taco Bell are found in many schools (Fuhrman, 2017). Incredibly, unsafe food is also found there. The Agricultural Marketing Service, which buys meat for schools, has served meat that "is only fit for pet food or compost" and that has had high levels of *E. coli* bacteria in it (Fuhrman, 2017, p. 37).

It is the government that is ultimately responsible for assuring healthy lunches for children. Most people agree that they are failing to do this. This is likely mostly the reason that food corporations have an enormous amount of power over what is offered in the nation's schools. Another reason is that the US FDA is still in the business of making sure enough of certain foods are sold in the United States, and some of this ends up in school lunches.

## Conclusion

In this chapter, you saw an extensive discussion of what experts say a healthy diet looks like, with the final word at the time of this writing coming from the Dietary Guidelines for Americans, 2015–2020. But, an analysis of what Americans actually eat shows that our eating patterns largely do not match government guidance on proper nutrition. Specifically, the typical American diet is too low in minerals, vitamins, whole grains, fruits, and vegetables, as well as too high in fat, saturated fat, cholesterol, processed grains, sugar, and salt. Average American eating patterns emerge from industrialized foods, so, to the degree such foods are being produced, marketed, and sold to others around the world, we will see similar eating patterns around the globe.

Americans do not also get the recommended daily amount of exercise either. And, though the concept of energy balance has been utilized in government guidelines for decades to promote exercise along with healthy eating, this concept has been latched on to by global food corporations to try to divert attention away from the harmful nature of their food products and to instead shift attention to increased exercise as the solution to problems of poor eating. Stated simply, this is a ploy of big food to deny its culpability for rising rates of obesity and diabetes both in the United States and around the world.

That school lunches are not even healthy speaks volumes to the challenges that remain in terms of making food healthier for people, including the most vulnerable of all, students enrolled in schools. While corporations are largely culpable for failing to assure healthy eating by Americans and others abroad, the government shares responsibility for failing to effectively regulate food in the name of public health and safety. The industry has used its immense power to influence even dietary guidance throughout our history, and this is likely one reason we will not see the government making statements such as "avoid ultraprocessed foods" that are both *not* healthy and demonstrably harmful to human health. It is past time for the government to break free from this power and use its own power to provide better recommendations for healthy eating and to encourage, if not force, food companies to do a better job of providing those foods to consumers.

# 4

# THE FOOD IS THE CRIME

## Putting the Food Back Into "Food Crime"

### Introduction

Even though some scholars are bringing attention to the subject matter of food crime, almost no one within that context studying food and its contents. In this chapter, I illustrate that it is the food itself that is the crime. That is, the great bulk of harm caused by "food crime" comes from the food itself. There is a lot about food that is disgusting; luckily for consumers, most of this is largely unknown!

In this chapter, I establish the reality of much food in the United States—ultra-processed food products that are high in salt, sugar, fat, and artificial additives. I examine each issue in turn, illustrating the potential harms of salt, sugar, fat, and additives.

This is not meant to suggest that food should be criminal, that any food should be a crime. Rather, there are substances in food (e.g., saturated fats, sugars and sugar substitutes such as high fructose corn syrup, salt, and chemical additives) that are found in foods in too high amounts to be considered healthy. Since this is known now, any failure to remove such substances from food products amounts to a form of negligence that could be used to hold food companies culpable for harms caused by these substances—outcomes such as obesity and diabetes. It is government agencies that are responsible for regulating food to make sure it is not dangerous. The review in this chapter shows they are failing to regulate products for too much salt, sugar, and fat content.

Based on the findings of this chapter, radical change will be needed to make foods, especially ultra-processed foods, more healthy. Additional focus is placed on other issues, including genetically modified organisms (GMOs), which I show in the book are not generally considered harmful to humans, as well as antimicrobials in animal feeds, which are!

### Processed Foods

Processed foods are essentially any food product that has been altered from the way it exists in nature (Constance et al., 2019). The NOVA classification scheme places foods into one of the following four categories:

DOI: 10.4324/9781003296454-4

1. Natural, unprocessed, or minimally processed
2. Processed culinary ingredients (e.g., oils, butter, sugar, and salt)
3. Foods processed in basic ways (e.g., bread, cheese, and canned or frozen goods)
4. Ultra-processed foods (e.g., packaged snacks and sodas)

*(Nestle, 2018, p. 213)*

Processed food, and especially ultra-processed food, is the result of the industrialization of food, discussed in Chapter 2. Constance et al. (2019, p. 92) answer the question of what is the relationship between industrial food and the quality of our food with "the dominant food system is characterized by an industrial diet of unhealthy *pseudofood* commodity chains centered on fats, sugars, starches, salt, and empty calories. This system leads to heart disease, diabetes, obesity" as well as many specific foodborne illnesses and deaths. Chapter 6 examines these outcomes.

Nestle (2018, p. 135) says this about processing: "Processing often removes vitamins, minerals, and fiber while adding salt, sugar, and trans fats, making foods junkier. Independent nutrition groups generally recommend diets containing foods that are minimally processed." Moss (2021) points out that most processed foods give us too many calories but deprive us of the nutrients we need to survive and thrive. The calorie load in processed foods—also known as *energy density*—is of great concern. Moss (2021, p. 94) gives the following foods as examples—"four-cheese pizza with a soda on the side; the Oreo Mega Stuff cookies; the extra-large bag of French fries. Our stomach as it evolved to love calories really loves these and tells the brain so." But, he points out, these foods tend not to stretch the stomach to signal us to stop eating, as they lack fiber and water, so, "By the time the gut sends the signal or *us* to stop, it's way too late." So, there is evidence that eating processed foods causes us to gain weight (Moss, 2021). Weight gain can be and often is associated with an increased risk of illness, as will be shown in Chapter 6.

There are many reasons why foods are processed. Some of these include not only added flavors and aromas but also increased shelf life and lower prices. Moss (2021, p. 48) adds that one of the major determinants of how much processed food is produced relates to the time it takes to produce them: "Saving production time translates into shaving expenses, which enables the processed food industry to lower prices, which in turn makes their products all the more desirable and rewarding." Later, he claims: "The processed food industry is . . . hooked on cheapness. The food manufacturers are fanatic about reducing their costs so they can lower their prices, knowing that lower prices will cause us to buy more" (p. 107).

Price is obviously something that is important to the consumer. Moss (2021, p. 106) points out: "Cheaper food means having to work less in order to part for that food, and thus we are drawn by instinct to grocery receipts and restaurant bills that are smaller." As such, he claims that when we are trying to decide whether to even put a product in our cart, "out first concern is for the price" (p. 107).

Time, or convenience of little time, is an advantage of processed food for the consumer, as well. Moss (2021, p. 49) notes that "there is speed in processed foods once they get into our hands. They unseal quickly, heat up in the microwave quickly, and, more important, excite the brain quickly when they reach the mouth." These "convenience foods" are sought out by us in large part because of how little time it takes to prepare (and eat) them. Americans now get about 75% of their calories from groceries that they do not need to cook but rather are ready to eat upon purchase or can merely be heated (Moss, 2021). Moss clarifies:

"Three-fourths of the groceries we buy, as measured in calories, are now processed, with most of this classified as *highly* processed food" (p. 119). He says that 68% are ready to eat and another 15% are ready to heat but do not have to be cooked. These foods also make it possible to eat food anytime, anywhere, a reality that has only existed with processed foods for the past several decades. As a result, snacking as a fourth meal became a new reality. So, it is the food environment itself that has changed over time in ways that tend to produce really bad health outcomes, including high rates of obesity and diabetes.

Another reason that processed foods are popular is that they are created to produce feelings of comfort. According to Herz (2018), there are at least two "central qualities" of comfort foods: (1) they are foods from childhood and (2) they are associated with home, nostalgia, and family. They also tend to have high levels of carbohydrates and sugars (i.e., processed foods). Carbs increase serotonin in the brain, yet a study of about 70,000 women found that more refined carbs were associated with a higher incidence of depression, whereas higher consumption of fruits, vegetables, and whole grains was associated with a lower prevalence of depression.

Comfort foods also cause endorphins to be released in the brain, a natural form of heroin that makes us feel good and blocks pain. Herz (2018, p. 238) notes that "the more common cause of endorphin release is the learned connections to comfort and pleasure that we have associated with certain dishes through our past experiences." She continues, writing that

> foods don't become comforting because of their starchy, sugary, or luscious qualities, or the mood-enhancing chemicals that they contain. It is by extrapolation . . . that comfort foods have these properties, and the comfort factor in many of these foods is then reinforced by their ingredients.
>
> *(p. 244)*

Take chocolate as one example; it will often "conjure memories and emotions connected to special treats, gifts, holidays, and romance" (p. 239). Chocolate also contains phenylalanine, which helps build dopamine in the brain, another neurotransmitter that produces feelings of pleasure, as well as anandamide, a relative of THC that can produce feelings of happiness and well-being. Comfort foods like chocolate are found to elicit nostalgia, "a sentimental yearning and wistful affection for one's past" (p. 244). Even foods like chicken noodle soup are found to bring about "unconscious thoughts of positive interpersonal relationships" (p. 242).

Studies show that people tend to eat, if not binge upon, comfort foods due to certain situations; one example is narrow losses in sports contests are found to increase such eating, whereas victories are found to decrease such eating. Herz (2018, p. 249) writes that

> the mild mania that comes from being victorious and also from falling in love—is an appetite suppressant. In these exuberant states the brain produces higher levels of norepinephrine (adrenaline), dopamine, and serotonin, so energy level is increased, the heart beats fasted, and we have less need for sleep and food.

Another factor that impacts eating is stress. Studies show that people who have undergone stressful situations tend to pick more unhealthy yet tasty foods. When this happens,

> the connectivity between the prefrontal cortex—where decision-making takes place—and areas of the brain where emotion, pleasure and taste are processed was especially

activated, while the connectivity between the prefrontal cortex and brain regions where self-control is regulated was underactive. In other words, stress made high calorie food more pleasurable and decreased the ability to make prudent decisions.

Ultra-processed food is made up of "industrial concoctions containing a multitude of additives: salt, sugar and oils combined with artificial flavors, colors, sweeteners, stabilizers and preservatives. Typically, they're subjected to multiple processing methods that transform their taste, texture and appearance into something not found in nature" (O'Connor, 2022). According to O'Connor: "Many ultra-processed foods are made in industrial machines that subject grains, corn and other raw ingredients to extremely high pressures and temperatures. This can destroy micronutrients and create new compounds that can be harmful, including carcinogens."

O'Connor (2022) provides us with examples, writing: "Think Frosted Flakes, Hot Pockets, doughnuts, hot dogs, cheese crackers and boxed macaroni and cheese." This author also points out they include "breakfast cereals, muffins, snack bars and sweetened yogurts. Soft drinks and energy drinks count, too."

O'Connor (2022) claims these foods make up to 60% of calories consumed in the United States and between 25% and 50% of calories consumed in other countries, including Brazil, Canada, England, France, Japan, and Lebanon! Why are such foods so popular? First, they are widely available and affordable, making them appealing to a wide range of people. Second, they "deliver potent combinations of fat, sugar, sodium and artificial flavors. They are what scientists call hyper-palatable: Irresistible, easy to overeat, and capable of hijacking the brain's reward system and provoking powerful cravings." This suggests food addiction, to be discussed in Chapter 5.

Recall the discussion of simplification and adulteration from Chapter 2. There has been a reduction in the complexity of whole foods and a reduction in the varietal diversity of foods in society due to processing, consistent with simplification. Constant efforts to increase the palatability of highly processed foods and increase shelf life, make processed foods cosmetically more appealing, and lower the manufacturer's cost to produce them are consistent with adulteration, and this results in diminished nutritional value in food products.

O'Connor discusses studies that eating ultra-processed foods is associated with earlier death from conditions such as heart disease. Part of this is likely due to the nature of the foods themselves, but part of it is also because studies show that, when people eat ultra-processed foods, they tend to eat more and thus consume more calories. The result is gaining more weight and body fat, two conditions associated with increased illness.

Studies, including two large studies of both men and women in the United States and Italy, show the dangers of ultra-processed foods, which tend to make up the bulk of our comfort foods. The first study found that eating a lot of ultra-processed foods significantly increases the risks of heart disease, colorectal cancer, and early death (LeMotte, 2022). Colorectal cancer is the third most diagnosed cancer in men in the United States. According to LeMotte, ultra-processed foods include "prepackaged soups, sauces, frozen pizza, ready-to-eat meals and pleasure foods such as hot dogs, sausages, french fries, sodas, store-bought cookies, cakes, candies, doughnuts, ice cream and many more." These foods tend to be high in added sugars and salts but low in dietary fiber; they also are full of chemical additives including artificial colors, flavors, and stabilizers. According to Fuhrman (2017, p. 36), French fries from fast-food restaurants can have up to

33 ingredients in them, "including sweeteners, salt, antifoaming agents, preservatives, and colorings."

In a new report on the studies, Marion Nestle is quoted saying: "Literally hundreds of studies link ultra-processed foods to obesity, cancer, cardiovascular disease, and overall mortality." And, according to LeMotte (2022):

> Processed and ultraprocessed meats, such as ham, bacon, salami, hotdogs, beef jerky and corned beef, have long been associated with a higher risk of bowel cancer in both men and women, according to the World Health Organization, American Cancer Society and the American Institute for Cancer Research.

In men, higher rates of bowel cancer were associated with higher consumption of ready-to-eat meats, poultry, and seafood, as well as SSBs. One of the authors explains why ultra-processed foods are so bad for us, saying that "ready-to-eat-or-heat industrial formulations that are made with ingredients extracted from foods or synthesized in laboratories, with little or no whole foods."

As for the issue of SSBs, Herz (2018, p. 218) notes one reason sugary drinks are problematic:

> Beverages in general aren't as filling as solid foods. In addition to bypassing a lot of sensory involvement as they swoosh down the throat, liquids reach our viscera and exit them more quickly than solids do and the glucose-sensing receptors in the gut don't recognize the calories from sodas as effectively as they do the calories from sandwiches because our glucose-sensing cells evolved to detect starchy legumes and fruit, not sugary drinks.

She continues:

> This is why sugar-sweetened beverages—or SSBs . . . —are so insidious, and why they have been implicated as a major offender in the obesity epidemic. Not only are we bad at estimating their caloric value, our bodies don't feel satiated from their calories though they add up just the same.

Herz notes that SSBs are deadly. While her estimate of deaths in the United States caused by SSBs is 2,500, another study published in the *Journal of the American Medical Association* (JAMA) put the number at more than 50,000! This was shown in Chapter 1.

The second study compared nutrient-poor foods like high-sugar and high-fat foods and ultra-processed foods and found that both increased the odds of chronic disease and early death from things such as cardiovascular disease. Ultra-processed foods were worse in terms of their impact on people's health, although nutrition-deficient foods tend to be ultra-processed, as well. A clinical trial from the National Institute of Health (NIH) found that people who eat ultra-processed foods tend to eat about 500 more calories in a setting than those who eat unprocessed foods, largely because they tended to eat at a faster rate. Moreover, they tended to gain about two pounds while on a diet of ultra-processed foods, compared with a two-pound loss for those on unprocessed foods. Again, weight gain is associated with increased risks of some illnesses. Figure 4.1 shows a snack aisle in a local

**FIGURE 4.1** Snack Aisle in a Grocery Store

*Source:* Photo by author

grocery store. Notice the enormous amount of potato chips, crackers, cookies, and so on, none of which are healthy.

Consider the possibility that ultra-processed foods are defective products. That is, though they are not intended to kill or injure people, they do that very thing, in the form of obesity, diabetes, heart disease, cancer, and more. The likelihood that such food products will ever be deemed defective is not high, yet it remains true that they are defective in at least that way.

Ultra-processed foods tend to be high in salt, sugar, and fat, each of which is discussed later in the chapter. As such, they are not healthy. These foods make up the bulk of foods consumed by children in the United States, accounting for as many as 67% of calories for kids and adolescents (Nunez, 2021). Interestingly, the FDA has announced new rules for front-of-packaging labels that will announce food as "healthy" as long as they contain meaningful amounts of certain food groups and limit the presence of others (Reilly, 2022). This is a positive step in the right direction to better inform consumers of better food choices.

While salt, sugar, fat, and other food ingredients are discussed separately in this chapter, it is important to note up front that they tend to have bigger effects on the brain and create more urges in us when they are combined, as well as combined in certain ways. We know this because of research conducted by major food companies, including brain scans of willing participants of all ages. Cohen (2014, p. 101), for example, notes that

> we get immediate positive feedback from sugar, chocolate, and fat, so we tend to view chocolate chip cookies in a very positive light and learn to desire them. Instant feedback is a wonderful behavioral learning technique when it comes to promoting eating, but is not very useful as a tool to support dieting.

A positive association between chocolate chip cookies and positive memories is "very hard to unlearn" (p. 101). The combination of sugar and fat, in particular, stands out. Moss (2013, p. xxxv), in describing scientific findings, writes: "They boosted one another to levels of allure that neither could reach alone."

Moss (2013, p. xxv) notes that salt, sugar, and fat are methodically studied and controlled by the food industry. They are, in essence, the main drugs in our food that can be manipulated to get us hooked and keep consuming their products (Robinson, 2022). Studying cravings and the "bliss point" of foods mathematically is evidence of this. And Moss gives us these examples of corporations experimenting with fat, salts, and sugars to prove the point:

> Scientists as Nestle are currently fiddling with distribution and shape of fat globules to affect their absorption rate and . . . their "mouthfeel." At Cargill, the world's leading supplier of salt, scientists are altering the physical shape of salt, pulverizing it into a fine powder to hit the taste buds faster and harder, improving what the company calls its "flavor burst." Sugar is being altered in myriad ways as well. The sweetest component of sugar, fructose, has been crystallized into an additive that boosts the allure of foods. Scientists have also created enhancers that amplify the sweetness of sugar to two hinder times its natural strength.

Who knew that such so much study and experimentation go into the foods that we (mostly) mindlessly buy in grocery stores and restaurants? The issue of food addiction is discussed in Chapter 5.

## Salt

Salt has an important place in our food. Herz (2018), for example, notes that salt helps our nerves and muscles in their correct functioning, as well as assists in the regulation of fluid balance. Stated simply, "if we don't consume enough salt we die" (p. 21).

But what is the real point of salt in products? According to Moss (2013, p. 281),

> salt is the great fixer. It corrects myriad problems that arise as a matter of course in the factory. Cornflakes, for example, taste metallic without it. Crackers are bitter and soggy and stick to the roof of your mouth. Ham turns so rubbery it can bounce. Some of salt's power has nothing to do with the food at all. In commercial bread making, salt keeps the huge, fast-spinning machinery from gumming up and the factory line from backing up: Salt slows down the rising process so that the overs can keep up with the pace.

Dozens of sodium-based compounds are also used in food manufacturing "to delay the onset of bacterial decay, to bind ingredients, and to blend mixtures that otherwise would come unglued, like the protein and fat molecules in processed cheese" (p. 282).

Yet, if you want to get a sense of whether salt is dangerous for us, consider that the Center for Science in the Public Interest (CSPI) published a report titled, "Salt: The Forgotten Killer." Moss (2013) notes that excessive salt is associated with high blood pressure, or hypertension, a major killer of humans. Even when people do not add salt to their foods, they may still get up to 75% of their salt intake from processed foods because "the salting of processed food had become a way to increase sales and consumption" (Moss, 2013, p. 270). Processed food manufacturers even found a way to increase cravings for salt when a natural craving for it does not exist in humans. Giving salt to young children, studies show, leads to cravings for salt and changes in preference for food that have more salt in them (Moss, 2013). It is food manufacturers who are, Moss concludes, "culpable" for this.

High salt intake is also associated with "coronary fibrosis"—scarring of the heart—which increases the risk of cardiac arrhythmia or irregular heartbeat. It is also associated with asthma, autoimmune disease, stomach cancer, osteoporosis, as well as kidney failure (Fuhrman, 2017, p. 35). Given these realities, people must be careful not to consume too much salt; yet, the salt content is so high in most ultra-processed foods, again raising the issue of defective products. Figure 4.2 shows the nutritional information for a bag of Dorito's chips. Notice the enormous amount of salt in them.

Should we limit the amount of salt in products? This seems unlikely given its importance for taste. Perhaps a better option is to assure labels that warn "High Salt Content," as in the case of Finland. According to Moss (2013, pp. 302–303), this effort, along with a public education campaign warning against the potential dangers of salt, led to an 80% decline in deaths from strokes and heart disease. Of course, any effort to achieve a goal like this in the United States would be objected to by the salt industry.

There are also industry groups, such as the Potato Chip and Snack Foods Association, who deny links between salt and deleterious health outcomes in the first place. This flies in the face of efforts by the FDA to educate consumers about the risks of high salt in their diet. And it makes you wonder why corporations such as Pepsi (which owns KFC and Taco Bell) would announce, in 2010, an effort to cut salt in its products by an average of 25%, as well as the amount of sugar? Why do this if salt (and sugar) are not harmful? In the United States, President Clinton sought to make school lunches healthier by reducing salt, sugar, and fat content. Again, why do this if these substances are not harmful?

Another option might be to take the sodium out of salt so that it is not as dangerous. The Cargill company produced a product called potassium chloride that had the same taste as salt but without sodium. But, in the United Kingdom, regulators decided that, since

**FIGURE 4.2** Nutritional Information in Dorito's Chips

*Source:* Photo by author

potassium can be linked to kidney problems, salt replacement should not be used in products (Moss, 2013). Cargill, according to Moss (2013, p. 286), has salts that "Are finely tuned bliss machines" that are specific in shape and size to each product to which they are added. So, perhaps Cargill already has the technology to make salt safer, rather than just less appealing and additive to us?

## Sugar

If any element of food gets really bad attention from food scholars, it is sugar. Yet, sugar was vital for our survival throughout history. Specifically, sweet foods were a sign of calories and were needed by our ancestors to survive and thrive. According to Herz (2018, p. 8): "Sweet taste lights up the same reward pathways in the brain as addictive drugs and alcohol, and triggers the release of dopamine." This issue of food addiction is discussed in Chapter 5. And, aside from the potential for addiction to be discussed later, it is a good thing, because being able to taste sweet saved us from starving to death throughout early history. Further, "the discovery of fire and cooking, which made carbohydrates such as tubers and starchy plants delicious and nutritious, is credited with leading to the evolution of the most complex creation on earth, the human brain" (p. 9). Herz notes that what fuels our brains is glucose, "and 20–25 percent of all the calories we consume and 60 percent of our blood glucose, goes toward keeping our brain powered up" (p. 9).

Herz (2018, p. 33) explains the relationship between innate desires and our survival:

We are innately programmed to love the tastes of sweet and salty because, respectively, they usually signal carbohydrates and proteins that we require to survive, and to reject the taste of bitter because most of the foods that taste bitter are toxic.

Ironically, in today's toxic food environment, "our love for these tastes is hazardous to our health, while not consuming enough bitter leafy greens can likewise threaten our lifespan."

Sweets have a direct impact on our metabolism because sweet taste receptors regulate the absorption of sugar into the bloodstream from the intestines and turn insulin on and off: "The taste receptors in our intestinal tract respond to the sweet we've swallowed by activating the release of insulin. When this balance goes away it leads to excess insulin and insulin resistance" which is the first step to metabolic syndrome,

a combination of conditions including type 2 diabetes, high cholesterol, and high blood pressure, all of which stem from an inadequate ability to store fat. When one's fat cells have reached their limit, excess fat is stored elsewhere, such as in the liver. With insulin resistance, the body is unable to absorb excess glucose. The body releases insulin in response to eating, as it should, but then doesn't do its job of helping to store glucose from the breakdown of carbohydrates, and glucose and insulin levels in the bloodstream remain chronically high.

*(Herz, 2018, p. 12)*

An extreme outcome including being overweight, diabetic or prediabetic, combined with high levels of triglycerides and cholesterol—metabolic syndrome—is a result of poor diet [including too much sugar and high-fructose corn syrup (HFCS)] and inactivity.

About one-third of Americans and one-fourth of people around the world suffer from insulin resistance, which significantly increases the risk of heart attack and stroke. So, more people die today from illnesses related to obesity than from even starvation. The producers of food share culpability for these outcomes. Consumption of refined starches and sugars is associated with a greater risk of inflammation, depression, and metabolic syndrome.

Yet, sugar has analgesic properties. Sugar increases tolerance for pain, reduces emotional agitation, and increases the release of endorphins—naturally occurring opiates in the body—to help us feel good. It also brightens our moods (Herz, 2018, pp. 16–17). Further, sugar can help us maintain self-control by increasing glucose levels in the brain. So, sugar, in itself, is not all bad. But a major problem with sugar is that it is in such a wide variety of food products, including ones that might surprise you—things like soup, sauces, and even meat products!

According to Herz (2018): "Since the 1950s, food companies have known that adding more sweetener makes a grape drink taste grapier and a cola drink taste more cola-ish." Even adding sugar to a tomato makes it taste better, or more accurately, disrupting the production of a protein responsible for how much sugar is in a tomato makes it not taste as good. To make tomatoes appear more uniform in stores, growers have been selecting those with a genetic mutation that results in this very outcome. Ketchup, a product from tomatoes, tastes good because it tends to have twice as much sugar per tablespoon as ice cream!

According to Moss (2021, p. 111):

Using an arsenal of more than sixty types of sugars—from corn syrup to concentrated fruit juice—[companies] marched around every aisle in the grocery store, sweetening products that didn't used to be sweet. The intent? To get us hooked on them, or, at the very least, to get us to buy them and keep buying them. Moss also reasons that this has created in us the expectation that all the food we eat should be sweet. It is food companies who are culpable for these realities.

One result of added sugars in our food products is that we are consuming more sugar over time. Walker (2019, p. 174) outlines the growth in the consumption of sugar in the United States over time. In the late 1700s, Americans consumed about four pounds a year, by 1850, this had increased to 20 pounds, and by 1994, it was 120 pounds. He says that today "the number-one food group consumed is added fats and oils, followed by processed flour and cereals, meat and eggs, and [then] added sugars."

Guidance from government agencies on avoiding added sugars has been consistent (though not consistently worded) across the years and agencies. Consider for example:

- The WHO advised to consume no more than 10% of calories from added sugars (2015).
- The US Dietary Guidelines advised to consume no more than 10% of calories from added sugars (2015).
- The US Dietary Guidelines advised to "avoid too much sugar" (1980, 1985).
- The US Dietary Guidelines advised to "reduce the intake of calories from solid fats and added sugars" (2010).

According to Nestle (2018, p. 45): "Sugars are today's food enemy number 1, toxic by some reports. How much is safe? None, claims science journalist Gary Taubes; sugar is responsible for obesity, type 2 diabetes, heart disease, stroke, gout, and Alzheimer's disease." To clarify, there should be no concern about sugars naturally occurring in foods; "the amounts are low, and the sugars are accompanied by vitamins, minerals, and fiber. In contrast, added sugars provide calories devoid of other nutrients." Yet, some clients with close ties to food and drink corporations claim that the causal relationship between sugar and chronic disease is unproven.

The major downside of sugar is that it is considered a major risk factor for disease and death (Minger, 2013), if not a poison to your body. Five pieces of evidence support this idea:

1. Our current intake of sugar is unprecedented in human history.
2. There is a strong correlation between sugar intake and heart disease deaths.
3. Increases in sugar intake parallel increases in cardiac mortality
4. Men with heart disease tend to eat a lot of sugar.
5. Sugar may raise triglycerides, insulin levels, blood glucose, and serum cholesterol.

*(Minger, 2013, p. 117)*

When you think of sugar, you probably think of sweet products such as candy, cake, ice cream, and other desserts. But, again, sugar is added by food producers into all kinds of foods. Take Prego spaghetti sauces for example: "The largest ingredient, after tomatoes, is sugar" (Moss, 2013, p. 37). Meanwhile, breakfast products such as Pop-Tarts have so much sugar in them that they "make no pretense of being anything other than cake for breakfast, or cookies at least" (Moss, 2013, p. 59). Figure 4.3 illustrates the nutritional information for a box of Pop-Tarts. Notice the enormous amount of sugar per package.

Of course, the same can be said of many, if not most, breakfast cereals, which are basically sugar in a bowl served with milk, and advertised directly to our kids with friendly cartoon characters such as "Tony the Tiger" who yells, about his Sugar Frosted Flakes (now called Frosted Flakes), "They're greeeeaaaat!" Moss (2013) characterizes many cereals sold by Post, General Mills, and Kellogg's (who control almost all of the cereal market) as candy for breakfast. Note that euphemisms for sugar are used in the industry, including honeyed, sugarcoated, sweet, syrupy, and candied (p. 152).

Sugar itself is a cancer promoter and is also associated with obesity, diabetes, and heart disease. Presumably, the more sugar a person eats, the worse the potential health outcomes can be. Part of the problem with sugar is that, "instead of being taken up by muscle cells all over the body, the fructose goes right into the liver and triggers the production of fats such as triglycerides and cholesterol," a process called "liogenesis" (Fuhrman, 2017, p. 30). Liver illness is the typically result of consuming too much sugar over an extended period of time.

One might thus ask why there is so much sugar in so many products we consume? Fuhrman (2017, p. 29) claims that "for more than fifty years, the soft drink industry and the sugar industry have provided millions of dollars of research funding to academic and government researchers to influence and cover up the health risks associated with consuming sugar." Big sugar is also one of the most significant food-related contributors to political campaigns and candidates every year. Fuhrman (2017, p. 173) notes that "the food and beverage industry spent $40 billion in lobbying Congress against regulations that would decrease the marketing of unhealthy foods to kids and promote soda taxes" just in 2010. Further:

The fast food industry spends more than $5 million a day advertising sugary cereals, junk food, and fast food to children; and it's working. It has been demonstrated that exposure to these advertisements induces children to eat higher amounts of sugar, fried food, and sweetened beverages.

**FIGURE 4.3**   Nutritional Information in Pop-Tarts

*Source:* Photo by author

Nestle (2013, p. 7) agrees, saying "It is anything but coincidental that kids prefer the very foods that are made and marketed to appeal to them." Figure 4.4 shows some of the types of characters used to promote sugary cereals to children.

**FIGURE 4.4** Cartoon Characters Used to Sell Cereal

*Source:* Photo by author

Nestle (2018) claims that food advertisers spend billions of dollars each year advertising directly to children. And why do they do this? Nestle offers three reasons:

1. To generate brand loyalty
2. To get kids to encourage their parents to buy the products
3. the third reason is the most insidious. It's to get them to believe that they know more about what they are supposed to eat than their parents do. Food marketing shifts the responsibility for deciding what children should eat from parents to the children themselves. As a result, meals become battlegrounds.

*(p. 77)*

Nestle adds that marketing to kids also subverts the parental authority over what their kids want to eat.

Fuhrman (2017) and others compare the tactics of Big Sugar to that of Big Tobacco—ironic given that large tobacco companies bought large food companies and controlled them for decades. Fuhrman (2017, p. 166) notes:

The current tactics of the sugar industry mimic how 'big tobacco' was able to define its practices and promote tobacco use for years and years. Key to its strategy was paying scientists to plant doubt in the minds of the public, thereby confusing them, and financially supporting political allies.

Then there is HFCS. According to Silbergeld (2016, p. 209), HFCS increased by 1,000% from 1970 to 2000, and "rapidly displaced refined sugar, which was disadvantaged by high tariffs, as well as other sources of sucrose." HFCS is really not much different than any other source of fructose in products, yet, it is

> a marker for poor-quality, nutritionally-depleted, processed industrial food full of empty calories and artificial ingredients. If you find "high fructose corn syrup" on the label, you can be sure it is not a whole, real, fresh food full of fiber, vitamins, minerals, phytonutrients. And antioxidants. Stay away if you want to stay healthy.

HFCS, first patented in Japan in 1971, is sweeter than cane sugar and also cheaper to produce; it also acts as a preservative, extending the shelf-life of products. So, we can understand its ubiquitous nature in products of all kinds, given the profit-motive of major food corporations. Moss (2013, pp. 4–5) describes how high fructose corn syrup is cheap and liquid, meaning it can be directly added to drinks and foods pretty easily.

High fructose corn syrup is also linked to insulin resistance, as it "interferes with the removal of other sugars from the blood, requiring the pancreas to produce more insulin to do so." Fuhrman (2017, p. 30) claims: "The explosion in the occurrence of obesity and type 2 diabetes in the past fifty years was caused (partially) by this high exposure to sugar and HFCS in fast food and soft drinks." It also contributes to high blood pressure through the promotion of plaque buildup (atherogenesis), the inhibition of an enzyme (endothelial nitric oxide synthase), and the increase of advanced glycation end products that age tissues and create nerve and kidney damage.

Fuhrman (2017, p. 32) lists the following as health-damaging effects of HFCS:

1. Obesity
2. High blood pressure
3. Diabetes
4. Blindness
5. Liver disease
6. Kidney disease
7. Premature aging
8. High triglycerides
9. High cholesterol
10. Heart attacks
11. Dementia
12. Strokes
13. Cancer

Seems like a defective product to me.

In spite of all of this evidence, the sugar industry denies any harm associated with its products. The industry—for example, the Sugar Association—uses "the playbook" (to be discussed in Chapter 9) to confront health claims related to its products. For example, in 2014, "The Union of Concerned Scientists (UCS) summarized the sugar industry's tactics for undermining policy: spread misinformation, deploy industry scientists, and influence academics" (Nestle, 2018, p. 46). You also have actions of the Corn Refiners Association denying harms associated with its products, most notably high fructose corn syrup or "corn sugar."

According to Nestle (2018, p. 122), the Sugar Association did all it could to cast doubt on the science. And the International Life Sciences Institute (ILSI) sponsored a study which concluded that "guidelines on dietary sugar do not meet criteria for trustworthy recommendations and are based on low-quality evidence" (p. 123). Imagine telling a body of experts who study nutrition for a living that their review of the evidence is not worthy of being taken seriously! And who are you going to believe, the nutrition experts or big sugar? Which one profits off of selling sugar and which one doesn't?

Experts also belong to organizations like the American Society for Nutrition (ASN), but here you must be cautious, given that the organization and its journal have corporate sponsors or "sustaining partners" (see Nestle, 2018, p. 128). In 2009, ASN formed a board of directors for its new program called "Smart Choices." According to Nestle (2018, p. 132): "This was a food-industry initiative in collaboration with leading nutritionists to put a stamp of approval on the front of packaged foods that met defined nutritional criteria." Nestle was asked to be on this board and declined. She writes: "Although Smart Choices seemed to be aimed at helping the public identify healthier food options, it seemed clear to me that its underlying purpose was to induce nutritionists to endorse highly processed 'junk foods' as healthy." Sure enough, the Smart Choices label first appeared on the front of a box of Froot Loops cereal, even though 44% of its calories come from added sugar—sugar in a box for breakfast is called a Healthy Choice!

Out of curiosity, I looked up the nutrition information for Froot Loops. For just one and one-third cups of the cereal (the listed serving size), the cereal has 150 calories (not including the milk). It has 1.5 g of fat, including 0.5 g of saturated fat, 210 mg of sodium, 34 g of carbohydrates, and 12 g of added sugars; 12g of sugar amounts to 24% of the recommended daily intake, so two servings of Froot Loops (just two and two-thirds cups) would amount to about half of one's recommended daily limit (Smart Label, 2022).

Many sugars are half sugar, just like many candy bars, and, on top of that, they are composed of corn and oats, which our body converts to sugar after we ingest it (Moss, 2021). Yet, cereal manufacturers use methods such as paying scientists to promote their foods. For example, Kellogg's pays nutritionists thousands of dollars a month each to promote their cereals on social media (Nestle, 2018). And groups like Cereal Partners Worldwide (made up of Nestle and General Mills) are working to promote sales of their products.

Nestle adds, about Smart Choices: "The program also appeared to be an industry-designed effort to head off a plan, then under consideration by the FDA, to regulate front-of-package food labels" (p. 132). The government did take action here, as both the FDA and USDA wrote to the Smart Choices folks and asked whether its log might encourage people to choose ultra-processed foods and refined grains like those in Froot Loops. By later that year, the Smart Choices program was at least ended. But this case provides an example perhaps of how far the food industry is willing to go to trick consumers into buying unhealthy foods by calling them healthy. Talk about misleading advertising! Food companies are culpable for this.

When the FDA proposed including information of food products about "added sugars," both consumer and public health groups supported the idea, but, of course, the industry did not. Surprisingly, neither did ASN. It said it has

concerns with the FDA's rationale for the inclusion of 'added sugars' on the food label. . . . Conclusions regarding added sugars remain elusive based on insufficient evidence regarding the effects of added sugars (beyond contribution of excess calories) on health outcomes . . . ASN recommends careful consideration of the totality of the scientific evidence, as well as consideration of compliance and other technical issues.

*(Nestle, 2018, p. 136)*

Read that again—"beyond contribution of excess calories"—as if that is not a good enough reason to inform people of how much added sugar they are being exposed to in products! By the way, as of 2016, information on added sugars does appear on food labels, after all. Stated simply, added sugars are unnecessary in food products other than to add to flavor and get you hooked on the products.

Meanwhile, sugar industry advertisements have claimed that sugar works to reduce appetite and increase energy, "sugar quenches fatigue," and even increases willpower to do things including to diet! Imagine that. Eat sugar and you will have more willpower to diet, which presumably means eating less sugar.

## Fat

Fat has "remarkable powers," so "the processed food industry relies on it like no other component" (Moss, 2013, p. 146). Sufficient fat in products allows them to stay on grocery store shelves longer, can be used to make hot dogs cheaper, and "is even the key determinant of the nutritional value of ground beef . . . depending on the percentage of fat . . . the levels of calcium, niacin, iron, and other elements in the meat go up or down" (p. 147). Moss continues, saying: "Fat also performs a range of culinary tricks for food manufacturers, thanks to another of its extraordinary powers. It can mask and convey other flavors in foods, all at the same time." Finally, fat has one more power that even sugar cannot match:

> Fat doesn't blast away at our mouths like sugar does; by and large, its allure is more surreptitious. . . . If sugar is the methamphetamine of processed food ingredients, with its high-speed, blunt assault on our brains, then fat is the opiate, a smooth operator whose effects are less obvious but no less powerful.
>
> *(pp. 147–148)*

Fat also provides "mouthfeel" to foods: "When it comes to fat, it detects the enthralling crunch in fried chicken. The velvet in melting chocolate and premium ice cream, the creaminess in cheese" (Moss, 2013, p. 154).

Yet, Herz (2018) notes that fat has the highest caloric density, meaning it has more calories than other substances such as sugar and twice as much as carbohydrates. She points out that "this was especially important in the time of our ancestors when a good calorie was hard to find." But, in today's food environment when calories are nearly everywhere at all times, our body cannot do a good job regulating substances we consume in large quantities, including but not limited to fat.

According to Herz, we are programmed to love fat, along with sugar and salt, and thus our bodies set us up to fail when it comes to regulating our weight. It is the contemporary food environment that makes it impossible for most people to manage their weight, due to their own biology.

Most dangerous to human health are trans fats and saturated fats, in that order. In 1961, the AHA suggested that people reduce their intake of saturated fats by replacing animal fats with vegetable oils, but the switch from butter to vegetable oil is not necessarily healthier. For example, even supposedly healthy olive oil has been found in some studies to decrease blood flow and increase inflammation.

It is clear that saturated fats found in foods such as meat and cheese are considered dangerous, so they could be considered defective products. One study, for example, found that a diet that replaced only 5% of calories from saturated fats was associated with a decline in death of 27% (Fuhrman, 2017, p. 52). Yet, another study found a 43% decline in death among people eating low-carb diets that were high in animal products. And still study found a 60% reduction in cardiovascular disease among people following this type of diet. Yet, still another study found a fourfold increase in cancer death and an 85% increase in death overall in those eating more animal products (Fuhrman, 2017, pp. 71–73). Consumption of animal proteins in amounts that top 10% of total calories are associated with increases in insulin-like growth factor-1, which can promote cancer and the spread of cancer: "Higher intake of animal products in the diet is associated with twelve different types of cancer" (Fuhrman, 2017, p. 70). Figure 4.5 shows the nutritional content of a sausage product. Notice the high levels of fat and salt in this product.

In 1910, Proctor and Gamble applied for a patent on hydrogenation in food products, and trans fats were born. Their "Crisco" product, derived from crystallized cottonseed oil, would be used as a shortening in cooking. According to Minger (2013, p. 153), Crisco was made from cottonseed waste, and thus, Proctor and Gamble "had pioneered what's now an American tradition: getting rid of agricultural waste products by feeding them to humans. The company had effectively bridged the gap between garbage and food." The company created a book, *The Story of Crisco*, and handed it out for free. It claimed that the product made foods more digestible, economical, and better tasting, and it "was presented as a panacea of sorts—healthier than lard, more economical than butter, and altogether in a category of its own" (Minger, 2013, p. 154). It was seen as "nothing short of a miracle. It came from plants; it was firm; it was tasty; it was cheap; it fried foods without smoking; and huzzah, it was even kosher and *parava*—usable with both milk and meat per Jewish law" (Minger, 2013, p. 155). But, of course, there is a downside:

> Crisco became the first ingredient to unleash unprecedented levels of linoleic acid—a polyunsaturated fat—into the American diet: trans fat resulting from partially hydrogenating oils and an astronomical intake of omega-6 fats—both now known to increase the risk of heart disease and cause inflammatory immune responses. It would be many decades before anyone realized what had gone so horribly wrong. In fact, the USDA would promote trans fats all the way up until 2005.
>
> *(Minger, 2013, p. 156)*

Margarine is in the same boat, filled with trans fats and thus really bad for you, yet, when discovered, the USDA reportedly suppressed the findings. According to Minger (2013, p. 158), this is because partially hydrogenated oils "were a sacred cow for food manufacturers: they were cheaper than animal fats, had a gloriously high melting point, prolonged the shelf life of whatever they touched, and provided just the right consistency to make foods profitably addictive." It was not until 2006 when the FDA took action, requiring

**FIGURE 4.5**  Nutritional Information in a Sausage Product

*Source:* Photo by author

food manufacturers to list the amount of trans fats in their products on nutrition labels. Earlier, the 1992 dietary guidelines did not mention that trans fats were under investigation as a potential health threat, and they even encouraged consumers to buy margarines with vegetable oils, thus encouraging Americans to eat trans fats.

Experts recommend limiting one's intake of fats, especially unhealthy fats like saturated and trans fats. Since at least the 1940s—when rates of heart disease fell during a time when low fat, low cholesterol, and rationed diets were popular due to the war—it is believed that too much fat is bad for you. Federal dietary guidelines suggest we should receive no more than 25–35% of our calories from fat. Yet: "Many soups, cookies, potato chips, cakes, pies, and frozen meals deliver half or more of their calories through fat." Still, "consumers won't identify these as fatty foods, which is great for sales. For some extra insurance on this, all the manufacturers have to do is add a little sugar" (Moss, 2013, p. 158). Interestingly, studies show that consumers can fairly accurately identify how much sugar is in a product but are unable to do this with fat. The danger of this is that even slight overeating is associated with weight gain, and when people do not know how much fat they are eating, they tend to eat more (Moss, 2013, p. 181). And adding sugar to fatty foods makes people tend to think they are getting less fat than they actually are.

One major source of fat in our diets is cheese, yet, in the United States, not much cheese is actually used in products. Instead, "cheese food, cheese product, and pasteurized processed American cheese" are terms used by federal regulators to describe the types of items added to many of our foods (Moss, 2013, pp. 162–163). According to Moss (2013), in 1970, Americans ate about 11 pounds of cheese a year; this increased to 18 pounds in 1980, 25 pounds in 1990, and 30 pounds in 2000. Here is one area where an easy change to manage or lose weight would be eliminating the amount of cheese one eats. Yet, companies add cheese to so many products because it "has proven to be a windfall for companies, driving up sales of cheese as well as the products that now use it to increase their allure" (Moss, 2013, p. 164). And this is problematic because adding cheese and thus more fat to products increases the odds that people will overeat.

Moss (2013, p. 214) explains, about saturated fat: "It is a primary cause of high cholesterol in the bloodstream, a waxy substance that lead to heart attacks and strokes." Meanwhile, the US government does not encourage people in its dietary guidelines to eat less meat and cheese, two major sources of saturated fat, likely because of the fact that the Department of Agriculture is in the business of promotion of their consumption. In fact, the 2010 guidelines even encouraged people to eat more cheese, not less! Further, "meat is touted throughout the report with the added assurance that neither it, nor milk products, have been specifically linked to obesity." The government calls these "important sources of nutrients in healthy eating patterns" (p. 217).

## Additives

According to Kaplan (2016, p. 264), the main classes of food ingredients per the FDA include "preservatives, sweeteners, color additives, flavors and spices, fat replacers, nutrients, emulsifiers, stabilizers and thickeners, pH control agents, dough strengthener and conditioners, firming agents, enzyme preparations, and gases." Many of these agents result in foods that are "bred for transportability and long shelf life" rather than taste and especially health (Nestle, 2013, p. 165).

In 1958, Congress defined two categories of food additives, one that needed to be proven as safe before they could be added to foods and other that have been used for some time and thus are "generally recognized as safe" (GRAS) (Nestle, 2018). As one example, consider maltodextrin. Maltodextrin is a substance added to many foods that does not taste sweet (yet has as many calories as sugar) that allows food producers to keep us coming back for more as it lights up the brain in the same way that sugar does.

Today, there are at least 10,000 substances that can be added to products that are considered GRAS. It is the companies that generally make this determination. Typically, a panel of three experts is retained to review evidence to make determinations about whether additives are safe, yet, sometimes it is just one expert who does the review; there is no oversight by the US FDA of this process. According to Walker (2019), about 1,000 of these substances have been added to foods without the knowledge of the US FDA.

Incredibly, the same food scientists used "to determine new additives can be deemed 'generally recognized as safe' [GRAS]" once worked for big tobacco companies to make the same determinations for additives for cigarettes. A review of 379 panels convened over a 17-year period found that three-quarters featured at least one of ten specific scientists and "at least four of the top 10 GRAS panel experts . . . had also served as scientific consultants for cigarette makers" (Center for Science in the Public Interest, 2022).

With regard to GRAS additives, Young and Quinn (2015) note: "Consumers regularly eat foods with added flavors, preservatives and other ingredients that are secretly added by companies and not reviewed for safety by regulators." That is, there is no notice to, or review by, the FDA, as noted earlier (Young & Quinn, 2015). Those who make these determinations may very well have conflicts of interests since they are paid by the industry to make decisions about the relative safety of product additives created by those companies. Young and Quinn (2015) report that a 2013 review by the Pew Charitable Trusts found that "financial conflicts of interest were ubiquitous."

Sometimes there are, allegedly, conflicts of interest when it comes to the designation of GRAS. Nestle (2018, p. 110) notes the example of the GRAS committee relying on materials provided by the Sugar Association. For example:

> The 1976 GRAS review concluded that other than the contribution made to dental caries, there is no clear evidence in the available information on sucrose that demonstrates a hazard to the public when used at the levels that are now current and in the manner now practiced.

To many, this might sound like there is no link between sugar and deleterious outcomes such as obesity, diabetes, and so on. Yet, research reported in this chapter suggests these links are real. Later, in 2013, an analysis of the GRAS review process "concludes that the industry ties of committee members not only threaten the integrity of the GRAS reviews by also the integrity of the FDA's entire scientific enterprise." Nestle adds that "without independent review of GRAS additives, it is difficult to be confident that the ones in use are safe" (p. 110). That is quite the claim, especially given, again, that GRAS means "generally regarded as safe"!

Walker (2019, pp. 180–181) provides this example of an additive that is considered unsafe, to show how difficult it can be to even get these things removed from products even when we know they are harmful: "Artificial trans fat, or partially hydrogenated oil, is made by adding hydrogen to solidify liquid vegetable oils." This harmful substance was exempted from exclusion in 1958 by Congress.

> Because trans fat is inexpensive, tasty, keeps foods from spoiling, and can be used multiple times in deep fryers, food providers love it. But for consumers, trans fat raises bad

cholesterol (LDL) and lowers good cholesterol (HDL), a one-two punch against good health. Despite decades of research consistently pointing out its dangers, FDA did nothing until a petition and lawsuit were filed in 2009.

Incredibly, it was not until 2015 that the FDA ruled

that trans fat be phased out over three years. This ruling, however, was not a prohibition but a statement that trans fat was not safe for consumption. Companies could still seek permission for specific uses. Also, usage and enforcement in non-interstate commerce like local restaurants was left to each state.

Many additives are artificial and come from labs. Moss (2021, pp. 104–105), for example, notes the presence of "flavor houses," chemical factories that create and perfect the flavors of foods. And he points out:

At the behest of the processed food industry, the flavor houses . . . make scents that mimic the char on meat for a tastier veggie burger, scents that stay dormant in a box until water is added, and scents that mask the undesirable smell that can arise in the making of processed foods. These potions are the unsung champions of modern-day food that comes packaged and able to sit on a shelf for months at a time without going bad, and they are, by design, quite secretive.

Incredibly, these chemicals do not have to be included on product labels but instead are noted as "natural and artificial flavors." This means "we can't know which chemicals are being used in the food we eat, though the brain sure does. The volatiles from these compounds strike the olfactory bulb with the singular goal of arousing our appetite." So, this is a devious way to get us to eat and keep eating. Food companies are culpable for these chemical additives to our food, as well as for any deleterious outcomes that result. Consumers cannot make informed decisions about which chemicals to avoid given we are not even entitled to know which ones are in our food!

## Antimicrobials

According to Silbergeld (2016, p. 89): "Growth promoting antibiotics or antimicrobials (GPAs) are what the FDA calls drugs that are added to feeds to increase growth rates of pigs, chickens, and other animals that are raised for human consumption." Adding antimicrobials to livestock feed supposedly accelerates the growth of the animals. Walker (2019, p. 194) says that, in 1925, it took 1,212 days to raise a chicken to 2.5 pounds, and today, one can grow a five-pound bird in just 50 days. According to Walker (2019), the FDA tried to stop the addition of antimicrobials to animal feed, but Congress would not take action. Incredibly: "The National Academy of Sciences, the World Health Organization, the Government Accountability Office, an FDA taskforce, and an FDA national advisory committee all advised against the use of antibiotics in feed." Yet, the FDA began permitting the use of antibiotics in animal food, and its use increased tenfold in ten years (Walker, 2019, p. 118).

Today, about 80% of all antibiotics used in the United States go to animals and agriculture. According to EcoNexus (2015), one-third of antibiotics sold in Germany are for animal production, and one-half are used that way in China: "In the US, where antibiotics are permitted to accelerate growth, eight times more antibiotics are used in factory farms than in hospitals." They are also used in "aquaculture, honeybees, and companion animals, and they are even sprayed on fruit trees like apples, pears, and peaches" (Walker, 2019, p. 193). When used on animals, the belief is that this allows large amounts of animals to be kept in confinement in tight spaces. So, chicken farmers went from 355 chickens per farm in 1950 to up to 500,000 now. Recall the discussion of confinement and concentration from Chapter 2.

Of course, antibiotics do not kill off all pathogens; some survive, and they become resistant to antibiotics. This impacts humans. Hauter (2012, p. 142) claims that antibiotic resistance is a threat to human health for we "come into contact with these strains through [our] food, the air, the water, and the soil." MRSA, a type of staph infection that is antibiotic resistant, kills more than 17,000 Americans a year (Hauter, 2012, p. 142). Incredibly, the European Union and WHO agree that antibiotic usage in animal feed should be banned.

It is the consequences in humans with which we must be most concerned: "The consequences are ever more antibiotic resistant bacteria and increasing numbers of people whose infections cannot be cured with antibiotics any more. The Word Health Organisation (WHO) says this is one of the most serious threats for human health." According to Walker (2019, p. 195), "at least two million people per year become infected with bacteria resistant to antibiotics; some 23,000 die as a result."

According to Silbergeld, these drugs have a major impact on human health, but the industry and government both fail to recognize it. FDA approval of these drugs is based on faulty interpretation of anecdotal historical evidence, she claims. But the reality is that the real rationale for adding these drugs to feeds is money, specifically to lower costs. Following the logic of "feed conversion efficiency," or using the least amount of feed needed to raise animals to the point of slaughter, the industry believes using drugs in feed reduces their overall costs. The industry also now claims that the drugs in feed prevent illness in animals, but in reality, the levels of antimicrobials in feed are so low; they are considered subtherapeutic. And evidence reviewed by Silbergeld shows no evidence whatsoever of their effectiveness: "Despite decades of hypotheses, no clear scientific rational exists to support GPA us in animal feeds" (p. 99). She adds: "There is no good basis for the approval of GPAs starting in 1947, and there has been no demonstration since that time that GPAs are necessary to industrialized food animal production" (p. 108).

Silbergeld (2016, p. 106) claims the continued use of GPAs in animal feed owes itself, at least in part, to the reality that it is much easier to start a practice in animal food production than it is to stop it, "even after a harm is evident." See? "Once something is being made and sold, there are economic interests in its perpetuation." Another reason is fear of liability: "Acknowledgment of the need to change is too often seized upon as tantamount to an admission of guilt for not doing it easier or doing it in the first place" (p. 107). Yet: "There is no question that antimicrobial use in food animal production is a major driver of resistance to treatment of infectious diseases." The millions of cases of illness and tens of thousands of deaths that occur as a result make the food industry culpable. If the use of antimicrobials in animal feed serves no purpose other than to reduce industry costs, profit-seeking comes at the expense of human health, and that is simply not a tenable situation. This practice should be stopped and, if it is not, both industry and government are responsible for the harms associated with its continuation.

## GMOs

GMOs are genetically modified organisms. According to Constance et al. (2019, p. 87): "The first commercial GMO activity occurred in the United States with the production of canola, corn, cotton, soybeans, and tomatoes in the 1990s." GMO technology, though, began in 1973, and the US Supreme Court ruled in 1980 that living organisms could be held as intellectual property, "which prompted rapid investment and growth of the industry" (p. 87). After the US Supreme Court ruled that inserting genetic material into a bacteria cell could be patented, Monsanto altered the DNA of soybeans to resist its own chemical product glyphosate (Roundup).

After this, the US Trade Patent Office ruled that "farmers could no longer save and replant seeds that companies like Monsanto had patented; outside research was also forbidden" (Walker, 2019, p. 129). This means that, even when Monsanto's seeds blow over onto their lands, they cannot be used by farmers unless they signed binding "technology use agreements" that require them

> to adhere to Monsanto's prescribed farming practices, which included granting the company access to their farm and all farm records. Should Monsanto determine that the farm was in violation of the agreement, the farmer agreed to pay all costs demanded by the company, including litigation expenses.
>
> *(p. 130)*

Incredibly, any study conducted by independent scientists was also a violation of federal law. In 2009, Monsanto granted partial access for research purposes.

The power of Monsanto can be seen with regard to the use of GMO seeds. When the EPA proposed using them on every other acre of land, to reduce the prevalence of resistant bugs, Monsanto objected and EPA backed off. They agreed to plant non-GMO seeds on every fifth acre (Walker, 2019, p. 197).

The first crop approved for commercialization was a tomato modified to delay ripening, in 1992 (Glenna & Tobin, 2019). A National Academy of Sciences report, titled, "Genetically Engineered Crops: Experiences and Prospects," contains a list of GMO crops. Today, products such as corn and soy are often GMO products, and "between 1996 and 2015, the percent of genetically modified corn and soy plants in the United States increased from less than 20 percent to 92 and 94 percent, respectively" (Gillon, 2019, p. 211). Odds are that, if you are eating corn or soy, it is a GMO product you are eating. Keep in mind, however, that corn is very useful in other products including sweeteners and processed foods, but it is also used as livestock feed, and it is turned into fuel for cars.

According to Clausen et al. (2019), 2015 saw the first FDA-approved genetically altered animal approved for food, in the form of salmon that grows twice as fast as normal salmon. Since less feed is required to grow the fish to full size, as well as less time, proponents claim the fish is more sustainable. Walker (2019, p. 188) explains the logic of GMO products, in this case, seeds: "farmers are planting seeds engineers to resist pesticides. By genetically altering the seeds' DNA to withstand compound and chemicals . . . farmers can blanket spray their fields, killing everything but the crops."

Walters (2019, pp. 270–271) outlines issues with GMOs, including:

- Risk and uncertainty regarding the development of GM crops
- Economic harm created through GM contamination via gene flow . . . or inadequate storage and transportation loss
- Environmental harm including the breeding of wild weed species . . . and the demise of beneficial insects
- Bt toxin persistence and unintended effects of soil biota
- Safety of long-term ingestion of GM foods
- Concentration of capital and privatization of genetic information and its influence on the control of agricultural production
- Deviant and illegal activities of corporate actors in controlling and monopolizing the production, distribution, and regulation of GM technologies, seeds, and plants

Notice there is no discussion of actual threats to human safety with regard to GMO foods. That is because, as of this writing, GMOs are considered safe for human consumption. The FDA, for example, says: "GMO foods are carefully studied before they are sold to the public to ensure they are as safe as the other foods we currently eat. These studies show that GMOs do not affect you differently than non-GMO foods."

## Conclusion

The conventional food system provides relatively affordable food for nearly all of the US population. It is a marvel worthy of careful study and consideration. Yet, a major criticism of the system is that a large majority of food products it produces are ultra-processed and contain none of the next-to-none nutritional value whatsoever. In fact, about 60% of calories consumed in the United States come from ultra-processed foods.

These foods, high in salt, sugar, fat, saturated and trans fats, plus additives that contain chemicals that consumers are not even entitled to know about (because they are phrased in products simply as "natural and artificial flavors") are quite literally dangerous. As shown in this book, nearly 400,000 Americans die from poor diet and nutrition every year. Yet, the foodstuff that is marketed to and sold to us starting as children is perfectly legal, in spite of the fact that much of it can be seen as defective.

The point of this chapter is not to argue that any food be criminalized. That is not what is meant by the phrase "the food IS the crime." Instead, the phrase is meant to call attention to the fact that there is significant harm associated with the foods we eat (and fail to eat), and there is culpability in the companies who market and sell these products to us. In essence, since defective products are generally understood to mean any product that kills or injures people that is not intended to kill or injure people (Robinson, 2020), many of the food products being marketed and sold to us can be seen as defective. Since many of them—the so-called "hyperpalatable" ultra-processed foods are addictive in nature (Robinson, 2022), one could argue that major food manufacturers are in the drug business. Seeing food products this way may be a stretch for some, but once one comes to grips with the reality that it is true, it is much easier being able to imagine food companies being held accountable for harms associated with their products.

A similar argument was made against big tobacco, although tobacco is obviously not a food product. The point is that, in spite of the enormous harms caused by tobacco use (roughly 480,000 deaths a year in the United States alone), tobacco remains legal. Yet, major tobacco companies have been held responsible for harms associated with their products.

The same could happen to major food companies, especially those who have been found to be targeting vulnerable children with their defective products.

I do not anticipate this happening any time soon. But, research like that presented in this chapter will go a long way to making the case that something needs to be done. The alternative is doing nothing and continuing to let hundreds of thousands of people die prematurely as a result. That would be a crime.

Importantly, if any entity is negligent here, it is the government. Stated simply, the government does little to regulate food for substances such as salt, sugar, fat, or additives, and it allows antimicrobials in animal feed, in spite of the clear harms associated with each of these substances. The issue of state-corporate criminality was introduced in Chapter 1. This chapter provides some pretty good examples of substances found in foodstuffs that amount to state-corporate crime in the sense that they are drastically harmful, included by corporations, and enabled by the government.

# 5

# FOOD ADDICTION

## Introduction

In this chapter, the issue of food addiction is examined. I begin the chapter with a differentiation of drug use and drug abuse, and I compare and contrast addiction with substance use disorder. I ask and answer why people use drugs in the first place and show the role the human brain plays in it. Then, I examine whether food is addictive, applying various definitions of addiction to food. The review of the evidence suggests that, yes, some foods may be considered addictive. In particular, ultra-processed and so-called "hyperpalatable foods" are found to meet many of the definitions of addiction. We have known for some time that eating can lead to disease!

Given that some foods can be considered drugs, this means the people who sell those foods can logically be seen as drug dealers. The implications of this for justice are profound, yet, virtually no one is addressing this reality in contemporary society. But, once we come to see certain types of foods as also drugs, it likely will change the way we react to their production and sales, or at least it should. It is important to keep in mind lessons learned in this chapter as you read the rest of the book, for one major potential harm of food is addiction.

## Drug Use versus Drug Misuse and Drug Abuse

*Drug use* refers to any ingestion of a drug into the body. A *drug* is generally understood to be any substance, that, when ingested into the body, produces a physiological change in the body. Although food would clearly fit this definition, scholars generally point out that the definition of drugs does *not* include food. For example, the Food, Drug, and Cosmetic Act (21 US Code Section 321) specifies:

> The term "drug" means (A) articles recognized in the official United States Pharmacopoeia, official Homoeopathic Pharmacopoeia of the United States, or official National Formulary, or any supplement to any of them; and (B) articles intended for use in the diagnosis, cure, mitigation, treatment, or prevention of disease in man or other animals; and (C) articles (other than food) intended to affect the structure or any function of the body of man or other animals; and (D) articles intended for use as a component

DOI: 10.4324/9781003296454-5

of any article specified in clause (A), (B), or (C). . . . A food, dietary ingredient, or dietary supplement for which a truthful and not misleading statement is made in accordance with section 343(r)(6) of this title is not a drug under clause (C) solely because the label or the labeling contains such a statement.

*(Cornell Law School, 2020)*

Interestingly, the federal law noted earlier states that foods and dietary supplements are *not* considered drugs even though research shows that, in some cases, they are.

An example of drug use is drinking a cup of coffee in the morning, having a beer after work, or drinking a glass of wine with dinner. *Drug misuse* refers to "improper or unhealthy use" rather than proper medical use, or overuse of a drug (including alcohol) rather than use in moderation. Drug misuse includes "repeated use of drugs to produce pleasure, alleviate stress, and/or alter or avoid reality. *It also includes using prescription drugs in ways other than prescribed or using someone else's prescription" (National Institute of Drug Abuse, NIDA, 2020, emphasis in original). One can envision the inclusion of food when it comes to misuse, given the realities of overeating, binge eating, obesity, and so forth.* In fact, even in the absence of physical dependence, Ziauddeen and Fletcher (2013, p. 21) suggest an alternative to the term food addiction—"food abuse or misuse." This issue is revisited later in the chapter.

## Why Do People Use Drugs?

According to the National Institute on Drug Abuse (NIDA), people use drugs:

- to feel good—feeling of pleasure, "high"
- to feel better—for example, relieving stress
- to do better—improving performance
- out of curiosity
- because of peer pressure

The "feeling" of course occurs in the brain. Hartney (2019) notes: "While the pharmacological mechanisms for each class of drug are different, the activation of the reward system [in the brain] is similar across substances in producing feelings of pleasure or euphoria, which is often referred to as a 'high.' " It is important to note that some foods are used by people to feel good or better and that they activate the same regions of the brain as drugs. This will be addressed later in the chapter.

### How the Brain Is Involved

The brain is involved in drug use because drugs "directly or indirectly target the brain's reward system by flooding the circuit with dopamine" (NIDA, 2020). Dopamine "is important for reward-related processes driving substance-seeking behavior" (Wiss & Brewerton, 2020, p. np). NIDA notes that

Dopamine is a neurotransmitter present in regions of the brain that regulate movement, emotion, cognition, motivation, and reinforcement of rewarding behaviors.

When activated at normal levels, this system rewards our natural behaviors. Overstimulating the system with drugs, however, produces effects which strongly reinforce the behavior of drug use, teaching the person to repeat it.

This same outcome occurs with food, especially those foods that taste good and consist largely of sugar, salt, and fat (Breslin, 2013; Drewnowski & Almiron-Roig, 2010; Jabr, 2016; Moss, 2013). Eördögh et al. (2016, p. np) confirm that "neurobiological circuits involved in the development of drug addiction also play a role in food consumption." Other scholars agree: "Drugs and food both exert a rewarding effect through the firing of dopamine neurons . . . resulting in the release of dopamine . . ." (Lindgren et al., 2018, p. 811). This issue will be the focus of much of this chapter.

In fact, all reward responses occur in the same areas of the brain, whether you are talking about drugs, food, or something else, including, for example, gambling. Thus, it makes sense the Diagnostic and Statistical Manual of Mental Disorders (DSM-V) includes not only drug use disorders but also Gambling Disorder and Internet Gaming Disorder. Each outcome may reflect a *Reward Deficiency Syndrome*, whereby low levels of dopamine are ultimately caused by excessive engagement with drugs, food, or gambling over time. This leads to greater use or participation to experience the desired effect due to increased *tolerance* over time, where more of a substance (or experience) is needed to achieve the desired effect. *Withdrawal*, a key sign of drug abuse, also happens with low dopamine; in this case, people seek out substances to deal with negative physical symptoms emanating from *not* being fulfilled. As will be shown later in the chapter, the same finding occurs with certain foods (Pursey et al., 2014).

### What Is Addiction?

There are numerous definitions of addiction available. The American Psychiatry Association (APA, 2020) defines *addiction* as a "brain disease that is manifested by compulsive substance use despite harmful consequence." APA adds that addiction involves an "intense focus on using a certain substance(s) . . . to the point that it takes over their life." That is, drug users will continue to use even knowing it will likely lead to problems in relationships at work, home, and/or school.

APA (2020) lays out the outcome of substance use disorders, as well as others:

- Social problems: Substance use causes failure to complete major tasks at work, school, or home; social, work, or leisure activities are given up or cut back because of substance use.
- Impaired control: A craving or strong urge to use the substance; desire or failed attempts to cut down or control substance use.
- Risky use: Substance is used in risky settings; continued use despite known problems.
- Drug effects: Tolerance (need for larger amounts to get the same effect); withdrawal symptoms (different for each substance).

*(APA, 2020)*

The American Society of Addictive Medicine (ASAM, 2020) defines addiction as

> a treatable, chronic medical disease involving complex interactions among brain circuits, genetics, the environment, and an individual's life experiences. People with addiction use substances or engage in behaviors that become compulsive and often continue despite harmful consequences.

Those harmful consequences include not only potential physical withdrawal and possible overdose but also interferences in relationships, as noted earlier.

NIDA (2020) defines addiction as a "chronic, relapsing disorder characterized by compulsive drug seeking, continued use despite harmful consequences, and long-lasting changes in the brain." NIDA goes on to point out that addiction "is considered both a complex brain disorder and a mental illness. Addiction is the most severe form of a full spectrum of substance use disorders, and is a medical illness caused by repeated misuse of a substance or substances" (NIDA, 2020).

Addiction is characterized by

> distorted thinking, behavior and body functions. Changes in the brain's wiring are what cause people to have intense cravings for the drug and make it hard to stop using the drug. Brain imaging studies show changes in the areas of the brain that relate to judgment, decision making, learning, memory and behavior control.
>
> *(APA, 2020)*

NIDA (2020) adds that addiction changes brain function in the areas of "natural inhibition and reward centers."

For food to be considered addictive by these definitions, it must have the potential to produce outcomes like those mentioned earlier. A review of the literature on food addiction, provided later in the chapter, shows that some foods do produce at least some of these outcomes.

## Addiction versus Substance Use Disorder

Addiction is not a diagnosis in the Diagnostic and Statistical Manual of Mental Disorders (DSM-5). Rather, the DSM includes "substance use disorder," a replacement of the previous categories (from DSM-IV) substance abuse and substance dependence. Substance use disorder has three subclassifications of mild, moderate, and severe based on the number of symptoms identified (two to three symptoms is considered "mild," four or five is considered "moderate," and six or more is "severe"). NIDA (2020) explains: "The new DSM describes a problematic pattern of use of an intoxicating substance leading to clinically significant impairment or distress with 10 or 11 diagnostic criteria (depending on the substance) occurring within a 12-month period."

The diagnostic criteria include the following:

1. The substance is often taken in larger amounts or over a longer period than was intended.
2. There is a persistent desire or unsuccessful effort to cut down or control the use of the substance.

3. A great deal of time is spent in activities necessary to obtain the substance, use the substance, or recover from its effects.
4. Craving, or a strong desire or urge to use the substance, occurs.
5. Recurrent use of the substance results in a failure to fulfill major role obligations at work, school, or home.
6. Use of the substance continues despite having persistent or recurrent social or interpersonal problems caused or exacerbated by the effects of its use.
7. Important social, occupational, or recreational activities are given up or reduced because of the use of the substance.
8. Use of the substance is recurrent in situations in which it is physically hazardous.
9. Use of the substance is continued despite knowledge of having a persistent or recurrent physical or psychological problem that is likely to have been caused or exacerbated by the substance.
10. Tolerance, as defined by either of the following:

   a. A need for markedly increased amounts of the substance to achieve intoxication or desired effect
   b. A markedly diminished effect with continued use of the same amount of the substance.

11. Withdrawal, as manifested by either of the following:

   a. The characteristic withdrawal syndrome for that substance (as specified in the DSM-5 for each substance)
   b. The use of a substance (or a closely related substance) to relieve or avoid withdrawal symptoms (Hartney, 2019).

According to Hartney (2019), the DSM acknowledges that substance use disorders can emerge from numerous classes of drugs, including alcohol, anxiolytics (i.e., anti-anxiety medicine), caffeine, cannabis, hallucinogens, hypnotics, inhalants, opioids, sedatives, stimulants (including amphetamines and cocaine), tobacco, and other or unknown substances. Hartney notes that "the use of other or unknown substances can also form the basis of a substance-related or addictive disorder." This could presumably include food. In fact, research into food addiction does show many of the aforementioned diagnostic criteria are in fact met when it comes to the use of some foods. This review is presented later in the chapter.

Table 5.1 illustrates the varying conceptions of addiction and substance use disorders. These are taken from the aforementioned definition, offered by APA, ASAM, and NIDA, and from information within the DSM. Later in the chapter, this table will be revisited to see which of these conceptions of addiction and substance use disorders apply to food.

Interestingly, a scale of food addiction—the Yale Food Addiction Scale (YFAS)—has been developed based on similar measures. This is a 25-point self-report questionnaire meant to discover the impact of addictive behaviors related to food, based on substance dependence criteria from the DSM. Table 5.2 shows this scale. The survey examines eating habits in the past year, and asks questions related to eating more than one plans, eating when no longer hungry, eating to the point of feeling ill, eating in spite of worrying about foods being consumed, various negative health outcomes resulting from eating, interferences in quality of life due to food, food cravings, as well as physical, emotional, and psychological outcomes consistent with withdrawal, and more. The YFAS allows us to get a sense of the prevalence of food addiction; those data are shared later in the chapter.

**TABLE 5.1** Indices of Addiction and Substance Use Disorders

_Social problems_—persistent or recurrent social or interpersonal problems caused by use; use interferes with work, school, or home; failure to sustain obligations; social, work, or leisure activities are given up or cut back (APA, DSM, and ASAM)

_Impaired control, compulsive use_—cravings or strong urges to use; desire to or failed attempts to reduce use; using more than intended (APA, DSM, ASAM, and NIDA)

_Risky use_—use in risky or hazardous settings; continued use despite known problems; continued use despite knowing one has a persistent or recurrent physical or psychological problem caused or exacerbated by the substance (APA, DSM, ASAM, and NIDA)

_Drug effects_—tolerance; withdrawal symptoms (APA and DSM))

_Brain disease_—causes physical problems in the brain; drive to use based on neurotransmitters (APA, ASAM, and NIDA)

_Time usage_—a large amount of time is spent in activities necessary to obtain the substance, use the substance, or recover from its effects (DSM)

_Source:_ <u>Key:</u>
_APA—American Psychological Association_
_DSM—Diagnostic and Statistical Manual for Mental Disorders_
_ASAM—American Society of Addictive Medicine_
_NIDA—National Institute on Drug Abuse_

**TABLE 5.2** Yale Food Addiction Scale, Version 2

This survey asks about your eating habits in the past year. People sometimes have difficulty controlling how much they eat of certain foods such as:

- Sweets like ice cream, chocolate, doughnuts, cookies, cake, and candy
- Starches like white bread, rolls, pasta, and rice
- Salty snacks like chips, pretzels, and crackers
- Fatty foods like steak, bacon, hamburgers, cheeseburgers, pizza, and French fries
- Sugary drinks like soda pop, lemonade, sports drinks, and energy drinks

When the following questions ask about "CERTAIN FOODS," please think of ANY foods or beverages similar to those listed in the food or beverage groups above or ANY OTHER foods you have had difficulty with in the past year

IN THE PAST 12 MONTHS: Never Less than monthly Once a month 2–3 times a month Once a week 2–3 times a week 4–6 times a week Every day

1. When I started to eat certain foods, I ate much more than planned.
   0     1     2     3     4     5     6     7
2. I continued to eat certain foods even though I was no longer hungry.
   0     1     2     3     4     5     6     7
3. I ate to the point where I felt physically ill
   0     1     2     3     4     5     6     7
4. I worried a lot about cutting down on certain types of food, but I ate them anyways.
   0     1     2     3     4     5     6     7
5. I spent a lot of time feeling sluggish or tired from overeating.
   0     1     2     3     4     5     6     7
6. I spent a lot of time eating certain foods throughout the day.
   0     1     2     3     4     5     6     7
7. When certain foods were not available, I went out of my way to get them. For example, I went to the store to get certain foods even though I had other things to eat at home.
   0     1     2     3     4     5     6     7

_(Continued)_

**TABLE 5.2** (Continued)

8. I ate certain foods so often or in such large amounts that I stopped doing other important things. These things may have been working or spending time with family or friends.

    0        1        2        3        4        5        6        7

9. I had problems with my family or friends because of how much I overate.

    0        1        2        3        4        5        6        7

10. I avoided work, school, or social activities because I was afraid I would overeat there.

    0        1        2        3        4        5        6        7

11. When I cut down on or stopped eating certain foods, I felt irritable, nervous, or sad.

    0        1        2        3        4        5        6        7

12. If I had physical symptoms because I hadn't eaten certain foods, I would eat those foods to feel better.

    0        1        2        3        4        5        6        7

13. If I had emotional problems because I hadn't eaten certain foods, I would eat those foods to feel better.

    0        1        2        3        4        5        6        7

14. When I cut down on or stopped eating certain foods, I had physical symptoms. For example, I had headaches or fatigue.

    0        1        2        3        4        5        6        7

15. When I cut down or stopped eating certain foods, I had strong cravings for them.

    0        1        2        3        4        5        6        7

16. My eating behavior caused me a lot of distress.

    0        1        2        3        4        5        6        7

17. I had significant problems in my life because of food and eating. These may have been problems with my daily routine, work, school, friends, family, or health.

    0        1        2        3        4        5        6        7

18. I felt so bad about overeating that I didn't do other important things. These things may have been working or spending time with family or friends.

    0        1        2        3        4        5        6        7

19. My overeating got in the way of me taking care of my family or doing household chores.

    0        1        2        3        4        5        6        7

20. I avoided work, school, or social functions because I could not eat certain foods there.

    0        1        2        3        4        5        6        7

21. I avoided social situations because people wouldn't approve of how much I ate.

    0        1        2        3        4        5        6        7

22. I kept eating in the same way even though my eating caused emotional problems.

    0        1        2        3        4        5        6        7

23. I kept eating the same way even though my eating caused physical problems.

    0        1        2        3        4        5        6        7

24. Eating the same amount of food did not give me as much enjoyment as it used to.

    0        1        2        3        4        5        6        7

25. I really wanted to cut down on or stop eating certain kinds of foods, but I just couldn't.

    0        1        2        3        4        5        6        7

26. I needed to eat more and more to get the feelings I wanted from eating. This included reducing negative emotions like sadness or increasing pleasure.

    0        1        2        3        4        5        6        7

27. I didn't do well at work or school because I was eating too much.

    0        1        2        3        4        5        6        7

28. I kept eating certain foods even though I knew it was physically dangerous. For example, I kept eating sweets even though I had diabetes. Or I kept eating fatty foods despite having heart disease.

    0        1        2        3        4        5        6        7

*(Continued)*

**TABLE 5.2** (Continued)

| | | | | | | | |
|---|---|---|---|---|---|---|---|

29. I had such strong urges to eat certain foods that I couldn't think of anything else.

 0  1  2  3  4  5  6  7

30. I had such intense cravings for certain foods that I felt like I had to eat them right away.

 0  1  2  3  4  5  6  7

31. I tried to cut down on or not eat certain kinds of food, but I wasn't successful.

 0  1  2  3  4  5  6  7

32. I tried and failed to cut down on or stop eating certain foods.

 0  1  2  3  4  5  6  7

33. I was so distracted by eating that I could have been hurt (e.g., when driving a car, crossing the street, and operating machinery).

 0  1  2  3  4  5  6  7

34. I was so distracted by thinking about food that I could have been hurt (e.g., when driving a car, crossing the street, and operating machinery).

 0  1  2  3  4  5  6  7

35. My friends or family were worried about how much I overate.

 0  1  2  3  4  5  6  7

*Source:* Food and Addiction Science & Treatment Lab (2020). Yale Food Addiction Scale. Downloaded from: https://fastlab.psych.lsa.umich.edu/yale-food-addiction-scale/

## Is Food Addictive?

Before considering whether food is addictive, it is first important to differentiate between two types of hunger. "Metabolic hunger" or "homeostatic hunger" is feeling a need to eat that is "driven by physiological necessity and is commonly identified as the rumbling of an empty stomach" (Jabr, 2016, p. np). This is regulated by the *hypothalamus* in the brain. After eating, hunger is normally suppressed by hormones produced by the hypothalamus as well as by the stomach. "Hedonic hunger" is understood to mean "a powerful desire for food in the absence of any need for it," as in the case of continuing to eat when you are already feeling full. It is "the yearning we experience when our stomach is full but our brain is still ravenous" (Jabr, 2016, p. np). It is hedonic hunger that is more relevant for this chapter, for it is eating to satisfy hedonic hunger that is most consistent with the idea of food addiction. Everyone eats; some people eat too much and/or too often.

The first scholar to introduce the term *food addiction* (FA) was Randolph (1956), who defined it as "a common pattern of symptoms descriptively similar to those of other addictive processes." More recently, Soto-Escageda et al. (2016, p. 175) claim: "There is physiological and behavioral evidence that some people may develop a true addiction to food." Yet, there is not universal agreement that FA is real. For example, a review of the evidence by Ziauddeen and Fletcher (2013, p. 19) concluded that FA is merely "a phenotypic description, one that is based on an overlap between certain eating behaviours and substance dependence." They explain:

> While work on animals supports the argument that the combination of high fat and high sugar, prevalent in modern processed foods, produces an addiction-like phenomenon in rodents . . . the FA concept in humans rests on a less well-explored extrapolation: namely that certain highly processed foods are addictive.

*(p. 20)*

Yet, Schulte and Gearhardt (2018, p. 112) disagree, writing that "studies in animals and humans have demonstrated biobehavioural indicators of addiction in response to foods high in fat and/or refined carbohydrates" making it "akin to a substance-use disorder." Additionally, Corwin and Grigson (2009: np) argue that "both behavioral and neurobiological evidence support the conclusion that food, under [certain] conditions . . . can induce and addiction-like state in subjects" (also see Pelchat, 2009).

According to Meule (2019), there are at least three major positions when it comes to the issue of FA:

1. Certain foods have addictive potential, consistent with substance use disorders (e.g., see Corsica & Pelchat, 2010; Ifland et al., 2015; Schulte et al., 2017).
2. No specific additive substances have been identified in food (e.g., see Hebebrand et al., 2014; Ruddock et al., 2017).
3. Food addiction is not a valid concept since it overlaps with at least one eating disorder—binge eating—and is thus unneeded (e.g., see Finlayson, 2017; Ifland et al., 2015; Long et al., 2015; Schulte et al., 2017).

Similarly, Fletcher (see Fletcher & Kenny, 2018, p. np) claims that "the addictive substance [in food] remains undiscovered"; there is significant overlap between food addiction and binge eating disorder, suggesting the concept is not valid; "there remains no convincing demonstration in humans that . . . neurobiological changes . . . underlie . . . food addiction behaviors"; and even convincing evidence from animal studies may not be applicable to humans. This chapter reviews the evidence and finds more support that food is addictive than that it is not.

Some have suggested the term "eating addiction" rather than FA (Jabr, 2016). For example, Novelle and Diéguez (2018, p. np) write:

Although eating behaviour cannot be considered "addictive" under normal circumstances, people can become "addicted" to this behaviour, similarly to how some people are addicted to drugs. The symptoms, cravings and causes of "eating addiction" are remarkably similar to those experienced by drug addicts, and both drug-seeking behaviour [and] eating addiction share the same neural pathway.

Schulte et al. (2017, p. np) agree, noting that

eating may be a behavioral addiction that can trigger an addictive-like response in susceptible individuals. One major rationale for the eating addiction framework is that the assessment of food addiction is based on behavioral indicators, such as consuming greater quantities of food than intended and eating certain foods despite negative consequences.

Lennerz and Lennerz (2018, p. 68) concur, writing that "food addiction may be a behavioral addiction, analogous to gambling disorder." They argue that "food addiction may be a behavioral addiction, analogous to a gambling disorder, which was recently included among addiction disorders in the DSM-5 catalog." Hunt (2020) agrees, calling food addiction "a complex condition that has similarities to other types of addiction, such as drugs, alcohol, shopping or gambling."

Other scholars also agree, claiming "animal and existing human data [are] consistent with the existence of addictive eating behavior . . . eating can become an addiction in . . . predisposed individuals under specific environmental circumstances" (Hebebrand et al., 2014, p. np).

Interestingly, one can be diagnosed with substance dependence even in the absence of physiological dependence—which requires evidence of tolerance and withdrawal—so being dependent or addicted to certain foods "can be diagnosed using entirely behavioral criteria" (Hebebrand et al., 2014, p. np). But, Morris et al. (2018, p. np) disagree, writing: "food addiction symptoms more closely resemble[] those of a substance use disorder due to the necessary consumption of a *substance* (food) and the inapplicability of certain behavioral criteria (e.g., monetary loss: DSM-5 criteria 1, 6, and 5)."

Still, given the foods regularly identified as relevant for addiction, perhaps the more specific concept of "refined food addiction" or "processed food addiction" would be more precise (Ifland et al., 2008, 2015). The types of foods that may be addictive are discussed later in the chapter.

## Food and the Elements of Addiction

As will be shown in the following, food addiction is consistent with *most* of the definitions of addiction and substance use disorders discussed earlier in the chapter. First, food changes the brain by altering neurotransmitter levels such as dopamine and serotonin. Further, the same areas involved in drug misuse and abuse and also involved in food misuse and abuse. Second, some foods produce intense cravings. Third, those foods can also produce tolerance. Fourth, discontinued use of some foods may lead to withdrawal (although this is less clear). Fifth, some people misuse food (e.g., compulsive eating) despite apparent harms to their well-being. This includes overeating even when people don't want to, suggestive of impaired control. Sixth, misuse of food leads to numerous social problems including enormous costs to health and other costs to society; some forms of food misuse also lead to significant interference in quality of life. Each of these issues is addressed in the following.

Before turning to these points, a meta-analysis of 52 studies published in 35 articles between 1999 and 2017 (20 articles on 22 studies were with humans and 15 articles on 30 studies were with animals) found support for "the following addiction characteristics in relation to food: brain reward dysfunction, preoccupation [with food], risky use, impaired control, tolerance/withdrawal, social impairment. . . . Each pre-defined criterion was supported by at least one study" (Gordon et al., 2018, p. 477). The most supported indicators of addiction were brain reward dysfunction (n=21 studies) and impaired control (n=12 studies), but risky use had the least support (n=1 study). Overall, 31 of the 25 articles and 47 of the 52 studies found support for the idea that certain foods are addictive, two studies more were mixed, and only three were unsupportive. The scholars conclude that "findings support food addiction as a unique construct consistent with criteria for other substance use disorder diagnoses" (Gordon et al., 2018, p. 477).

### Changes in the Brain

The foods we eat impact our brain chemistry. Studies show that eating certain types of foods under certain circumstances impacts the pleasure centers of the brain in the same way that drugs do. Research on both humans and animals shows highly palatable foods (i.e., foods that are highly pleasing or satisfying) impact the brain in similar ways to addictive drugs like heroin and cocaine. Specifically, "highly palatable foods trigger feel-good brain chemicals such as dopamine. Once people experience pleasure associated with increased dopamine transmission in the brain's reward pathway from eating certain foods, they quickly feel the need to eat again" (WebMD, 2020). Studies find the same areas of the brain involved in alcohol addiction and FA (De Ridder et al., 2016). More discussion on highly palatable foods follows later in the chapter.

An extensive review of the literature by Lennerz and Lennerz (2017, p. 64) focuses on the importance of brain changes related to food addiction. They note:

> Three lines of evidence support the concept of food addiction: (a) behavioral responses to certain foods are similar to substances of abuse; (b) food intake regulation and addiction rely on similar neurobiological circuits; (c) individuals suffering from obesity or addiction show similar neurochemical- and brain activation patterns.

With regard to the latter point, Leigh and Morris (2018, p. 37) claim: "There is a growing body of evidence that a subset of individuals with disordered eating display addiction-like behaviors in response to foods." Indeed, neuroimaging studies in obese subjects provide evidence of altered reward and tolerance. Once obese, many individuals meet the criteria for psychological dependence. Stress and dieting may also sensitize an individual to reward (Garber & Lustig, 2011, p. 146).

Lindgren and colleagues (2018, p. 817) add:

> Numerous functional and structural MRI studies have elucidated neurobiological correlates between drug addiction and obesity. Compared to healthy individuals, obese and drug addicted subjects show differences in reward and attention regions in response to cues and tasks as well as in a resting state.

Similarly, "there is considerable overlap in the behaviours associated to food addiction and binge eating disorder, and food addiction measures correlate highly with measures of binge eating." Drugs that are typically used to help with drug dependence also help address binge eating (Leigh & Morris, 2018, p. 31). According to Mayo Clinic (2018), binge eating is a disorder where people frequently eat more than they need and are unable to stop. Symptoms include eating a lot of food in a short amount of time, eating rapidly, eating when you are already full or not hungry, feeling that your eating is out of control, eating alone or in secret, and "feeling depressed, disgusted, ashamed, guilty or upset about your eating" (Mayo Clinic, 2018, p. np). People with binge eating disorder also frequently diet but typically do not lose weight.

Yet, it is not just with obesity or binge eating that similarities between food and drugs are seen. Even hedonic eating is similar to drug use and abuse. Zhang et al. (2011, p. 1149) provide a possible explanation for this, focused on the brain:

> Many of the brain changes reported for hedonic eating and obesity are also seen in various types of addictions. Most importantly, overeating and obesity may have an acquired drive similar to drug addiction with respect to motivation and incentive craving. In both

cases, the desire and continued satisfaction occur after early and repeated exposure to stimuli. The acquired drive for eating food and relative weakness of the satiety signal would cause an imbalance between the drive and hunger/reward centers in the brain and their regulation.

Indeed, research shows that "overconsumption of palatable foods triggers" a reduction in dopamine receptors "in the same way that drugs do" (Lerma-Cabrera et al., 2016, p. np).

One area of study in the brain is the *nucleus accumbens*. This is a small region just in front of the limbic system by the hypothalamus, and "neuroimaging studies show that our brain response is similar in the presence of food and drug abuse: increased cell activation" in the nucleus accumbens—"the brain's pleasure center" (Lindgren et al., 2018, p. np).

### Cravings

Fletcher (see Fletcher & Kenny, 2018, p. np) claims that "there are patterns of behavior and subjective experiences related to food consumption that bear a resemblance to [substance use disorders], most notably the strong urge to consume, which can become more powerful with abstinence and override personal desires to limit consumption." Part of the strong urge to consume comes from cravings related to past food experiences.

Even when foods do not immediately elicit cravings, over time with repeated exposure to foods, physical changes occur in the brain, and cravings are elicited that lead to food-seeking behaviors. Over time, habitual and compulsive consumption can result (Lennerz & Lennerz, 2018, p. 67). One possible outcome of this is obesity. That unintended, negative outcome of poor nutrition is discussed further later in the chapter.

All addictions include symptoms such as "craving, impaired control over the behavior, tolerance, withdrawal, and high rates of relapse." Lennerz and Lennerz (2018, p. np) also note that "commonly suspected problem foods share nutritive properties, suggesting a chemical or metabolic link rather than a mere behavioral phenomenon." That is, there are substances in the foods we consume that lead to cravings and ultimately addiction.

There appear to be several biological and psychological similarities between food addiction and drug dependence, including both craving and loss of control (Fortuna, 2012). This is partly due to the nature of some foods. According to Ziauddeen and Fletcher (2013, p. 24), certain foods impact brain wiring and behaviors "in ways that can be compared meaningfully with alterations produced by drugs of abuse." So, just as some drugs elicit cravings, so too will some foods. Some examples include sugar and fat (Hunt, 2020), to be discussed further later in the chapter.

### Tolerance

Research shown earlier illustrates that certain foods can produce effects consistent with tolerance. Pursey et al. (2014, p. 4581) explain *Reward Deficiency Syndrome*—discussed earlier—writing that

> an addiction to food could act in a similar way to other substance addictions, with repeated exposures to pleasurable food diminishing the dopamine brain response. This would

lead to larger quantities of food consumed in order to feel satisfied, subsequently perpetuating overeating.

This occurs when more of certain foods are needed over time to achieve the desired levels of satisfaction, caused by alterations to dopamine levels in the brain. One outcome is likely to be overeating, meaning eating more food than is needed due to an increased tolerance for certain foods (Pursey et al., 2014). Recall that neuroimaging studies in people suffering from obesity show evidence of altered reward in the brain, leading to possible tolerance (Garber & Lustig, 2011, p. 146).

The meta-analysis by Morris et al. (2018) reviewed a study that measured tolerance in bariatric surgery candidates who reported needing increasing amounts of food to reach satisfaction (Lent & Swencionis, 2012). It also reviewed a study of women craving carbohydrates who reported dispelling of negative moods upon drinking sweetened carbohydrate beverages; that effect diminished over time, consistent with tolerance (Spring et al., 2008). Further, another study found tolerance effects for high-fat, sweet foods as well as high-fat, savory foods (Markus et al., 2017).

### *Withdrawal*

Since food does produce cravings and tolerance, symptoms associated with the cessation of ingestion of certain foods may lead to withdrawal symptoms (Lennerz & Lennerz, 2018). Hunt (2020) lays out the relationships between cravings for certain foods, tolerance, and ultimately withdrawal: "Highly palatable foods often contain unnatural substances or higher-than-normal levels of natural substances that your body and brain can't process. This results in your body being flooded with 'feel-good' chemicals." In order for people to maintain or re-experience good feelings resulting from the ingestion of certain foods, people "will begin to crave highly palatable foods." Over time, a person's brain "will adjust its receptors to compensate for the rush of chemicals" and they will "eventually need to consume increasing quantities of highly palatable foods to get the same feel-good reaction," consistent with tolerance. Not eating those foods can lead to "withdrawal symptoms such as cravings, headaches, irritability, and restlessness" (Hunt, 2020).

Kenny (see Fletcher & Kenny, 2018) agrees, writing that evidence supports the idea that certain foods can produce withdrawal symptoms. Corwin and Grigson (2009, p. np) concur, claiming that "withdrawal from a high-fat diet leads to neurochemical responses comparable to those induced by withdrawal from drugs" (Corwin & Grigson, 2009, p. np; also see Lutter & Nestler, 2009).

The meta-analysis of studies by Morris et al. (2018) discovered that, among bariatric surgery candidates, those with the highest addictive personality scores reported feeling most anxious when not around food (Lent & Swencionis, 2012). This is consistent with withdrawal. Two studies of chocolate addicts determined that exposure to chocolate led to changes in anxiety and restlessness that are often seen in those suffering from substance addiction (Tuomisto et al., 1999). Another study found that people with at least one YFAS symptom reported some physiological effect of withdrawal from high-fat, savory foods, high-fat, sweet foods, low-fat, sugary foods, and low-fat, savory foods, in that order of magnitude (Markus et al., 2017). These findings are consistent with the idea of withdrawal.

### Compulsive Use Despite Harms, Lack of Impulse Control

According to Lerma-Cabrera et al. (2016, p. np): "The most common symptoms of food addiction are loss of control over consumption, continued use despite negative consequences, and inability to cut down despite the desire to do so." Generally speaking, the loss of control in eating is easily demonstrated by one simple fact: people "chronically eat some foods in amounts larger than needed for staying healthy" (Lerma-Cabrera et al., 2016, p. np).

Further, both hedonic and compulsive overeating are suggestive of spending too much time with food, consistent with one aspect of addiction (Jabr, 2016, p. np). When people eat too much and/or spend too much time with food, it is strongly supportive of a lack of impulse control.

Fletcher (see Fletcher & Kenny, 2018, p. np) also writes that substance use disorders are identified by "a manifestation of a behavioral abnormality that negatively impacts their life: specifically, the failure to control consummatory behavior despite repeated attempts to do so." This, too, is consistent with a lack of impulse control.

According to Kenny (see Fletcher & Kenny, 2018):

> Overweight individuals who experience real or perceived social, emotional, or health consequences because of their body weight will often express a desire to lose weight and will repeatedly attempt to do so, but limiting their food intake or the types of food over the prolonged time periods necessary to achieve and maintain a healthy body weight is notoriously difficult.

Further, even people who lose weight tend to gain it back, demonstrating "remarkably high rates of recidivism." Thus, "overweight individuals who are unable to exert control over their consummatory behavior, despite awareness of the negative consequences, demonstrate the same core failure to control consumption as those suffering from [substance use disorders]." Kenny cites an enormous amount of research on both adolescents and adults in support (see, e.g., Booth et al., 2008; Puhl et al., 2008; Saunders, 2001; Small et al., 2001; Stice et al., 2008, 2011; Yokum et al., 2014).

Individuals with FA are also more likely to be impulsive (Davis et al., 2011; Maxwell et al., 2020), similar to substance users. They tend to be characterized by "emotion dysregulation," or how well an individual is able to notice their own emotions and properly respond to them. This is likely to result in compulsive use, even with the knowledge that it can be harmful. Harfy et al. (2018, p. 368) note: "Individuals with poor emotion regulation abilities often have poorer decisions-making, i.e., a reduced ability to limit impulsive behaviors and adaptively handle unpleasant feelings." This is important because people who cannot control their emotions at times turn to drugs or food to make themselves feel better. One possible outcome is overeating and binge eating. Recall that drugs typically used to help with drug dependence also help people deal with binge eating (Leigh & Morris, 2018, p. 31).

This makes sense given the brain chemistry involved in both. Loss of control is logically caused by a reduction in *striatal* dopamine receptors, reducing metabolism in the *prefrontal and orbitofrontal cortex*, which normally exerts "inhibitory control over consumption" (Lerma-Cabrera et al., 2016, p. np).

The meta-analysis by Morris et al. (2018) examined studies that looked at the following indicators of poor impulse control: Substance use in larger amounts or over a longer period of time than intended, spending a great deal of time obtaining and using substances, and having cravings or strong urges to use substances. Here, studies found that people meeting FYAS measures of addiction, as well as obese people, had more cravings for food, hedonic eating, snacking on sweets (Davis et al., 2011, 2014; Tuomisto et al., 1999), and difficulty controlling their eating (Burmeister et al., 2013).

The meta-analysis by Morris et al. (2018) also looked at studies that examined the following measures of social impairment: Continued use despite social or interpersonal problems caused or exacerbated by use, and reductions in important social, occupational, or recreational activities due to use. Here, one study found that bariatric surgery candidates tended to choose spending time eating over other social and recreational activities, a behavior the authors referred to as maladaptive (Lent & Swencionis, 2012). Hardy et al. (2018, p. 367) explain: "Individuals with food addiction exhibit classic symptoms of addiction, such as a preoccupation with obtaining the desired substance, excessive ingestion of the substance and continued, excessive use, despite adverse biological consequences."

According to Jabr (2016, p. np), "people who are addicted to food will continue to eat despite negative consequences." This is consistent with one aspect of addiction. In fact, having trouble with stopping eating is measured in FA surveys with questions dealing with going out of one's way to obtain foods, food interfering with life activities, problems with normal daily functioning, and physiological withdrawal symptoms (Jabr, 2016, p. np).

### Social Problems

Social harms associated with food are enormous, and they rival those of drug abuse. Studies show the costs of drug abuse to society are about $740 billion per year, though this includes indirect costs such as losses of productivity. In 2016, drug overdoses killed more than 63,000 people. Another 88,000 died from excessive use of alcohol, and 480,000 died from illnesses caused by tobacco use (NIDA, 2020).

Poor diet is also responsible for very large financial losses as well as a large number of deaths. For example, diabetes was associated with $245 billion in losses (in 2008 dollars), while obesity was associated with $147 billion in losses (in 2008 dollars); both these figures also include indirect costs (Robinson & Tauscher, 2019). Additionally, 75% of all healthcare dollars spent in the United States are used to treat chronic diseases and medical conditions that are preventable, caused by things such as poor diet and lack of exercise (Nesheim et al., 2015).

In terms of death, poor dietary intake killed approximately 395,000 people in the United States in 2012 (Micha et al., 2017). This can be compared to only about 42,000 deaths caused by drug overdoses in the same year (Robinson & Tauscher, 2019). These data were first introduced in Chapter 1.

Additionally, FA can lead to numerous significant health outcomes. Outcomes of food addiction identified in studies include sluggishness and fatigue, declines in healthy functioning, efforts to avoid socializing with others, continuing to eat unhealthy food even after suffering from unhealthy outcomes, tolerance for foods in the form of not reducing negative emotions or increasing positive emotions with the same amount of food over time, and withdrawal symptoms after stopping eating certain foods (Arumugam et al., 2015). Such

**TABLE 5.3** Outcomes of Food Addiction

Physical effects
- Heart disease
- Diabetes
- Digestive problems
- Malnutrition
- Obesity
- Chronic fatigue
- Chronic pain
- Sleep disorders
- Reduced sex drive
- Headaches
- Lethargy
- Arthritis
- Stroke
- Kidney/liver disease
- Osteoporosis

Psychological effects
- Low self-esteem
- Depression
- Panic attacks
- Increased feelings of anxiety
- Feeling sad, hopeless, or in despair
- Increased irritability, especially if access to desired food is restricted
- Emotional detachment or numbness
- Suicidal ideation

Social effects
- Decreased performance at work or school
- Isolation from loved ones
- Division within family units
- Lack of enjoyment in hobbies or activities once enjoyed
- Avoidance of social events or functions
- Risk of jeopardizing finances or career

*Source:* https://www.eatingdisorderhope.com/information/food-addiction

outcomes are consistent with the idea that some foods are addictive, as these are similar outcomes in drug abuse. Table 5.3 illustrates other deleterious physical, psychological, and social outcomes.

## Which Foods May Be Addictive?

Now that the similarities between drug addiction and food addiction have been revealed, it is useful to identify the specific types of foods that may be associated with the outcomes consistent with addiction. Part of the overlap between food and drugs is due to what is in the foods people consume. Westwater et al. (2016, p. 56) claim that FA "is similar to substance addiction" and that "certain 'addictive agents' within food produce neurochemical effects in the brain similar to drugs of abuse."

According to Pursey et al. (2017, p. np): "Based on the current evidence base, highly processed, hyper-palatable foods with combinations of fat and sugar appear most likely to facilitate an addictive-like response. Total fat content and glycemic index also appear to be important factors in the addictive potential of foods." So, the ultra-processed foods discussed in Chapter 4, high in salt, sugar, and fat, are the culprits.

The YFAS specifically lists sweet foods including not only ice cream, chocolate, doughnuts, cookies, cake, candy, carbohydrates like white bread, rolls, pasta, crackers, chips, pretzels, French fries, and pizza but also steak, hamburgers, cheeseburgers, bacon, apples, bananas, broccoli, lettuce, strawberries, and even soda pop. And scholars suggest: "Energy dense foods, including sugary drinks like beverages, cakes, biscuits, and various salt and savoury snacks are the foods that are the most typically associated with reports of food craving and food addiction" (Ayaz et al., 2018, p. np).

Gearhardt et al. (2011, p. 1208) agree, writing: "Foods, particularly hyperpalatable ones, demonstrate similarities with addictive drugs. This is likely due to the nature of the food itself." For example, Pursey et al. (2017) write: "Functional neuroimaging studies have . . . revealed that pleasant smelling, looking, and tasting food has reinforcing characteristics similar to drugs of abuse."

Garber and Lustig (2011, p. 146) also agree that certain foods are more addictive than others: "Studies of food addiction have focused on highly palatable foods." In some individuals, palatable foods have palliative properties and can be viewed as a form of self-medication (Fortuna, 2012). According to Gordon et al. (2018, p. 477), "certain foods, particularly processed foods with added sweeteners and fats, demonstrate the greatest addictive potential." Other research supports this idea, finding that at least four out of the 11 DSM substance use disorder symptoms apply to highly palatable foods (Meule & Gearhardt, 2014). Processed foods and foods high in carbohydrates are the ones most identified as being most likely to lead to addictive-like eating among survey respondents (Schulte et al., 2015).

These foods include many fast foods. Fast food "has several other attributes that may increase its salience." It is high in fat, salt, and sugar, as well as other additives that might lead to dependence. Additionally, "fast food advertisements, restaurants and menus all provide environmental cues that may trigger addictive overeating . . . these findings support the role of fast food as a potentially addictive substance that is most likely to create dependence in vulnerable populations" (Garber & Lustig, 2011, p. 146).

The logic of why these foods are addictive is that they act on the same areas of the brain and in the same ways as addictive drugs. As noted by Hunt (2020, p. np):

Consuming "highly palatable" foods, or foods that are high in carbohydrates, fat, salt, sugar or artificial sweeteners, triggers the pleasure centers of the brain and releases "feel-good" chemicals such as dopamine and serotonin. These foods affect the same area of the brain as drugs, alcohol and behaviors such as shopping or gambling.

So-called "palatable foods" increase dopamine. For example: "High-glycemic-index carbohydrates elicit a rapid shift in blood glucose and insulin levels, akin to pharmokinetics of addictive substances. Similar to drugs of abuse, glucose and insulin signal to the mesolimbic system to modify dopamine concentration" (Lennerz & Lennerz, 2018, p. 64). Highly palatable foods impact the "mesolimbic dopaminergic circuit, the primary component of the reward system" in the brain (Leigh & Morris, 2018, p. 31). This system "is involved in a large number of behaviors including reward processing and motivated behavior"

(Lerma-Cabrera et al., 2016, p. np). These foods would include, for example, white breads, potatoes, rices, pastas, many cereals, and snack foods such as chips, pretzels, and crackers. Any drug of abuse tends to "increase the extracellular concentration of dopamine (DA) in the striatum and associated mesolimbic regions" (Lerma-Cabrera et al., 2016, p. np). Palatable foods also increase glucose levels which catalyze the absorption of tryptophan, and this is converted into serotonin, which can elevate mood (Fortuna, 2012).

Sweet, fatty, and salty foods also activate the same chemicals in the "reward circuits" of the brain that are linked to addictive drugs and gambling behaviors. The so-called "hunger hormone"—ghrelin—is released by the stomach, increasing dopamine in the brain's reward circuit with consumption of these foods, triggering the "feel good" effects we experience while and after eating these foods. Continued eating comes from a dependency on the good feelings people get from food (Jabr, 2016, p. np).

According to Moss (2021, p. 37), "food doesn't need harsh chemicals to throw us into a full-on crave. Salt, sugar, and fat do just fine. They get all the help they need from the brain, which is awash in chemical compounds that the brain conjures up all by itself." When we eat fat, for example, our brain chemistry responds:

> Dopamine gets all fired up when we eat fatty foods, and the brain areas that are activated by dopamine are responsible for why we crave deep-fried pickles and get much more enjoyment from eating them than from a plain sour spear. Just the fell of fat on the tongue, no swallowing required, lights up the reward and emotion centers of the brain, in particular the anterior cingulate cortex—a neuroanatomical bridge between the emotion limbic system and the analytical frontal love—and the amygdala, the central structure of the limbic system that governs our experience of emotion and emotional memory.
>
> *(p. 43)*

And, clearly, our emotions are involved heavily in what we eat. Herz (2018, p. 123) explains: "Odors trigger our most emotional and evocative memories." When it comes to fat, it is associated with positive emotions. The amygdala is also involved in the perception of odors.

And sugar and fat light up the brain more, in combination, than either does alone. Moss (202, p. 62) explains: "When we taste sugar, the taste buds on our tongue send the signal" to the brain.

> By contrast, the signal for fat gets transmitted by the trigeminal nerve that extends from the roof of the mouth to the brain. Food that has *both* sugar and fat will activate these two different paths, sending two separate alerts, and thus doubling the arousal of a brain that appears to place a high value on information for information's sake. Of course, ultra-processed foods tend to be high in both, as well as in salt.

Moss notes:

> The typical processed snack food has close to 24 percent fat and 57 percent sugar. Even savory foods like hot dogs, spaghetti sauce, bread, and frozen chicken dinners have been sweetened by the food manufacturers. And estimated three-fourths of our food contains added sugar, as well as loads of salt, which also adds to the thrill we get from fat.
>
> *(p. 62)*

Overeating highly palatable foods "saturates the brain with so much dopamine that it eventually adapts by desensitizing itself, reducing the number of cellular receptors that recognize and respond to the neurochemical" (Jabr, 2016, p. np). In essence, the brain demands more sugar and fat to reach the same level of pleasure as was once experienced with smaller amounts of food, consistent with tolerance. Overeating thus becomes a way of "recapturing . . . or maintaining . . . well-being" (Jabr, 2016, p. np).

Normally, leptin and insulin—produced by caloric consumption—suppress the release of dopamine. According to Lerma-Cabrera et al. (2016, p. np): "Leptin infusion into the tegmental ventral area, a reward system brain area, decreases food intake and inhibits the activity of dopamine neurons." Yet, as fatty tissue increases in the body, it does not respond to signals such as fullness and satisfaction; dopamine can override them, as well (Jabr, 2016, p. np).

Wiss and Brewerton (2020, p. np) write:

The mesolimbic dopaminergic circuit is clearly affected by both highly palatable foods and diet-induced obesity similar to exposure to drugs of abuse. Recent review articles have discussed highly processed foods (often high in glycemic index) as impacting neurohormonal and inflammatory signaling pathways in ways that create a vicious cycle of impulsivity, compulsivity, FA, and [eating disorders].

Consumption of fats and sugars can increase dopamine in the brain, "producing good mood effect" (Ayaz et al., 2018, p. np). Overconsumption of these foods, especially repeatedly, conditions the brain to expect high levels of dopamine, and not getting that will "ultimately promote depressive or anxious responses when those foods are no longer available or consumed" (Ayaz et al., 2018, p. np; also see Rodin et al., 1991).

Interestingly, in the short term, these foods are thought to relieve anxiety and temporarily lesson symptoms of depression among people with food addiction, who tend to have higher incidence of anxiety and depression (Benzerouk et al., 2018; Burrows et al., 2017; Fonseca et al., 2020; Linardon, 2018; Nolan & Jenkins, 2019; Spettigue et al., 2019; Tomiyama et al., 2011; Wiss & Brewerton, 2020). Perhaps this is why they are known as "feel good foods" or "comfort food" (Wiss & Brewerton, 2020). Still, low-quality diet is related to increased depression over the long term (Gomez-Donoso et al., 2019).

Lindgren et al. (2018, p. 811) concur with the idea that certain foods such as sugar are linked to food addiction: "Certain foods, especially those high in sugar and fat, act in a similar way to drugs, leading to compulsive food consumption and loss-of-control over food intake." Lennerz and Lennerz (2018, p. 69) assert: "Sugar elicits addiction-like craving, compulsive food seeking, and withdrawal in rats and has therefore been used in substance abuse models for some time." Sugar causes the "release of endogenous opioids in the [nucleus accumbens] and activates the dopaminergic reward system" (Lerma-Cabrera et al., 2016, p. np). It also increases the neurotransmitter acetylcholine, possibly explaining the sign of dependency of "increased intake of sugar after a period of abstinence" (Lerma-Cabrera et al., 2016, p. np).

Yet, at least one study concludes that sugar is not addictive. Westwater et al. (2016, p. 55) write: "We find little evidence to support sugar addiction in humans, and findings from the animal literature suggest addiction-like behaviors, such as bingeing, occur only in the context of intermittent access to sugar."

Studies on salt show it can lead to cravings, physical dependence, and tolerance (Soto-Escageda et al., 2016). Soto-Escageda et al. (2016, p. 180) conclude that "salt is categorically an addictive substance just as psychotropic drugs, as they share cerebral pathways that perpetuate their excessive consumption."

So, it may not be surprising to learn that people with high FA scores have significant brain responses to images of food similar to people with drug dependence who view images of drugs. These responses are found both in reward areas of the brain (i.e., *striatum, anterior cingulate cortex, dorsolateral prefrontal cortex,* and *amygdala*) and an area known to inhibit unhealthy behaviors (i.e., the *prefrontal cortical regions* and *medial orbitofrontal cortex*) (Gearhardt et al., 2011; Fletcher & Kenny, 2018; Holsen et al., 2005; Killgore et al., 2013; Lee & Dixon, 2017; Scholtz et al., 2014).

## Prevalence of Food Addiction

Studies vary on how prevalent FA is within the United States, but an average of studies is about 10–25% of the population (Leigh & Morris, 2018), with other scholars claiming 20% of the population (Pursey et al., 2014). Rates of FA are found to be highest among binge eaters (more than 50%), followed by obese people (nearly 25%), and then other members of the general population (about 11%) (Gordon et al., 2018). There is clearly a correlation between food addiction and binge eating, in particular (Leigh & Morris, 2018).

Studies also show that the likelihood of scoring high on the YFAS and thus showing symptoms of food addiction are related with the ingestion of certain foods, including more calories, fat, saturated fat, and trans fats, carbohydrates, as well as sugars (Schulte et al., 2018). This is consistent with the argument of scholars that hyperpalatable foods are the most problematic.

## Conclusion: What Does It All Mean?

While FA might not meet all the acceptable definitions of addiction that are used when considering drug addiction, there are clear parallels between drug addiction and FA. That is, some foods may be considered addictive in some very important ways, including that they produce changes in the brain (including in areas and ways related to drug addiction), produce cravings, lead to tolerance, lead to compulsive use in spite of harms and intentions to stop, lead to withdrawal under certain circumstances, and produce significant social problems that rival all illicit drugs combined.

Taken together, it is not unreasonable to conclude that some foods may be considered drugs and that their use may be considered drug use, misuse, and abuse. Table 5.4 revisits Table 5.1, with regard to which components of addiction are found with food. This review illustrates that most of the criteria for addiction are met when it comes to food, especially with regard to obesity and overeating (such as binge eating). Specifically, there is a high level of support for impaired control and compulsive use for obese individuals and people who overeat, a moderate level of support for social problems caused by food and interference in life from food issues, and a high level of support for drug effects on the brain and tolerance and withdrawal, especially for hyperpalatable foods.

Given this reality, the comparative harms (including illness, death, medical costs, and productivity losses) associated with licit and illicit drugs and food become increasingly

**TABLE 5.4** Degree of Support for Indices of Food Addiction

_Social problems_—moderate for obesity and overeating (persistent or recurrent social or interpersonal problems caused by use; use interferes with work, school, or home; failure to sustain obligations; social, work, or leisure activities are given up or cut back)

_Impaired control, compulsive use_—high for obesity and overeating (cravings or strong urges to use; desire to or failed attempts to reduce use; using more than intended)

_Risky use_—low (use in risky or hazardous settings); high for obesity and overeating (continued use despite known problems; continued use despite knowing one has a persistent or recurrent physical or psychological problem caused or exacerbated by the substance)

_Drug effects_—high for hyperpalatable foods

_Brain disease_—low, but high for brain changes

_Time usage_—low

relevant for academic disciplines such as criminology and related fields. This chapter demonstrates that the foods we consume lead to incredible financial and physical harms that rival those associated with both licit and illicit drugs. It is therefore not unreasonable to assert that we ought to reprioritize what we believe to be dangerous and worthy of the "criminal" label and to suggest that perhaps the "war on drugs" might be focused on at least some of the wrong things. That is to say, if the fundamental purpose of the criminal law is to protect us from harmful acts, particularly those committed against us by other people, one could at least argue that legislatures ought to reconsider what "drugs" are legal and which are illegal.

Of course, an obvious difference between food and drugs is that food is nourishing and thus has important value to humans in terms of functioning and surviving. So, no one will argue that food should be illegal. At the same time, drug use is considered normal as it has been found through nearly all societies in human history, and it also serves adaptive functions for humans (e.g., enjoyment, relaxation, socialization, religion, etc.) (see, e.g., Crocq, 2007; Lyman, 2016). The differences between drugs and food are thus smaller than they have been made out to be.

Further, many of the foods that we eat are actually quite dangerous, especially in the long term. If the purpose of the criminal law is to protect us from harms, government interventions into the conventional food system in the interest of public health are warranted. For example, Meule (2019, p. np) argues that "if certain foods have an addictive potential, policy regulations may be implemented to limit advertising, increase the price of, or restrict access to such foods, similar to alcohol and tobacco regulations" (also see Pomeranz & Roberto, 2014).

Finally, like with any drug, licit or illicit, the foods we consume of our own "free will" are also provided to us by "dealers" (in the case of the foods we eat, this is typically very large multi-national corporations that are the dealers). Food addiction and harmful outcomes resulting from it are partially due to the production and sale of addictive foods to consumers by those corporations, enabled by state, national, and international organizations including governments. This raises the issue of state-corporate criminality introduced in Chapter 1 (Kramer, 1990; Kramer & Michalowski, 1991; Kramer et al., 2002; Ross, 2017; Whtie, 2014). Leigh and Morris (2018, p. 31) note that it is "the ready availability of highly palatable foods . . . increases the incidence of hedonic, non-homeostatic feeding." That is, part of why there is so much hedonic eating of unhealthy and addictive foods is because they are available everywhere, heavily advertised, and highly affordable relative to healthier alternatives.

This raises the issue of culpability in the companies that produce these foods. Wiss and Brewerton (2020, p. 2937) argue: "Converging evidence from both animal and human studies have implicated hedonic eating as a driver of both binge eating and obesity." Humans have an evolutionary need for calorie-dense foods, so cravings for them can be seen as normal. Yet, this drive for certain foods is not healthy in contemporary society when calories are available everywhere and there is no need to hunt or fight in order to eat. The bulk of evidence shows that mere willpower is not enough to stop the overeating that leads to obesity and that change in the food environment is needed (Jabr, 2016, p. np). The highly palatable processed foods that are so highly sought after in contemporary society highly impact the brain and provide nutrients that do not occur naturally. These foods tend to be more addictive when they are blended to combine substances such as carbohydrates and fat, which are directly under the control of the corporations who manufacture our food (Fletcher & Kenny, 2018). Lerma-Cabrera et al. (2016, p. np) concur, writing: "Manufacturing industries have designed processed foods by adding sugar, salt, or fat, which can maximize the reinforcing properties of traditional foods (fruits, vegetables). The high palatability (hedonic value) that this kind of processed food offers, prompts subjects to eat more." Schrempf-Stirling and Phillips (2019, p. 123) note: "Food corporations are said to design foods in ways that maximise the addiction potential so that consumers buy more of their product." This suggests overeating is an expected outcome of the types of food being used to target consumers by food companies, again raising the issue of culpability.

According to White (2014, p. 835), there are also extensive government and corporate security mechanisms that have grown to protect "a platform of state, corporate, organized group wrongdoing and injustice." In the context of food, the goal of large corporations is to produce foods that sell, and natural as well as artificial flavors (sugars, salts, and fats) help them achieve wealth in spite of the costs, which include FA and the harmful outcomes associated with it. The government's approach seems to be just stand back and let them do it.

Surely, these realities are thus worthy of focus by criminologists who study culpable harmful acts, both illegal and legal. To the degree we study these realities, we are likely to promote the ideas:

1. The things we define as crimes in American society are not actually the most harmful acts committed with culpability.
2. The war on drugs does not target the substances that cause the most harm in society.

These are important lessons.

# 6
# HARMS ASSOCIATED WITH WHAT WE EAT

## Introduction

In this chapter, harms associated with the foods we eat are examined. I show, using data, the prevalence of health conditions such as obesity and diabetes, and I show how the foods people eat lead to these outcomes. I also outline other deleterious health outcomes associated with diet, including mental health outcomes such as depression. Further, I illustrate how the standard American diet, or SAD, is especially likely to produce negative health outcomes.

The federal government is widely aware of the realities discussed in this chapter. Yet, there is little regulation of food when it comes to outcomes like those mentioned earlier, so we find ourselves with epidemics of obesity and diabetes in the United States, as well as around the world.

I also revisit the issue of whether people have free will or self-control when it comes to the foods they eat but this time in the context of an analysis of obesity and diabetes. The question asked is whose fault is it that we are facing an epidemic of obesity and diabetes in the United States and around the globe? Does the fault lie at the feet of individual consumers through their bad dietary choices? Or are there structural sources of obesity and diabetes? The chapter also makes a point to identify what is a healthy diet based on sound science, in hopes that one day we will all move toward such a diet in order to reduce harms associated with food.

## SAD

Fuhrman (2017, p. 3) describes "excess-calories malnutrition" and "fast food malnutrition" as outcomes of the typical American diet or what he calls the "standard American diet" (SAD) (p. 9). Fuhrman (2017, p. 45) characterizes this reality: "The average American consumes an extravagant high-calorie, low-nutrient diet that stresses the brain with metabolic wastes while systematically depriving it of the micronutrients necessary to self-cleanse and undo the damage." One outcome of this is high levels of "metabolic wastes," or "toxins produced by our body that normally would be removed if phytonutrient exposure were

DOI: 10.4324/9781003296454-6

adequate" (p. 45). Basically, SAD is dangerous. While many commercial foods are fortified with vitamins and minerals—with only about 5% of our calories coming from produce and less than 10% containing fruits, vegetables, beans, seeds and nuts, and whole grains—we still tend to lack the nutrients we need to stay healthy. There are, by the way, significant geographical disparities when it comes to healthy and unhealthy eating, both within and outside of the United States.

Fast food is clearly part of the problem of SAD. Fuhrman (2017, p. 15) defines fast food this way:

1. It is served at commercial chair restaurants.
2. It is commercially made with artificial ingredients, processed grains, and high amounts of salt, sugar, and fat.
3. It has minimal nutrients.

He notes that fast food has the following six characteristics:

1. It is digested and absorbed slowly.
2. It contains multiple synthetic ingredients.
3. It is calorically dense.
4. It is nutritionally barren.
5. It is highly flavored.
6. It contains excessive salt and sugar.

*(p. 18)*

According to Chandler (2019), 80% of Americans eat fast food every month and 96% eat it at least once a year. Fuhrman (2017) claims that 28% of Americans eat fast food every week, while only 4% of Americans say they never eat fast food. Further, about one in three children between the ages of 2 and 14 years eats fast food every day This is due to fast food's "ubiquity and mass appeal" (p. 5). But, the industry has also engaged in "careful study and reflection of American desires and the national order—literal, figurative, and culinary" (p. 5).

Chandler (2019, p. 3) claims: "There are no inherited rights in America, but if one were to come close, it would involve mainlining sodium beneath the comforting fluorescence of an anonymous fast-food dining room or beneath the dome light of a car." Yes, people eat fast food not only because it tastes good, it is convenient and portable, it is relatively cheap and can be served with no waiting or reservations but also "because we have fond memories of wearing paper crowns at birthday parties, of Happy Meals and Frosties and road trips, and of bygone moments of innocence and intimacy with friends of parents or siblings." In part, we do it because "the places are familiar to us" and because fast-food restaurants do offer places for people to gather and socialize with friends or family. Consider the case of Dairy Queen.

Chandler claims will "probably always be associated with soft serve, idle hours, small-town communal innocence . . . wholesome Americana . . . enmeshed in that Rockwellian tapestry" (pp. 39–40).

Also, consider that fast-food companies "Are some of the most skillful cultivators of . . . collective loyalty out there" (Chandler, 2019, p. 167). They also "rely on intricate, genius gimircky and wacky marketing, jingles and slogans, baseline appeals to id and comfort, a hucketrism both grand and unflinchingly American" (pp. 175–176). Chandler quotes on CEO who says "We're in the temptation business" and that makes sense given that 60% of fast-food trips are made on impulse.

Since calories from fast food enter the bloodstream more quickly, there is a higher release of fat-storage hormones and a greater increase in dopamine in the brain. Likely outcomes include higher rates of insulin, more fat storage, weight gain, cravings, and addiction. As glucose builds up in the blood, the glycemic index increases, thereby increasing the risks of serious illness. Meanwhile, natural sugars from foods are accompanied by fiber and phyto-chemicals that slow the entry of sugar into the bloodstream: "The slower absorption rate allows more of the calories to be burned for energy, rather than stored as fat, and can delay and reduce the body signaling for calories (the feeling of hunger"; as such, the likelihood of overeating is reduced (p. 20).

Chandler (2019, p. 210) concludes that

fast food succeeds and has succeeded in large par because its appeal transcends nearly all demographic bounds. More than its innovation, imagination, convenience, value, or capacity to decode the national appetite, fast food's greatest virtue . . . is its creation of Americas most successful demographic gathering points: small matchbox chapels with practically no barrier to entry or belonging, regardless of race, age, class, gender, religion, or other.

You cannot separate the success of fast-food restaurants from our cars, roads, and high-ways. With the popularity of cars assured, the number of fast-food restaurants tripled between 1976 and 1986. Other factors that impacted the growth of fast food included a rise in dual-income households (meaning more money to spend on meals out), rising divorce rates (meaning convenience during meals would be more important), plus of course drive through windows and new products designed to be eaten more easily in cars. The creation of mini-vans and cupholders for cars also made eating and eating together in cars easier.

Chandler (2019, p. 119) notes:

One of the central complaints about the international expansion of fast-food chains (and along with the reach of processed-food peddling conglomerates) is that the footprint of US companies abroad is a form of cultural hegemony, a dynamic that upends traditional diets and foodways, kills national culinary traditions, and foists American dietary norms and values upon unsuspecting communities.

At the same time, fast-food restaurants have proven remarkably adaptable, making unique versions of their foods specific to individual countries. Examples include Burger King in France, with halal fare for Muslims there, as well as meat-free sandwiches in India, sausage hoagies in South Africa, and burgers topped with fifteen slices of bacon in Japan; Taco Bell in India also serves potatoes, chicken, and beans rather than beef; McDonald's in Spain offers gazpacho and macarons across Western Europe, as well as English muffins with vegemite in Australia and ground veal burgers topped with hash browns and mush-room sauce in Switzerland; Wendy's in Ecuador sells sweet honey hotcakes and savory fried plantains, as well as salmon burgers in the Republic of Georgia and cheeseburgers topped with bacon, ketchup, mayo, and a fried egg in Japan; Pizza Hut in Japan sold seafood piz-zas topped with shrimp, quid, tuna, and more; Domino's in Nigeria has pizza topped with jollof rice; Subway in Bulgaria sells roast-veal subs, and Popeye's in Turkey sell sesame onion rings; Hardee's in the Middle East sells cheeseburgers topped with cheesesteak and wrapped in tortillas; Pizza Huts there also sell pies with crusts stuffed with all sorts of meats (Chandler, 2019, pp. 124–131).

According to Chandler (2019, pp. 85–86):

> The economics of fast food aren't designed to benefit the surrounding communities much. The food is usually trucked in from centralized prep zones in far-flung places. More often than not, the majority of the money spent at larger chains wends its way out of the community and back to corporate offices and shareholder portfolios.

But even without fast foods, our standard food fare or SAD has been described as "a jam-boree of processed snacks, high fructose corn syrup, multivarious forms of corn and wheat, lab-concocted flavorings and chemicals, and other sketchy items that only vaguely resemble food" (Minger, 2013, p. 145). Minger (2013, p. 130) also concludes that SAD (or what she calls the "modern diet" and "typical Western diet") "tends to promote chronic, low-grade inflammation—a response triggered by damage to your body's tissues. The inflammation then cripples the body's ability to protect itself from other damage. Basically, that chronic inflam-mation sets the stage for disease." She notes that saturated fat can create insulin resistance and diabetes. So, too can excess consumption of refined carbohydrates, which "tend to spike blood sugar levels (and kick) off a process called *de novo* fatty acid synthesis or lipogenesis, and "enzymatic pathway your body uses to convert dietary carbohydrates into fat" (p. 131). The diet leads to your body being in a constant state of inflammation, "swimming in the sur-plus of saturated fatty acids it created from all those excessive carbohydrates" (pp. 131–132). Minger also blames both fat and sugar "which do not make good bedfellows" for the condi-tions that "magnetically attract chronic disease wherever it invades" (p. 132).

Cohen (2014) shows that 97% of Americans do not eat well (i.e., follow the Dietary Guidelines for Americans) or exercise enough (i.e., get 2.5 hours of moderate exercise a week). Part of this is owing to restaurants since less than 4% of meals there meet USDA guidelines for things such as fat, saturated fat, and sodium (restaurant meals also tend to be very high on calories due in part to very large serving sizes). With all the focus on diet and nutrition in this book, it cannot also be ignored that sedentary living poses a significant threat to one's well-being, as well. Minger (2013, p. 137) explains that not moving enough (i.e., not exercising enough) "induces rapid changes in how the body handles the food we eat, compromising our insulin sensitivity and glucose tolerance—both factors in conditions ranging from heart disease to diabetes, and influencing our ability to metabolize saturated fat and sugar." This reality was examined in Chapter 3.

## Illness

According to Walker (2019, p. 185): "Today, diets are dominated by refined carbohydrates, processed flour, loads of saturated fat, plenty of added salt, and minimal amounts of fiber. Consuming more than what our bodies require is tied to at least 30 different medical condi-tions and diseases." Some of these are discussed in the following.

First, it is important to point out that one of the major contributors to illness (and death) is overeating, something the conventional food system is aimed at assuring continues. According to Nestle (2002, p. 3):

> Overeating causes its own set of health problems; it deranges metabolism, makes people overweight, and increases the likelihood of 'chronic diseases'—coronary heart disease, certain cancers, diabetes, hypertension, stroke, and others—that now are leading causes of illness and death in any overfed population.

She argues that "many of the nutritional problems of Americans—not the least of them obesity—can be traced to the food industry's imperative to encourage people to *eat more* in order to generate sales and increase income in a highly competitive marketplace" (p. 4). This is all about satisfying stockholders (p. 21).

Second, it is also important to note that industrialization of food is to blame. It is simply an undeniable fact that

> the chronic diseases that now kill most of us can be traced directly to the industrialization of our food: the rise of highly processed foods and refined grains; the use of chemicals to raise plants and animals in huge monocultures; the superabundance of cheap calories of sugar and fat produced by modern agriculture; and the narrowing of the biological diversity of the human diet to a tiny handful of staple crops, notably wheat, corn, and soy.
>
> *(Pollan, 2008, p. 10)*

Pollan claims: "That such a diet makes people sick and fat we have known for a long time." Later, he shows that "people who eat the way we do in the West today suffer substantially higher rates of cancer, cardiovascular disease, diabetes, and obesity than people eating any number of different traditional diets."

According to Nesheim et al. (2015, p. 6), "unhealthy dietary patterns are identified as a risk factor in the etiology of several leading causes of mortality and morbidity. And it is important to point out at the outset that it is *overconsumption* that is the driving force behind illness." Nesheim et al. (2015, p. 90) note: "The primary diet-related risks to disease of the current food system are related to food overconsumption, and contribute to the etiology of several leading causes of mortality and morbidity, including cardiovascular disease (CVD), type 2 diabetes, cancer, and osteoporosis." That is, our health problems are "not related to nutrient inadequacy but mostly to inappropriate dietary patterns and overconsumption" (p. 116).

Fuhrman (2017) provides extensive evidence with regard to the deleterious health outcomes associated with poor diet and nutrition. They include weight gain, obesity, diabetes, heart disease, strokes, various forms of cancer (e.g., colon, breast, endometrial, lung, pancreatic, prostate), allergies, autoimmune diseases, autism, attention-deficit hyperactivity disorder, learning disabilities, memory loss, brain fog, Alzheimer's disease, poor school performance, anxiety, depression, mood swings, aggressive behavior, suicide, and even addiction.

### Obesity

With regard to obesity, it is defined as having 35 or more pounds of excess weight than a normal person. Moss (2021, p. 28) correctly notes that, in the United States, obesity began to rise in the 1970s, "climbing from 15 to 40 percent." Figure 6.1 shows increases in obesity in the United States among adults. Figure 6.2 shows the same trends for kids. Moss also shows that the problem is now global in nature, as there are at least

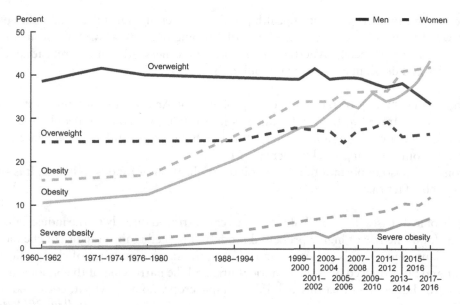

**FIGURE 6.1**   Trends in Adult Overweight, Obesity, and Extreme Obesity Among Men and Women Age 20–74, United States, 1960–2014

*Source:* https://stateofobesity.org/obesity-rates-trends-overview/

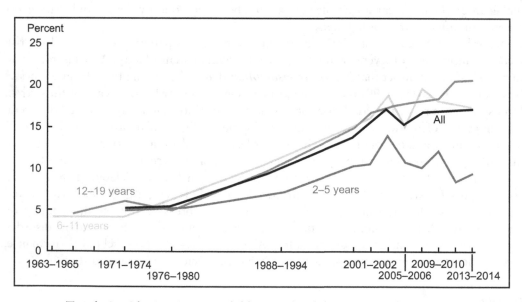

**FIGURE 6.2**   Trends in Obesity Among Children and Adolescents Aged 2–19 Years, by Age: United States, 1963–1965 to 2013–2014

*Source:* www.cdc.gov/nchs/data/hestat/obesity_child_13_14/obesity_child_13_14.htm

650 million obese people in the world; about 1.9 billion people worldwide are overweight (Coca & Barbosa, 2019, p. 347). Walker (2019, p. 216) notes that global obesity rates have tripled around the globe in just 40 years. This corresponds perfectly to changes in the food environment over the past four decades, which have been discussed throughout this book.

Within the United States, Herz (2018, p. 268) shows that obesity has more than doubled among children in the past 30 years and quadrupled among adolescents. Again, this corresponds to changes in the food environment over time. Herz claims that teenagers got about 17% of their calories from fast food and children ages 2–9 years got about 9% of their calories from fast food. Not surprisingly, diets high in fat and calories and low in fiber (like fast foods) are found to be associated with weight gain.

So, it is the food that is to blame. Yet, fast-food chains such as McDonald's deny this connection, and, in the case of McDonald's, it even created a film titled *540 Meals: Choices Make the Difference* to show that a person who ate nothing but its food for 90 days lost 37 pounds; the film ignored the fact that the subject in the film also exercised for 45 minutes a day, four to five times a week, and that he restricted his calories to 2,000 or less each day! Herz (2018, p. 269) notes that critics called this film "nothing more than a glorified infomercial that seems cynically calculated to get kids to eat even more fast food than they already do" since it was targeted at middle- and high-school-aged kids.

Interestingly, Herz (2018) notes a troubling correlation that may give us pause about organic foods:

In 1990, U.S. sales of organics were approximately one billion dollars annually. In 2014, organic sales were up to 39 billion dollars. In 1990, 11.4 percent of the U.S. population was obese, and by 2014 the percentage of obese U.S. adults had reached 34.9 percent. Just a coincidence? Probably not.

She presents examples of organic foods, including Oreo cookies, that are high in sugar, where consumers eat them more often and in greater numbers than non-organic, based on their false beliefs that they are healthier. Interestingly, some studies also show that the "fair trade" label (which has nothing to do with nutrition) causes some to believe food products are healthier, even though they are not.

The US Centers for Disease Control and Prevention (CDC, 2016) lists the health consequences of obesity. They include:

- All causes of death (mortality)
- High blood pressure (hypertension)
- High LDL cholesterol, low HDL cholesterol, or high levels of triglycerides (dyslipidemia)
- Type 2 diabetes
- Coronary heart disease
- Stroke
- Gallbladder disease
- Osteoarthritis (a breakdown of cartilage and bone within a joint)
- Sleep apnea and breathing problems
- Some cancers (endometrial, breast, colon, kidney, gallbladder, and liver)
- Low quality of life

- Mental illnesses such as clinical depression, anxiety, and other mental disorders
- Body pain and difficulty with physical functioning

Cohen (2014, p. 4) also lists outcomes of obesity, as well as food-related facts that pertain to weight gain. They include the following:

1. Obese people have a doubled risk of dying prematurely.
2. Eating too many trans fats increases the risk of coronary artery disease by 23%.
3. Not eating enough fiber increases the risk of colon cancer by 18%.
4. Drink one SSB a day increases the risk of diabetes by 83% in women.
5. Too much salt in the diet is thought to be responsible for 62% of all strokes.

Further, between one-quarter and one-third of all cancers are attributable to obesity, and being overweight or obese is also a strong risk factor for type 2 diabetes, heart disease, hypertension, joint and back pain, and a host of other medical problems (Cohen, 2014, p. 14).

Cohen (2014, p. 5) shows that, in as little as seven weeks of eating 40% too many calories in the diet, significant weight gain (more than 16 pounds) will occur, and waists will grow (more than three inches, and extra fat is deposited in organs rather than muscle; average fat cell size grew 54%). Further:

> Indicators of systemwide inflammation increased by 29–50 percent. Fasting blood sugar (and indicator of diabetes), cholesterol, and insulin resistance all increased. Blood pressure and heart rate went up and blood vessel functioning capacity was reduced by about 21 percent—altogether demonstrating a significant increase in factors associated with heart disease.
>
> *(pp. 5–6)*

### Diabetes

With regard to diabetes, it is a medical condition resulting from the interactive effects of poor nutrition, physical inactivity, and genetics. Diabetes causes blood glucose (i.e., sugar) levels to rise higher than normal (American Diabetes Association, 2017). Thus, the condition is also frequently referred to as *hyperglycemia*. Diabetes is a metabolic disorder, meaning it impacts the way the body uses food to help us grow and use energy (WebMD, 2017).

Under normal circumstances, the pancreas releases the hormone insulin so that the body can store and use sugar and fat from the foods we eat in order to generate the energy we need to live and move. The amount of insulin produced by the pancreas directly impacts glucose levels in the blood. Insulin is constantly released by the pancreas to regulate the amount of sugar in the blood. When glucose levels rise to a certain level, more insulin is released into the body to force more glucose into the body's cells, causing blood sugar levels to drop.

If glucose levels drop too far, the condition *hypoglycemia* occurs, creating a shaky feeling that signals it is time to eat. This also releases glucose into the blood stream that is normally stored in the liver. Diabetes occurs when the pancreas produces no insulin or very little insulin or the body does not appropriately respond to insulin (this is called *insulin resistance*), leading to higher levels of blood sugar.

There are three types of diabetes—gestational, type 1, and type 2. Gestational diabetes occurs during pregnancy. The primary difference between type 1 and type 2 diabetes is that the former has its roots in genetics and the latter is mostly caused by environmental or lifestyle factors (yet also impacted to some degree by genetics). Specifically, type 1 diabetes occurs when beta cells (the cells that produce insulin in the pancreas) are attacked by the immune system. This is produced by certain genes passed down from parent to child and is also thought to be impacted by viral exposure. The result is the production of no insulin, requiring insulin injections. According to the US Center for Disease Control and Prevention (CDC, 2017a), type 1 diabetes tends to appear quickly and be diagnosed in childhood, adolescence, or early adulthood. Historically, it was called "juvenile diabetes" and "insulin-dependent" diabetes. Only 5% of people with diabetes have type 1 diabetes.

Type 2 diabetes is far more common and develops over time, typically in adults. People with type 2 diabetes produce insulin, but the insulin is not at sufficient levels or the person has insulin resistance. Historically, type 2 diabetes was also called "adult-onset diabetes" because it only occurred in adults. Yet, a rapid rise in overweight and obese children has led to a surge in type 2 diabetes in children. I argue in this chapter that both corporate and government agencies share responsibility for this reality.

The major symptoms of diabetes include hunger, dry mouth and thirst, the need to urinate, fatigue, blurred vision and vision loss, itchy skin, slowly healing cuts, increased yeast infections, red or sore gums, and numbness, tingling, or pain in the extremities (American Diabetes Association, 2017; US Centers for Disease Control and Prevention (CDC), 2017a). Many of the symptoms of diabetes are relatively minor; others are quite severe and even life-threatening. Diabetes can cause macrovascular disease (damage to large blood vessels) and microvascular disease (damage to small blood vessels). Stroke and heart attack can be complications of macrovascular disease, and microvascular disease can lead to eye, nerve, and kidney problems.

Diabetes is also a risk factor for heart disease, and, according to the CDC, adults with diabetes have a 200–400% increase in risk of stroke or dying of heart disease. Diabetics can also develop neuropathy, which is nerve damage. The most common type of neuropathy is peripheral neuropathy, which causes numbness and pain in legs, feet, toes, and arms. High blood sugar can cause damage to the blood vessels in the retina, which can cause vision problems and even blindness. Additionally, the poor circulation of diabetics means they do not heal as well as non-diabetics due to the reduced flow of blood, oxygen, and nutrients (Dresden, 2017); this may lead to amputation of body parts in extreme cases (McDermott, 2016).

As noted earlier, type 2 diabetes makes up approximately 95% of all diabetes (Geiss et al., 2014). Given that type 2 diabetes is environmentally produced, nearly all diabetes can be prevented through environmental intervention. Yet, the prevalence of diabetes in America has been steadily increasing since 1958, according to the data from the National Health Interview Survey (NHIS) (Skyler & Oddo, 2002). There have been more and more cases diagnosed over the years, with prevalence increasing with age. Diabetes will only continue to grow as people continue to become obese and live sedentary lifestyles (Skyler & Oddo, 2002).

Figure 6.3 shows trends over time in the incidence and prevalence of diabetes in the United States. This figure demonstrates that diabetes began to rise around 1990, roughly one decade later than the increases in obesity; this makes sense given that obesity raises the risk of diabetes over time.

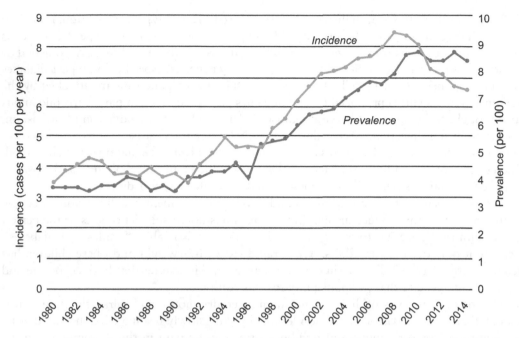

**FIGURE 6.3**  Trends in Incidence and Prevalence of Diagnosed Diabetes Among Adults Aged 20–79, United States, 1980–2014

*Source:* www.cdc.gov/chronicdisease/resources/publications/aag/diabetes.htm

By 2015, 30.3 million Americans (9.4% of the US population) had diabetes. Additionally, more than 84 million Americans had prediabetes, "a condition that if not treated often leads to type 2 diabetes within five years" (US Centers for Disease Control and Prevention, 2017b). According to the US Centers for Disease Control and Prevention (CDC, 2017a):

> Prediabetes is a serious health condition where blood sugar levels are higher than normal, but not high enough yet to be diagnosed as diabetes. Prediabetes increases your risk for type 2 diabetes, heart disease, and stroke. It has been estimated that the prevalence of diabetes in the United States will increase by 165% between 2000 and 2050.
>
> *(Narayan et al., 2003)*

For comparison purposes, there were five million victims of violent crime in 2015, 1.8 million of which were victims of serious violent crime. Of those, only 658,040 were physically injured (Truman & Morgan, 2018).

The National Diabetes Statistics Report found that, in 2015, there were approximately 1.5 million new cases of diabetes among Americans aged 18 years old and older (US Centers for Disease Control and Prevention, 2017b). The mean age at which diabetes was diagnosed has also decreased over the years (Koopman et al., 2005). From 1988 to 1994, the mean age was 52 years, and from 1999 to 2000, the mean age decreased to 46 years (Koopman et al., 2005). This change could indicate an earlier detection of type 2 diabetes or an earlier onset of the disease (Koopman et al., 2005).

There has also been evidence that type 2 diabetes has been diagnosed more in adolescents (Koopman et al., 2005). Data analyzed from the SEARCH for Diabetes in Youth study show that there is a relative 4.8% increase in youth with type 2 diabetes annually after adjustments are made for ethnicity, age, and sex (Mayer-Davis et al., 2017). It was also estimated that there were 3,800 cases of type 2 diabetes diagnosed among youth annually from 2002 to 2003, while the number grew to 5,300 cases per year in 2011–2012 (US Centers for Disease Control and Prevention, 2017c; Mayer-Davis et al., 2017).

According to the American Diabetes Association, "Diabetes was the seventh leading cause of death in the United States in 2015 based on the 79,535 death certificates in which diabetes was listed as the underlying cause of death" (Statistics About Diabetes, 2017). Diabetes was also listed as a cause of death in a total of 252,806 death certificates. It is also possible that diabetes is being underreported as a cause of death in the United States; studies have shown that only about 35–40% of diabetics had the illness listed on their death certificate after they died, and only 10–15% had diabetes recorded as the underlying source of death (Statistics about Diabetes, 2017). For comparative purposes, there were 15,696 acts of murder and nonnegligent manslaughter in the United States in 2015 (FBI, 2019).

### Mental Health and Other Conditions

Even mental health is impacted by nutrition, and this makes sense given that neurotransmitter levels in the brain are built from the foods we eat (Robinson & Beaver, 2020). For example, depression is associated with deficiencies in EPA and DHA fatty acids, commonly found in fish oils as well as vegan forms through algae. Deficiencies in EPA and DHA are related to:

1. Lower intelligence
2. Poor school performance
3. Depression and suicide
4. Memory loss and cognitive decline
5. Brain shrinkage and dementia
6. Antisocial behavior including crime.

*(Fuhrman, 2017, pp. 56, 64)*

The crime links are likely interesting to criminologists. Fuhrman (2017) shows that crime rates are correlated with rates of diabetes in areas of the country, and that rises in the consumption of linoleic acid from increased use of food oils are associated with increased murder rates. He also shows, at the individual level, that excesses of omega-6 fats and deficiencies of omega-3 fats are associated with low levels of serotonin in the brain, a source of violent behavior, as are high levels of soda consumption, which is also correlated with withdrawn behavior and poor attention. The impact of food on the brain should also be of interest to criminologists, as well as the rest of us, since lack of nutrients can damage the brain, junk food can shrink the brain, and people can become desensitized to dopamine by eating too many processed foods (which can lead to tolerance, indicative of addiction—see Chapter 5): "The amount of scientific evidence is irrefutable that junk food, candy, soft drinks, fast food, and commercial baked goods damage the brain" (Fuhrman, 2017, p. 62).

### Death Associated With Diet

And then there is the issue of death. The WHO reports that "degenerative and chronic diseases such as heart disease, cancers, stroke, and diabetes cause 70 percent of deaths worldwide. They account for 37 percent of deaths in low-income countries and up to 88 percent in high-income countries" (Winson & Choi, 2019, p. 39). These are largely considered "lifestyle diseases" because they result from poor diet and inactivity (Chenoweth & Leutzinger, 2006).

About 16 million people around the world die each year from diabetes and cancer alone, and these lifestyle diseases can largely be prevented by making changes to the foods we eat and by making sure people exercise more. One question is what is the best way to achieve these results? Countries around the globe have taken steps, including Hungary which imposed heavy taxes on unhealthy foods and the United Kingdom which introduced a 20% tax on SSBs in 2017. Such change is unlikely in the United States due to the immense power of key actors in the conventional food system here and the unwillingness of the government to do anything about it.

As for the United States, a 2004 study by the US Centers for Disease Control and Prevention (CDC) claimed that poor diet and inactivity lead to about 400,000 deaths per year, but this estimate was downgraded to about 112,000 later by the CDC. Moss (2021, p. 29) claims that obesity leads to about 300,000 deaths each year in the United States, and costs us about $300 billion in health-care costs. And a study published in the JAMA found that poor diet leads to 395,000 deaths a year (Micha et al., 2017), as shown in Chapter 1. So, a safe estimate of deaths attributed to food is in the hundreds of thousands.

Recall from Chapter 1 that the aforementioned study examining data from 2012 found that just over 700,000 people died that year from a "cardiometabolic disease" (CMS). This included more than 500,000 from heart disease (including almost 375,000 from coronary heart disease, nearly 130,000 from stroke, and just under 70,000 from diabetes) (Micha et al., 2017). Of the 700,000 CMS deaths, about 395,000 (about 56%) were attributed to the foods that people ate (Micha et al., 2017). Recall that roughly 66,500 of the deaths identified in the study were due to high sodium intake. Another 58,000 deaths were caused by eating too much processed meat. Many other deaths were attributable to the absence of certain foods in the diet, including 59,000 deaths due to a lack of nuts and seeds in the diet, 55,000 deaths due to too little omega-3 fats, 53,000 deaths from too few vegetables, and 52,500 deaths from too few fruits. Finally, about 52,000 deaths were from too many SSBs (Micha et al., 2017).

Recall from Chapter 1 that only about 42,000 people died from illicit drugs in 2012 (National Institute on Drug Abuse, 2018). Stated simply, this means that, in 2012, the foods we ate (and did not eat) killed about ten times more people than all illegal drugs combined. Further, SSBs are more deadly than all illegal drugs combined! Note that many of these SSBs are also served daily in our children's schools.

With regard to fast food, in particular, Fuhrman (2017) shows that eating fast food two or three times a week increases the risks of fatal coronary heart disease by 50%; eating fast food four or more times per week increases the risks by 80%, and eating fast food just once a week increases the odds by 20%. This is partially due to the build-up of "toxic metabolites" or

waste products produced by the body that can cause inflammation and disease if allowed to accumulate. Consuming a diet low in micronutrients and phytochemicals [such as fast and processed foods] results in inflammation, oxidative stress, and the accumulation of these toxic metabolites, including free radicals.

*(p. 24)*

Another outcome is "toxic hunger" which consists of "headaches, light-headedness, irritability, and fatigue" (p. 26). Fuhrman claims that toxic hunger feeds overeating, whereas normal hunger does not.

Incredibly, when faced with criticisms over the unhealthy food products they produce, manufacturers try "to spin themselves as 'part of the solution' when it comes to solving this public health crisis" (Simon, 2006, p. xv). And, although many major food manufacturers do indeed offer healthier versions of many of their products, one could see health-related claims from junk food manufacturers as simply public relations (PR), typical of corporations of all types (Robinson & Murphy, 2008). As explained by Simon (2006, p. 8), "food makers initiate only marginal, largely cosmetic changes to their traditional business practices—and then greatly exaggerate the benefits of these efforts for maximum PR effect." After all, these companies make their living off of unhealthy food, and they seem not concerned at all with the health consequences to their own consumers that result from their unhealthy products. Instead, large companies such as Kraft, McDonald's, and General Mills "have launched massive PR campaigns aimed at convincing policymakers—and indeed all Americans—that they are committed to making substantive changes that will rebound to the benefit of the nation's health. The reality, though is far different."

According to Simon (2006, p. xv), the PR campaigns are designed to achieve two goals— "shoring up positive public images and deflecting the threat of government regulations and lawsuits." Meanwhile, Simon notes, major corporate lobbyists are "conducting business as usual" so that "virtually every commonsense health policy, such as getting soda out of schools, has either been blocked, delayed, or significantly compromised" (p. xvi). According to Simon:

A food lobbyist's main goal is to provide any distraction, misdirection, or obfuscation possible to avoid talking about corporate accountability. The goal of all industry rhetoric, activities, and lobbying is to keep government out of their business while maintaining the media focus on individual behavior change as the true solution to America's health problems.

*(p. xvi)*

Responsible actions on the part of food companies are discussed in Chapter 9.

### Health-Care Costs

Not counting the value of the human lives lost to diet and inactivity, there are other significant financial health-care costs associated with the foods we eat. According to Nesheim et al. (2015, p. 91), 75% of all health-care dollars spent in the United States are used "to treat preventable chronic diseases and conditions." Specifically, "the burden of cardiovascular disease in the United States is more than $300 billion each year, including the cost of health care services, medications, and lost productivity." This figure includes the roughly $245 billion lost to diabetes.

Some of the indirect costs of diabetes include missed days at work, reduced productivity at work, and an inability to work due to health conditions (Petersen, 2013). Type 2 diabetes also causes other significant economic burdens. In 2007, the total approximate cost of diagnosed diabetes in the United States was $147 billion. This number increased to $245 billion in 2012, $176 billion due to direct medical costs and $69 billion due to reduced productivity. Medical costs include "hospital inpatient care (43% of the total medical cost), prescription medications to treat the complications of diabetes (18%), antidiabetic agents and diabetes supplies (12%), physician office visits (9%), and nursing/residential facility stays (8%)" (Petersen, 2013). For comparison purposes, the direct costs to victims of street crime were between $12.5 billion and $13.1 billion in 2008 (Robinson & Tauscher, 2019).

It was estimated that the lifetime cost of treating type 2 diabetes and its complications in men diagnosed at ages 25–44 was $124,700. The lifetime cost for men diagnosed between 45 and 54 was $106,200, for men diagnosed between 55 and 64, the cost was $84,000, and for men above the age of 65, the cost was $54,700. The lifetime cost for women was found to be slightly higher. Women diagnosed at 25–44 years had a lifetime medical cost of $130,800. Women who were diagnosed between the ages of 45–54 had a cost of $110,400, women diagnosed at ages 55–64 had a cost of $85,500, while women diagnosed over the age of 65 had $56,600 in medical costs (Zhuo et al., 2013).

According to US Centers for Disease Control and Prevention (CDC, 2016), obesity is also financially costly to the United States: "The estimated annual medical cost of obesity in the U.S. was $147 billion in 2008 U.S. dollars; the medical costs for people who are obese were $1,429 higher than those of normal weight." Even though these medical conditions contribute to the nation's GDP (sick people go to doctors and hospitals and spend money), these costs are significant higher than the direct costs of street crime in the United States (Robinson, 2015).

US Centers for Disease Control and Prevention (2018a) notes:

Obesity and its associated health problems have a significant economic impact on the U.S. health care system. Medical costs associated with overweight and obesity may involve direct and indirect costs. Direct medical costs may include preventive, diagnostic, and treatment services related to obesity. Indirect costs relate to morbidity and mortality costs including productivity. Productivity measures include "absenteeism" (costs due to employees being absent from work for obesity-related health reasons) and 'presenteeism' (decreased productivity of employees while at work) as well as premature mortality and disability.

US Centers for Disease Control and Prevention (2018a) tabs lost productivity due to absenteeism at between $3.38 billion ($79 per obese individual) and $6.38 billion ($132 per obese individual).

The deaths of 300,000 Americans per year from obesity-related conditions would also produce potential losses of another $17 billion in salaries annually, using the average salary of $56,000 for all workers (300,000 multiplied by $56,000). The number of deaths from diabetes would be $3.9 billion (70,000 deaths multiplied by $56,000).

## A Healthier Diet

Meanwhile, healthier diets are found to reduce the incidence of outcomes like obesity and diabetes, thereby reducing illness as well as mortality. Healthier diets are also rich in such substances and are associated with less hyperactive behavior, less moodiness, more happiness, higher levels of intelligence, better school performance, less illness and death, as well as less antisocial and aggressive behaviors. Studies even show that dietary changes within schools and even correctional facilities are associated with lower rates of disciplinary infractions and violence. Fuhrman (2017, p. 78) promotes what he calls the "nutritarian diet," " 'the gold standard' eating style to maximize health and longevity." It involves food focused on nutritional value above and beyond all else, including foods high in fruits, vegetables, whole grains, nuts, and seeds. This diet would result in the following:

1. Prevention as well as reversal of high blood pressure
2. Prevention and even remission of type 2 diabetes
3. Prevention and reversal of heart disease
4. Prevention and reversal of many types of cancer
5. Prevention of autism
6. Improvements in health and happiness.

*(Fuhrman, 2017, pp. 94–94)*

An example of a healthy diet that is associated with a lower incidence of illness is illustrated by the Oslo Diet-Heart Study. It increased the intake of fish and shellfish, omega-3 fatty acids, beans, nuts, and fruits and vegetables, as well as vitamins A and D, calcium, and whole grains. It also reduced sugar consumption, trans fats, and whole milk products. Vegetarianism and veganism also tend to be healthier, although vegetarians and vegans also tend to live healthier lives than most meat-eaters; this does not necessarily lower mortality rusks (Minger, 2013). Incredibly, about half of all grains consumed in the United States are of the refined variety, something linked to insulin resistance and diabetes.

A review of the Paleo diet, Mediterranean diet, and plant-based diet found each to provide some notable health benefits, but Minger (2013, p. 225) notes that each of these diets excludes industrially processed vegetable oils, refined grains and flours, refined sugars, chemical preservatives, and "lab-produced anythings . . . nearly anything coming in a crinkly tinfoil package, a microwave tray, or a McDonald's takeout bag," and all include tubers, low glycemic fruits, and non-starchy vegetables. So, all of them "kick modern, heavily processed fare to the curb." They are all also dense in nutrients associated with greater health. Based on her extensive review, Minger (2013, p. 242) recommends eliminating one's intake of refined grains, sugars, and high-omega-6 vegetable oils, securing a source of fat-soluble vitamins A, D, E, and K, stocking your diet with nutrient-dense items, and limiting animal muscle meat consumption. It appears that many harms associated with SAD can be eliminated by following such a diet. But why does not the conventional food system stop producing such high numbers of industrialized, ultra-processed foods and replace them with foods that are part of a healthier diet? It is culpable for failing to do this. Further, government agencies are culpable for failing to achieve high production of healthier foods so they can be more accessible to and affordable for Americans.

## Obesity, Diabetes, and Choice

As noted in Chapter 4, the dominant view, at least in America, of health, diet-related conditions such as obesity and diabetes, is that people choose what to eat, and thus, they are solely responsible for deleterious outcomes which they suffer. Cohen (2014, p. 6), for example, writes: "The conventional wisdom is that obesity is the expression of individual weakness, gluttony, and/or lack of personal responsibility," or basically a crisis of poor self-control. Clearly, many choices we make about what we eat, how much, as well as what we do not eat are "not ideal" (Dogget & Egan, 2016, p. 109). Kukla (2018, p. 593) agrees, writing that

> the dominant mode of food messaging holds individuals accountable for making "life-style" choices that keep them thin and healthy, and shames those who fail to do so. Given this is such an important topic, it is worthy of revisiting here, as we discussed health implications of obesity and diabetes in this chapter.

In essence, we blame poor self-control for problems like obesity. A lot is actually known about self-control: It is poor early in life, and people learn to develop it in intimate social groups such as families. It is also fairly stable across a person's life because it is at least partially heritable, it is visible in brain scans (Robinson & Beaver, 2020), and early measures of low self-control are associated with later weight gain (Cohen, 2014). Finally, when it comes to eating, people often fail to exercise self-control when they should, for example, in the presence of too much food. So, it appears we have less control over our choices than we like to believe.

As noted in Chapter 4, consider our failures with dieting as evidence of our weak self-control. According to Nestle (2013, p. 41):

> Dieting . . . is rarely effective to long-term weight loss. Most people who lose weight regain it over time, mainly because dieting fights physiology. Although a great many hormones and factors are involved in regulation of hunger, satiety, and body weight, nearly all of them stimulate people to eat; barely any tell you when to stop.

So, we seem to be fighting our own bodies when we try to lose weight, but we are also fighting evolution, because

> humans evolved from populations who had to hunt or gather. When food intake drops below the amount required for physiological needs, metabolism slows to conserve energy and you need fewer calories. Once you do lose weight, you need even fewer calories to maintain your smaller body size. Dieters must counter their own physiology as well as the "eat more" food environment. Under these circumstances, personal responsibility doesn't stand a chance.

In spite of the myth of free will, the truth is much more complicated. Like virtually every behavioral and medical condition, the causes of obesity and type 2 diabetes are complex and involve a combination and interaction of various factors, including genetics and environmental factors. Whereas many Americans believe in the idea of free will and thus likely hold that we are solely responsible for how much we weigh and how (un)healthy we are, research shows that weight and health are impacted by many variables, including genetics/epigenetics, as well as the foods and drinks we consume, and the amount of physical activity in which we engage. The latter two factors are impacted by community-level factors, advertising by businesses, as well as other decisions of elites in the conventional food system. Due to poor diet and nutrition promoted and even celebrated by companies that comprise the food industry—with the assistance of the state—Americans suffer from heart disease, many forms of cancer, strokes, type 2 diabetes, numerous other health conditions, diminished quality of life, and even death, many of which are produced by type 2 diabetes and obesity, its antecedent.

The CDC notes that both genetics and behavior contribute to the problem of being overweight or obese. In addition, where people live directly impacts their weight, based on factors such as the availability of quality food and access to exercise opportunities. The behaviors that influence weight "include eating high-calorie, low-nutrient foods and beverages, not getting enough physical activity, sedentary activities such as watching television or other screen devices, medication use, and sleep routines" (US Centers for Disease Control and Prevention, 2018b). Conversely, eating healthy foods, remaining physically active, and getting enough sleep are associated with healthier weights as well as reductions in "chronic diseases such as type 2 diabetes, cancer, and heart disease." Access to low-quality and high-quality foods and the availability of exercise opportunities are heavily dependent on the actions of both food companies and government agencies.

There is simply no question that one's diet is associated with one's risk of developing outcomes such as type 2 diabetes. According to Sami et al. (2017, p. 65): "Dietary habits and sedentary lifestyle are the major factors for rapidly rising incidence of [type 2 diabetes] among developing countries." And since Americans really very heavily on the conventional food system—controlled by only a handful of large, multi-national conglomerates that provide us with nearly all the food we consume at home, school, work, and on the road—there is also no question that the conventional food system plays a major role in the epidemic of type 2 diabetes in the United States as well as in the rest of the world.

Published studies show, for example, that diets high in sugar and low in fruits and vegetables (like the typical US diet) are associated with higher a likelihood of developing type 2 diabetes (Crosta, 2008). These studies are longitudinal in nature and focused on tens of thousands of individuals and find that greater consumption of sweet drinks such as soda and fruit juices and other sugary drinks like Kool-Aid lead to diabetes, although part of this effect is moderated by BMI (since obesity also contributes to diabetes) (Palmer et al., 2008). According to Crosta (2008),

> drinking two or more soft drinks each day was associated with a 24% increase in diabetes risk and drinking two or more fruit drinks each day was associated with a 31% increase in diabetes risk compared to women who had less than one soft drink or fruit drink per month, respectively. There was no association noted between type 2 diabetes risk and diet soft drinks, grapefruit juice, or orange juice.

Similarly, a meta-analysis of nine cohort studies published in 11 journals found

> a relationship between sugars-sweetened beverages [SSBs] and the incidence of type 2 diabetes. The link between sugar consumption and diabetes is both direct and indirect—with sugars-sweetened beverages being directly linked to the incidence of type 2 diabetes, and equally sugar consumption leading to obesity, one of the main risk factors for type 2 diabetes.
>
> *(Action on Sugar, 2017)*

Another meta-analysis of data from eight studies on SSBs and the risk of type 2 diabetes found that individuals who consumed one to two servings per day had a 26% greater chance of developing type 2 diabetes than individuals who consumed none or less than one serving per month (Malik et al., 2010).

Sugar is undeniably an addictive substance that humans naturally crave. Yet, it is global companies that decide how much sugar to add to food products, advertise those products to consumers (including directly to children), and sell those products. It is government agencies that allow and even encourage it (Moss, 2013; Taubes, 2021). The same findings occur when it comes to carbohydrates, particularly refined carbohydrates, which raise glucose levels in the blood. Of course, there are different kinds of carbohydrates and they impact the body differently. For example, there are simple sugars that are found in processed foods and offer few or no nutrients (e.g., vitamins and fiber), and there are complex carbohydrates that are in their "whole food form" and also provide additional nutrients. Simple sugars raise glucose and contribute to diabetes, and complex carbohydrates generally do not because they can help stabilize sugar in the blood by slowing down the absorption of glucose. Thus, dietary experts suggest people eat complex carbohydrates such as brown rice rather than white rice, whole wheat foods rather than processed wheat foods (including pastas), other natural foods like steel-cut oatmeal and quinoa, and of course, fruits and non-starchy vegetables, lentils, and beans. White foods like bread, pasta, potatoes, flour, cookies and pastries, and sugar (including fruit juices and sodas) should be eaten only minimally or in moderation. Yet, the latter foods are highly addictive and advertised, as well as much more available than the former. Global food corporations and government agencies meant to regulate share responsibility for this reality.

The American Diabetes Association (ADA, 2017) claims the idea that excess sugar consumption causes diabetes is a myth. Yet, it writes: "Being overweight does increase your risk for developing type 2 diabetes, and a diet high in calories from any source contributes to weight gain. Research has shown that drinking sugary drinks is linked to type 2 diabetes." And it recommends "that people should avoid intake of sugar-sweetened beverages to help prevent diabetes." This includes regular soda, fruit punch and fruity drinks, energy and sports drinks, sweet tea, and other sugary drinks. According to the ADA (2017), these drinks "will raise blood glucose . . . the main sugar found in the blood and the body's main source of energy. Also called blood sugar . . . and can provide several hundred calories in just one serving!" The ADA promotes healthy eating, which includes foods that increase glucose such as sweets, carbohydrates, and fruits, meaning they should not be avoided but instead eaten in moderation.

A multi-national study (n=175 countries) over a decade found mere availability of sugar was associated with higher levels of diabetes, even after controlling for many potentially confounding variables. The authors write

> every 150 kcal/person/day increase in sugar availability (about one can of soda/day) was associated with increased diabetes prevalence by 1.1% (p <0.001) after testing for potential selection biases and controlling for other food types (including fibers, meats, fruits, oils, cereals), total calories, overweight and obesity, period-effects, and several socioeconomic variables such as aging, urbanization and income.
>
> *(Basu et al., 2013)*

Moreover, they found that there was no other food type which "yielded significant individual associations with diabetes prevalence after controlling for obesity and other confounders." Additionally, "declines in sugar exposure correlated with significant subsequent declines in diabetes rates independently of other socioeconomic, dietary and obesity prevalence changes." The conclusion of the study is that "[d]ifferences in sugar availability statistically explain variations in diabetes prevalence rates at a population level that are not explained by physical activity, overweight or obesity."

Another study focused on thousands of people over many years found that higher levels of vitamin C in the diet (from higher consumption of fruits and vegetables) lowered the risk of developing diabetes (Harding et al., 2008). According to Crosta (2008):

> Compared with men and women in the bottom quintile [fifth] of plasma vitamin C, the odds of developing diabetes was 62 percent lower for those in the top quintile of plasma vitamin C. A weaker inverse association between fruit and vegetable consumption and diabetes risk was observed.

A study of thousands of people over six years compared dietary patterns between two groups.

> The pattern labeled "prudent" was characterized by higher consumption of fruits and vegetables, and the pattern labeled "conservative" was characterized by consumption of butter, potatoes, and whole milk—and the prudent diet was associated with lower occurrences of diabetes while the conservative diet was associated with higher occurrences of diabetes.
>
> *(Montonen et al., 2005)*

Still another study found that consumption of lower-fat foods was associated with a slightly lower incidence of diabetes (Tinker et al., 2008). Similarly, a recent meta-analysis of studies found general evidence that high-fat diets, and diets high in fried foods, red meat, and sweets, are associated with higher levels of type 2 diabetes (Sami et al., 2017). This is in part due to the link between saturated fats which serve to activate immune cells and produce inflammatory proteins such as interleukin-1beta (Wen et al., 2011).

Not only do high-fat foods increase the odds of diabetes, but so too do processed foods. When foods are processed, the fiber from the bran and germ is removed, as are a large majority of the food's nutrients. The absence of fiber allows the food to be digested more rapidly, quickly elevating sugar levels in the blood. As noted earlier, global food companies, enabled by the government, produce, advertise, and sell the foods that lead to increases in diabetes and other harmful medical conditions.

In type 2 diabetics, elevated levels of glycated hemoglobin (HbA1c), a protein in red blood cells that combined with glucose in the blood, have also been considered one of the leading risk factors for developing microvascular and macrovascular complications. Improvement in the elevated HbA1c level is achievable through diet management (Sami et al., 2017).

Nutritional intake does not only impact diabetes by increasing glucose in the blood. The foods and drinks we consume also impact blood flow and can cause inflammation. Markers of homeostasis (which occurs when blood flow stops) and inflammation (an autoimmune response to injury or damage to tissues) are found to be associated with higher likelihood of developing type 2 diabetes. A study by Liese et al. (2009) was able to determine that certain foods were to blame for this relationship:

> High intake of the food groups red meat, low-fiber bread and cereal, dried beans, fried potatoes, tomato vegetables, eggs, cheese, and cottage cheese and low intake of wine characterized the pattern, which was positively associated with both biomarkers. With increasing pattern score, the odds of diabetes increased significantly.

The authors specifically found that the odds of diabetes increased three to four times with that type of dietary intake. This follows earlier an earlier study (Schulze et al., 2005) which found two to three times higher rates of diabetes resulting from diets high in processed meats, refined grains, and sugary drinks but low in wine, coffee, and vegetables, and another study (Heidemann et al., 2005) which found two to five times higher risks of diabetes among people who ate a lot of processed meats, red meats, chicken, refined-grain breads, beer, legumes, and sugary drinks but few fruits.

Additionally, chemicals added to foods to make them taste better play a role in the development of diabetes. For example, a study of four generations of lab mice (whose DNA and central nervous systems are very similar to humans) found that mice fed foods with advanced glycation endproducts tended to have lower levels of antioxidants but greater levels of inflammation, body fat, and importantly, more insulin resistance (Cai et al., 2012).

Interestingly, the factors reviewed earlier that lead to diabetes also tend to lead to weight gain and obesity. Consider what it takes to actually first lose weight and then to keep it off, as well as to be healthy, when you live in environments that are quite literally packed with food. Here you would have to

> resist food multiple times per day, rule out many options, and make large trade-offs every day of the week, every week of the year, for many years. In the current environment, with temptations everywhere, the mental effort needed to lose weight is enormous.
>
> *(Cohen, 2014, p. 33)*

The reality of the situation is that, often, we choose the easy and the convenient, which tends to be unhealthy for us. This is easy to understand why when you consider the concept of "food swamps" in which we all live (p. 80).

Simon (2006, p. xi) notes the ubiquity of unhealthy food in America, writing that we "take for granted the mind-boggling amount of unhealthy food currently available anywhere, anytime, from entire grocery store aisles devoted to chips, soda, and ice cream to the endless fast-food chains found of every block and at every highway exit." She continues: "Nowadays you can't even walk into a bookstore without being tempted by mocha lattes and megamuffins." She thus describes our food environment as "toxic"—filled with fat, hormones, and additives—caused by "overzealous corporate marketing strategies" (p. 1). Simon adds that "clever manufacturers have made processed food artificially stimulating by isolating particular chemicals that cause pleasure reactions, creating new 'foods' that don't exist in nature and ensuring that we stay hooked." The issue of food addiction was discussed in Chapter 5.

Simon (2006, p. 1) also points out that much of the fat, sugar, and salt we eat is "hidden," "either in processed foods or by restaurants when we eat out." Additionally, "about 90 percent of salt intake comes from food processing, preparation, and flavoring. Only 10 percent is intrinsic to the food itself." So, as more processed foods appeared on the scene, the amount of calories consumed by Americans grew 25% from 1970 to 2000. It is food companies that are culpable for this reality.

Voluminous evidence repeatedly supports the conclusion that food choices are elaborately determined, and that the causal relationships between eating practices, weight, and health are endlessly complex (Nestle, 2013, p. 41). The complex set of factors that impact eating includes genetic, social, situational, psychological, as well as structural factors such as demographics, education, region, and race. In fact, experts claim: "So much of our eating is unthinking, and, so far as we think about what we eat, that thought is typically not careful, thorough deliberation about whether we may eat this or that" (Barnhill et al., 2018, p. 21; also see Doggett & Egan, 2016). Our knowledge about what and how to eat is constrained and limited and obstacles outside of us are created to eating healthy foods.

Cohen (2014, p. 7) claims that obesity is the result of at least two factors:

1. Immutable aspects of human nature, namely the fundamental limits of self-control, the inflexible decision-making strategy of the brain's noncognitive system, and the automatic and unconscious way that we are hardwired to eat; and
2. A completely transformed food environment . . . all the food-related elements of our surroundings, including grocery stores, restaurants, prices, portion sizes, availability, marketing, and advertising.

With regard to the first point about unconscious eating, it is important to note that humans are hardwired to eat, and perhaps even to eat unhealthily, at least to a point. Cohen (2014, p. 8) claims:

We are biologically designed to overeat when presented with the opportunity (that is, to eat more than we need in order to ensure our survival); most of us have a shockingly limited capacity to deliberately and consistently regulate our eating behaviors; our eating behaviors are not a matter of thoughtful, mindful decision-making, but instead happen automatically, without our full awareness.

According to Cohen (2014), our noncognitive system is in control about 95% of the time. It takes the cognitive, rational part of the brain to control our behaviors. "Mental contamination" occurs when our emotions, judgments, and behaviors are impacted by unconscious mental processing. Here, things such as background music and other things that impact our moods are known to impact what and how we eat. These are things that "prime" our behavior, often in the interests of the corporations that have control over them. When we act impulsively, we tend to make poorer food choices. When we are distracted, whether it be by watching TV, playing on our phones, or just not focusing on what we are eating, we tend to eat too much. So, there is a relationship between watching more TV and being obese. Part of this is owing to the fact that many TV programs "are accompanied by repeated commercials for high-calorie junk foods and soft drinks" (Nestle, 2013, p. 44). Poverty also produces stressors that are also likely to lead to poor food selections.

Another issue to consider is that our food selections are heavily influenced by our physical senses (as shown in Chapter 4), and those "were designed to work automatically and reflexively" (Cohen, 2014, p. 45). This means we will typically react to our senses before we even think about it. Consider the sense of smell, which, when it comes to food, will trigger the flow of digestive juices and the watering of the mouth. Then, consider sight: Our eyes pay more attention to foods than nonfood items and especially to so-called vice foods that are high in sugar and fat. In fact, Cohen (2014, p. 47) claims: "Conscious awareness of our behavior appears to be activated *after* we begin an action in a secondary, indirect way, almost as an afterthought." When it comes to food, research shows that many decisions about food "can be so quick that no real conscious effort or direction seems to be involved" (Cohen, 2014, p. 49). Even buying decisions can be made in less than one second!

And sometimes, our behaviors can be triggered by subliminal triggers, something known about and studied by the food industry. Consider the comments by Moss (2013, p. 346) with regard to the things that trigger our food purchases:

Some of the tricks being used to trick us are subtle, and awareness is key: the gentle canned music; the in-store bakery aromas; the soft drink coolers by the checkout lanes; the placement of some of the most profitable but worst-for-you foods at eye level, with healthier staples like whole wheat flour of plain oats on the lowest shelf and the fresh fruits and vegetables way off on one side of the store.

But he claims that there is nothing subtle about the products themselves:

They are knowingly designed—*engineered* is the better word—to maximize their allure. Their packaging is tailored to excite our kids. Their advertising uses every psychological trick to overcome any logical argument we might have for passing the product by. Their taste is so powerful, we remember it from the last time we walked down the aisle and succumbed, snatching them up. And above all else, their formulas are calculated and perfected by scientists who know very well what they are doing. The most crucial point to know is that there is nothing accidental in the grocery store. All of this is done with a purpose.

Moss (2013, p. 4) notes that companies "have on staff cadres of scientists who specialize in the senses, and the companies use their knowledge to put sugar to work for them in countless ways." Surely, this makes food companies at least partially culpable for our poor eating; this issue is revisited in Chapter 9.

Subliminal messages often involve food but do not have to; images and scenes that feature happiness and other positive emotions can trigger purchasing and eating behaviors. It is important to note that food companies know that palatable foods "whets our appetite more than seeing fruits and vegetables" (Cohen, 2014, p. 96). Food companies have even used eye-tracking software to determine that the more people pay attention to a particular target, the more likely they are to buy it (pp. 87–88).

Nestle (2013, p. 15) notes that people are induced "to eat more food than they need or necessarily want, not least because 'eat more' stimuli are largely invisible or not consciously perceived"—that is, are subliminal. We see the use of subliminal triggers in the placement of products in grocery stores, for example (Lorr, 2020). Cohen (2014, p. 65) explains, saying

> there are now hundreds of studies that document how simple changes in the design, content, format, and layout of stores, displays, advertisements, packaging, and images make significant differences in how people automatically respond. Those in the food industry . . . have applied these techniques to their products and sales campaigns. They have increased the frequency and location of messages about food and dramatically altered food accessibility, availability, and variety. Most of the time, we just don't notice. We simply adapt.

Taken together, all of this information would seem to strongly suggest that our eating behaviors tend to be more automatic than rational.

Cohen reasons that we are basically evolutionarily hardwired to eat as much as possible, but in the current food environment, this tends to lead to really bad outcomes. That environment features cheaper foods (especially high-calorie foods with high levels of fat and sugar), more accessible food, and vast cues to eat that appear in "sophisticated and ubiquitous" advertisements (Cohen, 2014, p. 71). The ads, by the way, are shown to work to make people eat more, such as in the case of children watching a TV show with ads for food who ate 45% more than kids who did not see those ads (p. 95).

The American Psychological Association (APA)'s Task Force on Advertising to Children stated:

> Considerable research has examined advertising's cumulative effect on children's habits. Studies have documented that a high percentage of advertisements targeting children feature candy, fast foods, and snacks and that exposure to such advertising increases consumption of these products. . . . Several studies have found strong association between increases in advertising for nonnutritious foods and rates of childhood obesity. . . . We believe the accumulation of evidence on this topic is compelling enough to warrant regulatory action by the government to protect the interests of children, and therefore offer a recommendation that restrictions be placed on advertising to children too young to recognize advertising's persuasive intent.
>
> *(Simon, 2006, p. 248)*

A 2005 Institute of Medicine report, summarizing hundreds of studies, found "marketing promotes preferences for high-calorie, low-marketing foods and beverages, and encourages children to request and consume these products" (p. 249). Food companies obviously disagree, and thus they take numerous measures to defeat efforts to do this very thing. These effects are discussed in Chapter 9.

Broadcast airwaves are regulated by the Federal Communications Commission, and the FTC regulates advertisements. Both are clearly failing to effectively regulate advertising of unhealthy foods to Americans, including children. Here is a clear example of failing government regulation. According to Simon (2006, p. 256): "Corporate advertising falls under what's called 'commercial speech,' a category that actually enjoys less constitutional protection than the broader category of 'free speech,' the king normally associated with individuals and the press." Of course, as long as corporations are treated as individual persons, their speech will be protected like that of people.

Incredibly, trade associations hold an annual meeting titled "Kid Power" where they feature the latest ways to appeal to children through advertising. The way people of the world feel about this can likely be gleaned from the fact that 85% of 73 countries reviewed by the WHO imposed some form of restrictions on advertising to children.

Why does all of this matter? You may have heard the saying, "the best predictor of future behavior is past behavior." This is true with regard to eating; the way people eat as children tends to be a good predictor of how they eat as adults (Simon, 2006). That children see as many as 40,000 commercials a year, many of them related to food, puts parents who want their children to eat healthy food at a major disadvantage. Add on to that, that many food manufacturers use cartoon characters and superhero figures to target children with their products. Incredibly, the food industry has its own oversight body—the Children's Advertising Review Unit (CARU), that once simply encouraged children to eat breakfast (in most cases, sugary cereal breakfasts featuring the products the manufacturers produce), based on the premise that breakfast is the most important meal of the day. Yet, CARU did challenge Kraft on one of its Lunchable campaigns that asserted some leftover meals were less healthy than their own products. Still, critics and lawmakers alike, generally view CARU as an ineffective form of self-regulation. Neither is the Alliance for American Advertising, comprised of Kraft, General Mills, and Kellogg's.

Kessler (2009) suggests that being constantly exposed to food so often and in so many places makes people not only eat more as "conditioned overeaters" but also crave high-calories foods. So, we eat too much and we simultaneously eat the wrong things. This appears to be largely out of our control.

Further, we seem to be impacted by norms for eating created by the food system itself, as in the case when we eat more just because more food is available to us. According to Cohen (2014, p. 83),

> Restaurants . . . have . . . figured out how to get their customers to buy (and eat) more than they intend, not only by serving larger portion sizes but also by altering the way they present and bundle their products. The ubiquitous combo meal is no accident: It has been finely developed and bred for success

to feed on the idea that getting more for less is automatically perceived as a bargain to us. Add on to that the fact that people seem to be pretty bad at judging portion size and it is easy to understand why many people overeat. Just as we tend to eat more when food is served in larger containers, we eat more when portion sizes are larger, and those have grown significantly in restaurants over time, along with the number of calories per meal. Simon (2006, p. 197) describes portion sizes in restaurants as "out of control" and says they "serve two to three times more than what is considered a standard serving size."

Cohen (2014, p. 53) writes: "There is no doubt that the increased portion sizes bear responsibility for a significant part of the obesity epidemic" especially when about half of our food dollars are spent on meals away from home. Studies show that when presented with a wider variety of foods (e.g., a buffet), people tend to eat more. Cohen (2014, p. 60) writes: "Variety delays satiation and encourages and prepares our body in many ways to continue to accept and crave food." Food companies know this, and they are responsible for the stunning array of products on shelves in grocery stores and other food stores. According to Cohen: "The food industry is very aware of people's attraction to variety, and as a consequence it introduces more than ten thousand new processed food products annually into US markets" (p. 60). Many of these products offer "empty calories"—calories with no nutritional value (Nestle, 2013, p. 155).

The cost to roll out a new product is about $1.5 million, yet tens of thousands of new products are rolled out each year, many of which are so-called "line extensions" that offer new versions or flavors of commonly-known products (Moss, 2013, p. 26). Using the craft called "optimization," companies "alter [color, smell, packaging, and taste] ever so slightly in making dozens and dozens of new versions, each just a bit different from the next." Clearly, food companies know the importance of variety to shoppers, especially given the automatic nature of much of our food shopping.

The grocery store operates in some ways as an assembly line, where shoppers move systematically, with shopping cart in hand (which was invented in 1937), and without much thought, aisle by aisle (Lorr, 2020). Lorr (2020, p. 12) characterizes grocery stores as being in the "business of desire"— that is, tapping into if not exploiting people's wants when it comes to food products. Lorr also characterizes people's food preferences:

> People are skittish and insane when it comes to their food. They not only want, they demand, through buying power, completely impossible unsustainable opposites—low price and high quality; immediate availability and customized differentiation—and then react apoplectically to the often ingenious, if Frankensteinian, solutions industrial food creates to bridge the gap.
>
> *(pp. 76–77)*

Interestingly, careful, thoughtful, conscious behavior consumes glucose, leading to a greater need for sugar for energy: "The more the brain uses up the available sugar and the more difficult it is to continue to exercise self-control. The lower our blood sugar is, the more likely we are to choose foods high in sugar and fat" (Cohen, 2014, p. 31). When we are worn out by the stress of the day, our brains are literally drained of energy, and in these conditions, we are more likely to make poorer food choices. Later she adds that "eat available food"—"a survival imperative passed on through evolution"—is basically "etched into our DNA" (p. 50). And still later, she claims: "When too much food is available, our DNA has us wired to eat more than we need" (p. 195).

The second point from Cohen about a transformed food environment raises the issue of moral and perhaps even legal culpability for actors in the conventional food system for outcomes such as obesity and diabetes. Nestle (2013, p. 37) writes: "Beginning in the early 1980s, food became more widely available. Fast-food places proliferated. The sizes of food portions increased. People began to eat outside the home more often and to snack more frequently." So, Cohen (2014, p. 8) adds that "the modern food environment practically assaults us wherever we go." She notes that the number of restaurants more than doubled since 1977, and the number of vending machines has also grown enormously; you see them nearly everywhere you go.

This environment has been described as *obesogenic*, which

> used to describe social, environmental, and cultural factors that are thought to tend to cause the "illness" of obesity, such as inadequate exercise or lack of access to exercise facilities; the suburbanization of societies and reliance on automobiles; the wide

availability and advertising of high-calorie foods; lack of access—through poverty or poor quality of local shopping facilities—to a wide enough range of 'healthy' foods; low interest in or understanding of food and nutrition; lack of consumer understanding of positive food choices due to poor or misleading food labeling; the globalization of food cultures leading to an incomprehensible "dietary cacophony"; the demise of "home cooking" and "the family meal."

In essence, the term means we all live in an environment where obesity is likely to keep occurring, regardless of efforts by individuals to try to maintain or lose weight.

And to that, I would add that we now see food nearly everywhere, even in places we do not expect it to be. As one personal example, today, I went to two stores for items I needed for an art project—Lowe's Hardware and Michael's art supply store. At the cash register at each location, there were numerous types of snacks available, all high in salt, sugar, and fat, and none at all healthy in any way. This is the case, according to Cohen (2014, p. 133), at more than 40% of all retail businesses (including 96% of pharmacies, 55% of hardware stores, 55% of automobile sales and repair outlets, 22% of furniture stores, and 16% of apparel stores). Why am I being offered "food" (if you can even call it that) at these places when my only goal is to purchase my items and leave? These are meant to promote what Cohen (2014, p. 75) refers to as "impulse buys" and she correctly points out that they tend to be "cookies, candies, chips, and sodas." Cohen argues that

> displaying candies, chips, and soda at the cash register increases our odds of buying them . . . because our ability to resist them is likely to be low at the moment we are forced to encounter them. Much of today's strategic marketing pushes us to make poor choices by confronting us when we are unprepared to make a thoughtful choice.
>
> *(p. 89)*

Cohen then adds: "Advertising, too, has become increasingly sophisticated and insidious, so much so that we might not even recognize it as advertising." The snacks I was surrounded by when I made my purchase (of paint, by the way), can be seen also as advertisements for the corporations selling them, especially given the evidence that we begin to recognize and automatically react to brands and their symbols by the age of two years. Logos standard in image, color, size, and so on tend to offer "a clear, concise, and compelling message." According to Cohen (2014, p. 98):

> Advertisers spend a great deal of time and money to make sure that the bran works: they create mock designs and logos and conduct focus groups to see how people react to them. Once they have a brand that grabs our attention and makes us think positive thoughts, they develop a strategy to disseminate that brand image. To do so, they make great pains to increase exposure to the brand, so it is burned into the consumer's brain and becomes the brand of choice.

Nestle (2013, p. 76) claims that "research shows that children presented identical foods with and without advertised brand names greatly prefer the advertised products." This may be explained by "mere exposure conditioning"—"simply seeing a product . . . influences the likelihood that a customer will choose it" (Cohen, 2014, p. 99).

Cohen (2014, p. 8) believes that the "food environment is the largest determinant of our behavior." Yet, I did not buy any of these products, so it is of course possible to resist. Cohen (2014, p. 76) also discusses the role of "product placements" in TV shows, movies, and Internet products, and she shows that children are targeted this way; they tend to see about 65 messages from TV every day, half of which are for foods. She cites a study by the National Academies' Institute of Medicine that found food advertising does impact children's choices as well as their health. Incredibly, research also shows that unhealthy food ads are more targeted toward people of color, as in shows with African American characters (p. 77).

It turns out that rates of obesity in a community and a country correlate with many other factors, including the degree of healthy and unhealthy foods within the food environment, poverty, social inequality, as well as technological and lifestyle changes over time (Barnhill et al., 2018). Nestle (2013, p. 23) notes, for example: "The highest prevalence of overweight and obesity is observed among the poor." Incredibly, food insecurity is a significant predictor of overweight, particularly in low-income groups." Part of this can be seen as structural in nature, since Supplemental Nutrition Assistance Program (SNAP) recipients receive what is meant to be supplemental and temporary benefits, and so recipients come to rely on the "cheapest sources of calories—the snacks, fast food, and sugar-sweetened sodas pejoratively called 'junk foods,' or more politely, 'foods of minimal nutritional value'" (Nestle, 2013, p. 23).

Taken together, all the aforementioned information helps us understand why obesity rates began to rise in the 1980s and type 2 diabetes started to increase in the 1990s. It is not a change in genetics or degree of self-control in the population that accounts for the rise—it is not choice or free will—but rather very specific and measurable changes in the food environment, under the control of major actors in the conventional food system.

If hardwired cognitive limitations make it hard if not impossible for us not to eat too much, the answer to addressing obesity seems to be to make changes to our food environments. Nestle (2013) writes, for example: "It works much better to change the food environment to make the healthier choices easier." This is especially true given that food intake seems to matter more for outcomes such as weight management than exercise. Still, the food environment "encourages people to eat more often, in more places, and in larger amounts, than is good for maintaining a healthy weight" (p. xix). Yet: "The beneficiaries of the current situation argue that their right to sell whatever they want, however they want, is a matter of personal liberty—that it is up to individuals to make better choices if they really care about their health." This idea is deeply engrained in Americans and will be quite difficult to challenge, yet this places the burden for avoiding unhealthy foods and gaining weight entirely on us as individuals even though "America produces twice the calories that people need-about 3,800 calories per person per day" (p. 119).

Cohen adds that she does not believe global food corporations aim to make people fat, but rather their motivation is simply profit. Take Coca-Cola, for example. Through its "Foreign Department," the goal is to get Coke in the hands of everyone (this is the company's "ubiquity strategy") not to make them fat, but instead to make them happy. The problem is the consumption of sugary soda is linked to weight gain and "our bodies are less aware of excess intake when the calories are liquid" (Moss, 2013). Packaging of some of these products, such as in the 7-Eleven 64-Double Gulp, comes with a stunning 44 teaspoons of sugar (p. 99)!

Nestle (2013, p. 12) writes:

> The overabundance of calories [produced daily per capita by food corporations] forces the food industry to be highly competitive, but other changes in the early 1980s required even more competition. Shareholders began to pressure corporations to reward them with higher returns on investment. Food companies not only had to compete for sales against 3,900 calories a day, but now had to *increase* sales and report growth in profits to Wall Street every 90 days.

This is a major reason why food began appearing everywhere: "bookstores, libraries, and stores selling clothing, business supplies, cosmetics, or drugs" (p. 13). In this way, eating too much is not an issue of personal weakness or moral failure but rather is "an unavoidable response to today's 'eat more' food environment," all arranged for corporate profit (p. 14). As a criminologist, I take great issue with profiting off of deleterious health outcomes, as well as human suffering in the form of illness and death.

Cohen (2014, p. 117) uses candy bars to make the point that governments cannot tell you not to ever eat candy bars if that is what you choose to do, but governments can protect you from marketing efforts that force you to be confronted by candy bars every day everywhere you go! She uses the regulatory approach to pornography as an instructive approach to how we might deal with foods:

> We don't ban it, but we restrict its availability. It is not on every street corner, at every cash register, or in vending machines. If it were, it would be impossible to ignore. If we had to see pornography everywhere we went, no doubt our ability to think about anything other than sex would be challenged. Food can be a little like pornography.

She also discusses regulatory approaches to alcohol, writing: "How many people think we should sell alcohol from vending machines, display it at every cash register from bookstores to hardware stores, serve it at meetings at work, allow people to routinely drink on the job, or sell it to children?" If we don't do this with alcohol, perhaps it's time to stop doing it with food, especially products that we call food but really aren't.

## Conclusion

In this chapter, you saw that the standard American diet, or SAD, is quite literally, dangerous. Common outcomes associated with the foods we eat include illness from conditions such as obesity and type 2 diabetes, each of which contributes significantly to both deaths and enormous financial costs. Hundreds of thousands of Americans die from these conditions alone, and this excludes deaths from other diet-related conditions such as heart disease and hypertension. Diet is also associated with a wider range of poor health outcomes, including in mental health.

A healthier diet is possible, yet, it is up to the conventional food system generally and food corporations, in particular, to produce these foods in larger amounts so that they will be available to and affordable for consumers. This should also be a goal of US regulatory agencies—to force food corporations to cut back on the production of unhealthy ultra-processed foods in the public interest and to force the production of healthier alternatives.

Yet, it appears that the government prefers a hands-off approach to regulating food companies when it comes to issues of public safety. As such, the government is culpable through its negligence for outcomes such as obesity and diabetes.

Consumers clearly play a role in this, but it is not just up to consumers to make better choices. In this chapter, you saw that the notion of consumer choice, or free will, is largely a myth. Further, it is actually quite hard (and more expensive) for consumers to choose healthier foods when they are literally surrounded by unhealthy choices everywhere they go, especially when you add in the reality of food addiction, established earlier. Surely, if we hold drug dealers accountable for their actions, we can do the same for food corporations; they are the ones responsible for the widespread availability of poor food choices.

# 7

# HARMS ASSOCIATED WITH THE SYSTEM

## Introduction

In this chapter, harms associated with the conventional food system itself are identified and analyzed. Many topics are discussed, including harms to animals (i.e., animal death and animal suffering) and harm to workers (i.e., hazardous working conditions). Potential harms to consumers are also identified, but this time in the context of pathogens found in food. Other issues include the contradictory realities of food insecurity (where people do not have enough food to eat) and food waste (where a large portion of produce food is just thrown away or rots away in fields). I hold these to be costs of the system itself, even though there is much blame on the part of consumers when it comes to wasting food.

Perhaps the most important issue identified in this chapter is the issue of environmental damage caused by the conventional food system. Agriculture is one of the leading causes of greenhouse gas emissions, for example, and the implications of climate change for people around the globe are now enormous. Leading food scholars are clearly aware of harms associated with the system itself, the focus of this chapter. While it is in the interest of global food companies to reduce their "harm footprint" here, not enough is being done or quickly enough in the conventional food system to reduce harms in this area.

## Killing and Eating Animals

Perhaps the first real harm of the conventional food system is that we kill a lot of animals. Yet, humans have been killing and eating animals for more than a million years. In that sense, we might view killing and eating animals as normal, for it has been happening for even much longer than homo sapiens have walked the Earth. What has changed over time are the methods and breadth of this killing and consumption. Today, Americans eat nearly 300 pounds of meat every year!

More than 70 billion land animals are farmed for the purposes of producing food (i.e., meat) every year. More than 120 million hogs are slaughtered each year in the United

DOI: 10.4324/9781003296454-7

States. Further, according to People for the Ethical Treatment of Animals (PETA, 2021), each year:

- 9 billion chickens are killed;
- 245 million turkeys are killed, including 46 million at Thanksgiving and 22 million at Christmas;
- 31 million ducks are killed;
- 30 million cows die in meat and milk production;
- Tens of billions of fish and shellfish are killed.

It is expected that livestock production will double between 2000 and 2050, so we are nearly halfway there (Aramark, 2018). Generally speaking, food animal production is a messy, dangerous, and deadly business. But it is clearly also not good for animals who are held in conditions of concentrated confinement (more on this later in the chapter).

For now, something needs to be said about the ethics of eating animals in the first place, much less the mass killing of animals for food for humans. Some animals we eat, perhaps most notably pigs, are as intelligent if not more so than animals we keep as pets. Just because they taste good does not mean we should necessarily be killing them. But, even if we decide as a people that their lives are not worth celebrating other than as food, it is undeniably true that their deaths are harmful, at least to them. As such, this can be viewed as a cost of the conventional food system itself.

Lorr (2020) refers to much of our foods as simply "Consumer Packaged Goods" (CPG) rather than food (p. 8). Another common term, at least in grocery stores, is "Shop Keeping Units" (SKUs), and the average store has roughly 32,000 of them, while the largest stores have more than 120,000 SKUs (Lorr, 2020). Lorr (2020) notes the following types of grocery store niches: Convenience, traditional supermarkets, specialty, natural, gourmet, and mash-ups. Lorr (2020) describes the process of factory killing of animals such as chickens, being processed at 140 birds per minute, as the moment when animals become food rather than living beings. Then, when they are shipped to stores, they become product or merchandise rather than food; they become SKUs. When the customer comes to the store and selects the SKUs, they become food again, or at least they will when they are prepared and eaten. What a strange reality.

## Animal Welfare

According to Lakhani et al. (2021), at least 1.6 billion animals lived on 25,000 factory farms in 2017. Even before their deaths, these birds, pigs, cows, and other animals we eat

> are confined in densely stocked indoor facilities. A single Industrial Food Animal Production (IFAP) facility, for example, typically houses over 5,000 hogs, or over 100,000 laying hens. Beef and dairy cattle spend at least part of their lives in similarly large-scale indoor or outdoor feeding operations. IFAP facilities are designed to maximize the amount of meat, milk, or eggs from a particular breed of animal in the shortest amount of time and at the lowest cost.
>
> *(Johns Hopkins Center for a Livable Future, 2021)*

That is Chickenization or McDonaldization of food, as explained in Chapter 2. Figure 7.1 shows the inside of an IFAP.

Animals are largely raised indoors because this

eliminates weather from the equation, giving producers greater control over temperature, lighting, and other factors that influence productivity. Recall the discussion of confinement and concentration of animals from Chapter 2, indicative of industrialized animal food production. Specialized confinement facilities maximize the number of animals per area, and allow for automated feeding, watering, ventilation, and waste removal.

*(Johns Hopkins Center for a Livable Future, 2021)*

The majority of animals in IFAP

rarely if ever see the outdoors, let alone feel sunshine, experience day or night, or walk on grass. They are unable to perform many of their natural behaviors; pigs raised on concrete are unable to root in the earth with their powerful nose discs, chickens are unable to dust bathe, cattle are fed unnatural diets of grain that cause painful liver abscesses. In most egg production facilities, hens are confined to "battery cages" so small they are unable to spread their wings. Female pigs (sows) used for breeding

**FIGURE 7.1**    Inside of an Industrial Food Animal Production (IFAP) Facility

*Source:* Credit: Barbara Barbosa

are confined to "gestation crates" in which they are unable to turn around or groom themselves.

*(Johns Hopkins Center for a Livable Future, 2021)*

All of these realities would likely disturb Americans if they knew them, but these facts are largely unknown to most of us.

According to FoodPrint (2021), dairy cows are milked for 10–12 months a year: "Average US milk production is about 17,000 pounds per cow annually, though herds with averages of up to 24,000 pounds per cow are not unusual." They spend their lives making milk for humans and are killed years before their normal life span of 15–20 years (Humane League, 2021).

Moss (2013, p. 168) describes the reality of life for dairy cows (not even the ones we kill to eat!):

They were moved into gigantic sheds where artificial lighting extended their workday. This industrialization, along with a heartier diet of corn and added fats, transformed the American dairy cow into a prodigious producer. Where each animal used to give barely a gallon and a half of milk each day, the modern cow puts out more than six gallons each. Six gallons of full-fat milk.

Moss (2013, p. 168) points out the clear link between corporate interests and the US government here, noting:

Since the 1930s, the federal government has viewed milk as vital to the nation's health, and thus, it has labored to ensure that dairies never go under. It subsidized the industry by setting price supports and used taxpayer money to buy any and all surplus dairy products.

As cows began producing more milk than could be drunk, "all that unwanted milk and extracted milkfat" was turned into cheese (p. 169)! Here is thus an example of how government action would lead to greater consumption of a product high in calories and calories from fat, which would ultimately lead to weight gain and its affiliated health outcomes. The government's actions here as quite strange, as it bought up so much milk and milk fat that it "began secreting it away in caverns and a vast, abandoned limestone mine" (p. 169).

To sell more cheese, companies such as Kraft began campaigns like its "Real Women" campaign to have people promote their products. And so Paula Deen was born, an American icon of food, a chef, and cook book author. Ironically, Deen would develop type 2 diabetes, a condition caused by poor diet and inactivity.

Beef cattle are typically slaughtered at 12–16 months of age, and grass-fed beef cattle are killed between 22 and 28 months (FoodPrint, 2021). Often, cows are fed grains they cannot digest, leading to gastrointestinal problems. Male dairy calves are not useful for milk and so they are sold for veal or beef production, and nearly "half of female calves are raised as replacement milk cows, as the older ones slow down; the rest are sold for veal or beef" (FoodPrint, 2021).

As for chickens, layers (egg birds) and broilers (meat birds) both live lives of misery. Hens begin laying at 18–20 weeks of age and lay an egg almost every day of the week, in battery cages, so small they are unable to even spread their wings (FoodPrint, 2021). Forced molting (e.g., denying food and water) is used to achieve more egg production. Male chicks are fed into industrial grinders. Chickens tend to be slaughtered after a year even though their life span is 6–12 years (FoodPrint, 2021). The same is true for pigs. Figure 7.2 shows another example of overcrowded conditions, this time for chickens.

Then, add on debeaking chickens, and tail removal on cows and pigs (without anesthesia) to prevent animals from hurting other animals, and animals literally living in their own feces and urine. According to FoodPrint:

Animal behaviors, like pigs rooting in the dirt or chickens taking dust baths, are stifled when animals live in cages or in houses with metal floors; in some cases, such as with veal calves and nursing pigs, animals are unable to turn around. Animals also have to live in

**FIGURE 7.2** Another Example of Overcrowded Conditions, This Time for Chickens
*Source:* Credit: Mark Stebnicki

or on top of their own excrement, breathing in the toxic fumes, and in perpetual stress from the crowded conditions.

*(FoodPrint, 2021)*

If Americans knew these realities, could we reasonably expect them to eat less meat? This is unknown, but eating less meat would be associated with less harm produced by the conventional food system. Currently, we have a conventional food system built on the immense and unimaginable suffering of animals.

## Factory Farming/CAFOs

CAFOs are used to feed animals in large numbers, often in allegedly inhumane conditions. Linnekin (2016, p. 64), for example, described cows being kept inside a CAFO and having access to a small pasture through a single door but "CAFO cows spent nearly all of their time confined inside and rarely, if ever, actually *accessed* that pasture." He also describes the large numbers of birds living in poultry-based CAFOs are often responsible for the outbreak of things such as avian influenza.

Hauter (2012, p. 158) describes the typical CAFO:

Today, many thousands [of cows] are housed in acres of steel pens, each corralling around two hundred cattle that are fed from a mechanized feed delivery system or a trough that can be filled by a tractor. The cattle stand without shelter, shade, or grass and sleep on their own waste, which forms a concrete-like surface that is dusty when it's dry and sewer-like if it rains. The cattle are squeezed together tightly so that no calories are lost to unnecessary movement.

For pigs the picture is much the same:

Thousands of pigs are packed together tightly in giant warehouses, where they generate tons of liquid and sold waste, posing health hazards to the surrounding communities and degrading he environment. The feces and urine produced in hog factories—containing ammonia, methane, hydrogen sulfide, cyanide, phosphorous, nitrates, heavy metals, antibiotics, and other drugs—fall through slatted floors and into a catchment pit under the pens. Giant exhaust fans pump the toxic fumes out of the warehouse-like buildings, twenty-four hours a day, to prevent the hogs from dying. The accumulated waste in then pumped into enormous lagoons that can cover six to seven and a half acres and hold as much as 45 million gallons of wastewater. Leaking and flooding lagoons pollute local waterways, and the fumes from the waste spread on neighboring fields choke and sicken the local community.

*(p. 170)*

For chickens, the situation is just as bad:

Thousands of birds are crowded together for their brief lives in extremely crowded conditions in the filthy warehouses where they live in their own waste. At the end of their short lives, the birds are roughly loaded in tractor trailers, denied food and water, and sent to processing facilities where they are hung by their feet and slaughtered by the truckload.

*(p. 206)*

According to Constance (2019, p. 88), CAFOs are "directly owned by or contractually linked to major agribusiness integrators." And they are "the preferred model of intensive, industrialized animal production, including dairy, veal, beef, eggs, broiler chickens, turkeys, and hogs." "Yet, whereas CAFOs are very efficient regarding production, they are problematic regarding the environmental, community, public health, and animal welfare impacts" (p. 88). As an example of an environmental issue, EcoNexus (2015, p. 6) claims: "Two-thirds of nitrous oxide emissions, which remain more than 100 years in the atmosphere, and are particularly damaging to the climate, originate from concentrate feed-based industrialized livestock farming." The issue of environmental degradation is discussed later in this chapter.

Martin (2016, p. 203) defines factory farming as "production that employs high stocking density, which deprives animals of the opportunity for even minimal movement and causes many animals to engage in self-mutilating behavior and to attack nearby animals." It involves "debeaking chickens and clipping pigs" teeth without the use of anesthesia. It also involves breeding "for extremely fast growth rates and high meat productivity" in unhealthy animals that often die of illness or injury even before slaughtering.

These conditions also can be quite unhealthy for workers and are ultimately bad for the farmers who service the slaughterhouses. They "must accept increasingly demand contracts that often lasts for only the life of the flock . . . along with decreasing net income" (Martin, 2016, p. 203). In fact, it is alleged that many contract farmers "exist in a condition tantamount to indentured servitude" (p. 204). Then, conditions in meat packing plants are unsafe and restrictive of worker rights.

Laws across the country are meant to protect CAFOs and slaughterhouses, for example, by making it illegal to film and share footage of what actually goes on in these facilities, something Linnekin (2016) describes as animal cruelty. This is an example of the government not only failing to regulate industry but also going further to protect it. Footage filmed by independent filmmakers has, nevertheless, depicted unbearably horrendous conditions of animals in confinement. Images shown in this chapter confirm this reality.

## Change Is Coming?

The first ever food industry-led initiative aiming to advance animal welfare—the Global Coalition for Animal Welfare—was initiated in 2018. Included in the efforts were Aramark, Compass Group, Elior Group, IKEA Food Services, Nestle, Sodexo, and Unilever. These companies serve nearly 4 billion customers a day and earn about $165 billion in revenues every year (Aramark, 2018). We will see what, if anything, will come from this.

It is clear, though, that change is possible. One example comes from the People for the Ethical Treatment of Animals (2022):

Following extensive discussions, PETA persuaded the food and beverage companies marked with an asterisk (*) below to stop conducting or funding deadly experiments on animals in order to establish health claims for the marketing of products or ingredients.

Now, they'll no longer take part in animal tests unless they're required by law or government regulators (and PETA is working on that, too).

Companies that do not test on animals, including those that PETA persuaded to stop, are shown in Table 7.1.

TABLE 7.1  Food and Beverage Companies That Do Not Test on Animals

Accolade Wines
Adagio Teas
Agropur Dairy Cooperative
AGV Products Corp.*
Alfred Ritter GmbH & Co. KG
Amy's Kitchen, Inc.
Ankerkraut GmbH
Apeejay Surrendra Group (Typhoo Tea only)
Arbor Crest Wineries & Nursery, Inc.
Arcor S.A.I.C.
Asahi Group Holdings, Ltd.*
Australian Fruit Tea Company Pty Ltd.
B&G Foods, Inc.
Bacardi-Martini, Inc.
Ball Corporation*
Barilla SpA*
Barry Callebaut*
Bettys & Taylors of Harrogate Ltd.
Beyond Meat
Bigelow Tea
Bimbo Bakeries USA
Black Star Farms, LLC.
Blommer Chocolate Company*
Bonduelle Group
Borealis Foods
Brown-Forman Corporation
Campbell Soup Company
Chobani Global Holdings, LLC
The Coca-Cola Company*
Constellation Brands, Inc.*
Daintree Tea
Dean Foods
Del Monte Pacific, Ltd.
Diaspora Tea & Herb Co (doing business as Rishi Tea & Botanicals)
Dole Food Company*
Don Lee Farms (Goodman Food Products, Inc.)
Praeger's Sensible Foods
E. & J. Gallo Winery
Eclipse Foods
Ensuiko Sugar Refining Co., Ltd.*
Ethicoco
Ezaki Glico Co., Ltd.*

*(Continued)*

**TABLE 7.1** (Continued)

Field Roast
Flowers Foods, Inc.
Follow Your Heart
Fortnum & Mason
The Frauxmagerie Ltd.
Fuji Oil Holdings Inc.*
Gathered Foods
General Mills Inc.*
Givaudan
Grape King Bio*
Green Monday Group/OmniFoods
Grupo Peñaflor S.A.
Hain Celestial Group
Hari Har Chai
Heineken N.V.
The Hershey Company*
HOT EARTH GmbH
House Foods Group Inc.*
Ingredion Incorporated*
Intelligentsia Coffee, Inc.
ITO EN, Ltd.*
J&J Snack Foods Corp.
James White Drinks
JINKA Foods
Kellogg Company*
Keurig Dr Pepper Inc.
Kewpie Corporation*
Kikkoman Corporation*
KIND
Kipster
Kirin Holdings Co., Ltd.*
Koala Tea Company Pty Ltd.
Korea Yakult*
Kuleana
Lactalis American Group, Inc.
Lakewood Organic Juices
Lancaster Colony Corporation
Lian Hwa Foods Corp.*
Lightlife Foods, Inc.
Lindt & Sprüngli
May Wah with Lily's Vegan Pantry
McCain Foods Limited*
McCormick & Company, Inc.
Megmilk Snow Brand Co., Ltd.*
Meiji Co., Ltd.*
Microbio Co.*
Mitch's Vegan Jerky
Molson Coors Brewing Company
Monde Nissin

(*Continued*)

**TABLE 7.1**  (Continued)

Monogram Foods Solutions, LLC
Morinaga & Co., Ltd.*
Nagase & Co., Ltd.*
Next Gen Foods
NH Foods Ltd.*
Nippon Suisan Kaisha, Ltd.*
Nissin Foods Holdings Co., Ltd.*
Nitto Beverage Co., Ltd.
Nuts For Cheese
Oatly Group
Ocean Spray Cranberries, Inc.*
The Original Ceylon Tea Company
OSI Group
Ostfriesische Tee Gesellschaft GmbH & Co. KG
PepsiCo, Inc.*
Pernod Ricard*
Pinnacle Foods Inc.
POM Wonderful LLC*
Primeval Foods
Primo Water Corporation
Rebellyous Foods
Reily Foods Company
Reine Vegan Cuisine
Rich Products Corporation
Riken Vitamin Co., Ltd.*
Robertet SA*
Sanderson Farms, Inc.
Sapporo Holdings Ltd.*
Saputo Inc.
Satake Corporation*
Semper AB
Seneca Foods Corporation
Sensient Technologies Corporation*
Shiok Meats Pte. Ltd.
Simply Eggless
Standard Foods Group*
Stash Tea Company
Strand Tea Company
Strauss Group*
Sugar Creek Packing Co.
Sunshine Burger & Specialty Food Co, LLC
Suntory Holdings Limited*
Sweet Earth Enlightened Foods
Swilled Dog Hard Cider*
Swire Coca-Cola Taiwan*
Takasago International Corporation*
TeeGschwendner
Das Teehaus
Teekanne GmbH & KG

(Continued)

**TABLE 7.1** (Continued)

Tesco PLC (tea products only)
T. Hasegawa Co.*
Tofurky
Toyo Suisan Kaisha, Ltd.*
True Blue Holdings, LLC
Twinings North America
Uni-President*
Unilever*
Vegano Foods LLC
Viña Concha y Toro S.A.
Vitalon Foods Group*
Welch Foods Inc.*
Weston Foods (Canada) Inc.
Whispering Pines Tea Company
Wholesome Savour/OsomeFood
Wildtype
Yakult Co., Ltd.*
Yakult Honsha Co., Ltd.*

*Source:* PETA persuaded the food and beverage companies marked with an asterisk (*) below to stop conducting or funding deadly experiments on animals in order to establish health claims for the marketing of products or ingredients. https://www.peta.org/features/victories-food-drink-companies-refuse-animal-tests/

PETA (2022) notes:

Prior to contacting the companies, PETA uncovered disturbing documents showing that thousands of animals were cut into, tormented, and killed during cruel laboratory experiments for decades—all so that companies could attempt to make marketing claims about products ranging from Ramen noodles to candy bars and from breakfast cereals to liquor. Marketers aimed to boost product sales with suggestions of health-promoting ingredients, and they turned to experiments on animals for proof—even though the scientific evidence shows that animal testing is ineffective and fails to lead to human treatments.

PETA discusses very specific examples of mistreatment of animals by major food and beverage companies in experiments they conducted, none of which were required by law:

In these food- and beverage-industry experiments, animals were restrained in tubes, hung by their tails, forced to run on treadmills, and made to stand on hot plates; force-fed and starved; injected with chemicals, drugs, alcohol, and cancer cells; made to swim until they were exhausted and inhale smoke; cut apart; made to endure the exposure of their nerves and electrocuted; given facial lacerations; infected with harmful bacteria and viruses; inflicted with erectile dysfunction; and killed by suffocation or neck-breaking, after which they were dissected.

Aside from experimentation on animals, there is the issue of animal welfare of these animals that either produce food or become it. PETA (2021) notes:

When they've grown large enough to slaughter or their bodies have been worn out from producing milk or eggs, animals raised for food are crowded onto trucks and transported for miles through all weather extremes, typically without food or water. At the slaughter-house, those who survived the transport will have their throats slit, often while they're still conscious. Many remain conscious when they're plunged into the scalding-hot water of the defeathering or hair-removal tanks or while their bodies are being skinned or hacked apart.

## Pathogens

Slaughter and processing of animals for food are messy processes. Contamination is very common. Silbergeld (2016, p. 177) explains:

The conveyor belts, knives, instruments, and other surfaces and equipment are quickly contaminated by those 30 percent of carcasses still carrying pathogens. This is what was found in a study conducted by the French food safety authority that concluded that cross-contamination increases over the workday at poultry slaughter plants due to con-taminated equipment, work surfaces, chillers, and process water.

Then you have the issue of there being multiple shifts at the same processing plant, where there is typically no clean-up in-between shifts. Dipping tanks and chillers tend to be only cleaned once a day, also. A study that examined interventions to reduce *Salmonella* infections in processing plants in the United States and other countries found "the disin-fection bath accumulates pathogens and other contaminants such that, relatively quickly, carcasses that go into the bath are just as or more contaminated coming out of the bath as they were prior to being put in it" (Silbergeld, 2016, p. 178). Silbergeld concludes that "it is *by design* that there is no effective control of foodborne pathogens or worker exposures in any system that is managed in this way by HACCP" (pp. 179–180). So, she says, "it is not surprising that up to 95 percent of poultry products purchased in stores end up testing positive for *Campylobacter* after they have been cut and packaged at the plant" (p. 179).

HACCP, or Hazard Analysis and Critical Control Points, is called by the FDA to be "an effective and rational means of assuring food safety from harvest to consumption." Based on basic principles of hazard analysis—CCP identification, establishing critical limits, monitoring procedures, corrective actions, verification procedures, and record-keeping and documentation—the system aims to identify threats pose by pathogens and eliminate them as quickly and safely as possible. According to the Institute of Medicine (IOM) and National Research Council (NRC) (1998): "It is widely accepted by the scientific community that use of HACCP systems in food production, processing, distribution, and preparation is the best known approach to enhancing the safety of foods." This program uses "a systematic approach to identify microbiological, chemical, and physical hazards in the food supply, and establish critical control points that eliminate or control such hazards." IOM and NRC note:

Implementation of HACCP is the responsibility of food producers, processors, distribu-tors, and consumers. The role of government is to ensure that HACCP programs are properly implemented throughout the food supply continuum by evaluation of HACCP plans and inspection of records indicating monitoring of critical control points.

Silbergeld (2016) characterizes HACCP as ineffective and controlled completely by industry. If so, this is yet another failure of government regulation, this time a serious threat to human safety. Hauter (2012, p. 129) agrees that HACCP systems "put the industry in the driver's seat on food safety and meant that giant food companies could move processing to the developed world." She claims that HAACP is a weaker system than previous regulations, also noting that it "reduced the role of USDA inspectors and created a new paperwork function. Inspectors in processing were told not to stop the line for contamination, but to wait until the meat product reached the end of line, where a 'treatment' would take care of the problem." She notes the following as examples of those treatments: ammonia, chlorine, trisodium phosphate, and irradiation.

According to Silbergeld (2016, p. 174):

In 2010, the USDA and the industry jointly claimed that through application of HACCP the average prevalence of *Campylobacter* on poultry products had been lowered to less than 30 percent of products tested. Falling into line, the National Institute for Occupational Safety and Health, the research arm of OSHA and part of CDC, relied on these data to conclude that risks of worker exposures to these pathogens were also well controlled. But this did not make sense: at the same time as this pronouncement came out of the USDA, we were finding that between 60 and 90 percent of consumer poultry products were contaminated with *Campylobacter* when we conducted studies of consumer products, buying chicken breasts and thighs in packages sealed at processing plants by major producers such as Tyson and Perdue. And my lab wasn't the only one; shortly after our publication, FDA scientists reported similar findings, but without the freedom to name names. In 2010, Consumers Union reported that 62 percent of chickens tested by an outside laboratory were contaminated by *Campylobacter*. This is a global problem for the industry. In May 2015, a government survey in the United Kingdom reported that over 70 percent of poultry products purchased at retail stores were contaminated by this same pathogen.

Still, US officials would have us believe our food system is safe. According to a joint statement by the FDA, USDA, and INSERT (NOAA): "The US enjoys one of the world's safest food supplies." Yet, to at least some degree, this must be due to chance. For Marion Nestle describes our national approach to regulating food for safety "breathtaking in its irrationality" (Walker, 2019, p. 155). Walker himself says "it is cobbled together using twelve to fifteen different agencies, thirty to thirty-five separate laws, and more than fifty interagency agreements" (p. 155). He discusses the USDA's FSIS and FDA, noting that the former oversees 10–20% of the food supply while receiving 60% of the budget, while the FDA oversees 80–90% of the food with less money and staff. Walker notes: "What Congress has never done is set out to design a food safety system that matches the complexity and scope of the globally expanding modern food system. Their approach has always been a patchwork" (p. 155).

According to Walker (2019, p. 156), the FDA is

to provide oversight and technical support, ensure compliance, strengthen the global food-safety system, enhance protection of public health, deliver training, serve as a repository of science and expertise, provide leadership for innovation and action, ensure that firms are consistently implementing effective prevention systems, enhance partnerships with states and other government counterparts, build robust data integration and analysis systems along with information-sharing mechanisms, significant expand its inspection and surveillance tools to include a wider range of inspection including sampling, testing, and other data-collection, activities, and—as always—respond when food-related problems and outbreaks emerged.

*(pp. 156–157)*

Walker notes that the FSIS mandate includes "assuring that meat and meat food products are wholesome, not adulterated, and properly marked, labeled, and packaged" (Walker, 2019, p. 158). Nowhere in this language are there any words about safety. A food standard created in 1994 considered *E. coli* to be an adulterant, making its sale illegal, but only in ground beef. One wonders how much of it is even being looked for in other meat products, given the speed of their production lines—10,500 chickens, 1,300 hogs, and 390 cattle per hour (Walker, 2019). A USDA Office of Inspector General report, from 2013, found that we do not actually know if product safety has been improved as a result of measures meant to reduce *E. coli* in meat products.

With regard to our food safety system, Silbergeld (2016, p. 192) claims:

This is not a food safety policy that protects the public and improves food safety. This is a food safety policy that protects the industry and accepts unsafe food, shifting responsibility to the victim without attention to the origin of repeated exposures to the hazard.

Later, she explains: "We are at the same point in ensuring food safety as we were at the moment of the revolution in agricultural production in the early nineteenth century" (p. 195). She continues:

We do not have an agency dedicated to food safety whose authority extends back to the farm. Agencies dedicated to food safety were set up in France, the United Kingdom, and elsewhere in the EU during a brief moment of activism following outbreaks of zoonotic diseases, such as avian and swine influenzas and mad cow disease.

No such centralized agency exists in the United States. Further: "Most of our programs in food safety are voluntary and nontransparent. These voluntary programs are run by the industry with little public access, and the stated goal of most of these programs is not to protect consumers but to ensure the productivity of the industrial sector" (Silbergeld, 2016, p. 195).

The reality of food safety standards is that they tend to be created by industry. One reason corporate food chains create their own safety standards is "food is often produced in one country and consumed in another, making it hard for any single national government to monitor the product along the whole supply chain" (Bloom, 2019, p. 136). Silbergeld (2016, p. 5) calls the state of regulation for food safety "nonexistent" and blames deregulation of business over a three-decades-long period for this reality. Further, she suggests that the food industry largely relies on self-policing to function.

One example is the Global GAP standards (Good Agricultural Products), created in 1997: "The rise in importance of Global Gap over national food safety standards is an illustration of the shift in governance from the *nation-state* (government-based regulations) to the *market* (corporate-based regulations)." Bloom (2019, p. 137) writes:

> The rise in power of corporations in a global economy (and the corresponding decline in the power of the nation-state) to determine regulations that affect who can participate in markets, who benefits, who bears risk, and who loses out is referred to as neoliberalism.

At least three big policy changes are a part of neoliberalism. First, we see a reduction in regulations. Second, we see the removal of barriers to trade. Third, we see a shift in responsibility for things such as safety from the government to private entities (Konefal & Hatanaka, 2019). Consolidation of business means companies such as Walmart and Unilever have great control over not only the food people eat but also how much it is produced: "Today, large corporations control nearly all aspects of food and agriculture, including agricultural inputs (e.g., seeds, fertilizers, and pesticides), processing (e.g., grain elevators and manufactured foods), and retail (e.g., supermarkets and restaurant chains)" (Hatanaka, 2019, p. 5). Suppliers to large supermarkets must abide by certain product standards, putting those stores in charge of much of our food system: "Increasingly, these private standards cover nearly all aspects of agricultural production and food processing, including product quality and safety, packaging, shelf life, traceability, and on-farm practices, such as sustainability and animal welfare" (p. 9). Consistent with both Chickenization and McDonaldization, one result of this is widespread uniformity of foods across varied locations.

Ransom (2019, p. 149) says the founding of the WTO is seen as evidence of the

> full realization of neoliberal policies. It deals with the rules of trade between nations, and confronts barriers to that trade such a tariffs, rules of origin, and nontariff regulations. In concrete terms, neoliberal policies allow companies to now import and export products or portions of products across multiple countries.

Products will also be shipped to numerous countries just as part of the processing process. All of this is made possible by neoliberalism. And, by the way, there are now hundreds of Regional Trade Agreements, such as NAFTA, in effect to assist with the trade of goods across the globe (Ransom, 2019).

A study by the University of Florida Institute of Food and Agricultural Sciences identified numerous sources of food contamination. According to Silbergeld (2016, p. 163):

> The first six "causes" are all related to workers: sick employees, employees with unwashed hands or contaminated gloves, employees with open cuts and scrapes, employees who touch their faces and mouths with their hands, employees who improperly dispose of hygienic items . . . and employees who do not wash their hands after using the restroom. Yet, it is the nature of the workplace itself that causes the contamination—i.e., the workplace contaminates the workers and not the other way around! Even the WHO has expressed concerns that the industrialization of agriculture and food production through integration and consolidation has created an environment which foodborne diseases can

emerge and spread rapidly through farms and food processing, increasing the likelihood of larger, even if fewer, outbreaks of foodborne illness.

*(Silbergeld, 2016, p. 203)*

Silbergeld (2016, p. 116) estimates that pathogens can spread up to one football field away from concentration areas due to large fans that are required to blow air to keep confined animals alive. The evidence is clear that pathogens from animals are not contained because "food animals are held in buildings that are open sources that provide little to no barriers to the release of what is inside or the entrance of what is outside." So pathogens end up in the soil, in the air, and in rivers and streams. The *carrying capacity* of a river is defined as the ability of that river "to absorb and utilize inputs without sustaining damage" (Silbergeld, 2016, p. 150). We often exceed these limits, and it should always be a goal to not do this when producing food products.

Silbergeld (2016, p. 116) notes that every chicken produces about ten pounds of waste, each hog about one ton of it. And she claims "there is little regulation of animal waste management." Silbergeld (2016, p. 117) reminds us: "What the animals excrete is feces, along with the drugs and other feed additives they have eaten. They also excrete bacteria and other pathogens in their waste, along with the genes that make bacteria resistant to antibiotics." So, the stakes are clear. What we have, really, is not waste management, but rather,

> waste holding and dumping. . . . There are no regulations and no monitoring. There are *recommendations* from the [US Department of Agriculture] about handling animals wastes, such as holding times before disposal on land, and some guidance from some states as to how often and where wastes can be applied to land. But neither reduces the hazards present in the waste.
>
> *(p. 119)*

Yet, both the industry and USDA "continue to insist that their operations are 'biocontained' (keeping biological agents and biotoxins inside) and 'biosecure' (keeping these same hazards outside)" (p. 115). There is no confirmation of this claim by the government.

A description of how animal holding facilities are (not) cleaned out is pretty stunning:

> The usual practice in the industry is to remove only the top layer of litter—known as the "crust"—which becomes caked with dried liquid waste and water. This is not a complete cleanout. Based on growers' reports, a complete cleanout may occur once every twelve to eighteen months, unless there is a disease outbreak, which would necessitate an earlier cleanout. . . . There is no decontamination of this waste, and growers usually store the house wastes in an open-sided shed until its eventual disposal on land, either by the grower or by someone else after transfer off the farm.

Another means of holding wastes is in so-called "lagoons." Animal agriculture produces immense waste problems in the form of lagoons, "where it threatens to overrun or leech into water" (Barnhill et al., 2018, p. 7). Here, when lagoons overflow, "the wastes are transferred into trucks that spray the liquid wastes on fields in most instances, but not for fertilizing. At all stages of holding and disposal, the odors and irritant aerosols from this process can be staggering in intensity" (Silbergeld, 2016, p. 118). According to Silbergeld: "There are few regulations on the placement or management of these impoundments. They do not have to be lined to prevent leakage into the ground, and they are often poorly managed, such that heavy rains can cause spills and overflows," as in the case of when Hurricane Floyd hit North Carolina in 1999. The outcome is clearly land pollution, not to mention water pollution.

Silbergeld suggests that animal wastes are simply not regulated, but she notes that there is an "array of hazardous waste management laws implemented by the [Environmental Protection Agency]" yet, the

> EPA clearly assigns responsibility for downstream management to the industry that produces the wastes. But this concept of corporate responsibility has not been adopted for animal wastes, once again because of the general refusal—in which we are all complicit—to recognize that agriculture has become industrialized and to accept the convenient divisions that have been create by the integrated structure. The integrators bear no responsibility for animal wastes. Individual farmers are responsible for meeting any state regulations . . . or the short-lived EPA regulations related to obtaining permits for releasing animal wastes into surface waters (not soils).
>
> *(Silbergeld, 2016, p. 187)*

Though individual farmers have been sued for not complying with regulations, courts have generally ruled that the corporations who actually produce the food products that lead to such large amounts of waste have no responsibility for that which is produced by farmers working under contract with them. According to Silbergeld (2016, p. 120), neither the USDA nor the FDA have authority to actually impose agricultural waste management:

> Neither has stepped up their obligation to regulate. For the USDA, the basis for its silence on waste management is its persistent allegiance to the long-gone traditions of small farms producing limited amounts of wastes that can be locally recycled for improving soil quality for agricultural benefit.

Again, this is indicative of regulatory failure by the US government.

So, rather than actually regulating waste, the USDA "only *suggests* that wastes should be composted 180 days prior to land disposal in areas where crops are grown for human consumption" (Silbergeld, 2016, p. 121). Silbergeld claims: "Composting as practiced by the poultry industry basically consists of piling wastes, sometimes under a roof, until it is hauled away. Not much happens in terms of reducing hazards in the waste." Meanwhile,

> The FDA has proposed but not issued regulations to require standards and treatment to reduce microbial contamination of animal wastes (including documentation of composting; limits on land application in fields where crops are produced for human consumption; rules on storage, transport, and handling of wastes; and recordkeeping).
>
> *(p. 122)*

All of this is a major risk to workers, neighbors, and even drivers who transport animal products as well as other drivers whose cars are exposed to the pollutants being expelled by animals in open truck beds. Inaction is produced by politics and self-interest (i.e., profit), both of which are seen as more important than science. So, there is lots of silence within the industry and government with regard to these issues.

Why this should be so concerning is the entire food system is porous, meaning no one is safe from it:

> Food animal production takes place within a set of environments or ecosystems, starting with the farm and ending with us in our world. All of these systems are highly porous; they are in contact with each other through the movement of animals and humans, transport systems, air, and water. We enhance this movement by our failure to manage wastes from food animal production and by the globalization of our food supply.
>
> *(Silbergeld, 2016, p. 141)*

Silbergeld (2016, p. 133) notes that

> certain strains of *E. coli* are highly virulent and can cause life-threatening human disease, including diarrhea . . . and urinary tract infections, particularly in women and children that can progress to kidney failure. . . .. Some of these virulent communities of *E. coli* are also resistant to multiple antimicrobials.

A major outbreak of *E. coli* occurred in the United States in 2013, attributed to Rich Products Corporation, "a company that has repeatedly made the FDA list of food recalls. Rich Products demonstrates how 'too big to fail' plays out in the world of today's food industry." It had to recall millions of its products, all produced at the same plant, including some pizzas intended for school lunches! A couple of years earlier, in 2011, National Beef Packing Company recalled about 60,000 pounds of beef contaminated with *E. coli*. Such major newsworthy outbreaks of food safety issues have caught our attention over the years. Consider the shocking case of Jack in the Box restaurants when *E. coli*-infected meat sickened more than 700 people in four states and killed four children! The outcomes were preventable because the chains ignored state laws about cooking temperatures as well as internal suggestions from the company to cook patties longer.

Hauter (2012, p. 123) claims that

> the industrialized food system is responsible for its proliferation. Cattle spend then last three to four months crowded together in megasize feedlots, where they wallow in their own waste. Because they arrive at the slaughter facility covered in fecal matter, from the first step of killing the animal, throughout processing, decal bacteria are dispersed in the meat product.

And it is hamburgers that this is most problematic, for "it is ground in enormous batches that contain parts from thousands of cows that originated in feedlots in multiple countries."

Foodborne outbreaks seem to be increasing in the United States. Walker (2019), for example, counts 8 multi-state outbreaks in 2002 versus 38 in 2016. According to Walker (2019, p. 154), "between twenty-nine to seventy-one million foodborne illnesses occur annually. . . . Of those infected, 63,000 to 216,000 require hospitalization. Between 1,500 and 3,000 ultimately die." In 2012, data show that there were "831 foodborne illness outbreaks, 14,972 illnesses, 794 hospitalizations, and 23 deaths" (Nesheim et al., 2015, p. 101). Yet, CDC data from 2000 to 2008 suggest "47.8 million illnesses, 127,839 hospitalizations, and 3,037 deaths related to foodborne illness occur every year in the United States" (Nesheim et al., 2015, p. 102). This means one in every six people becomes ill from the foods they eat every year!

Linnekin (2016, p. 27) estimates that pathogens in food kill about 3,000 Americans a year and make ill nearly 50 million more. These numbers have declined over time as food has generally gotten safer, yet, he notes that the FDA fails to regulate food preparation and handling at restaurants, hospitals, cruise ships, and individual homes where viruses lurk (this is the jurisdiction of states, counties, and cities, as well as individual families). "The FDA also doesn't regulate beef, pork, poultry, and other meats that are responsible for another 22 percent of foodborne illnesses . . . that's the job of the USDA's Food Safety and Inspection Service" (FSIS).

Regulatory efforts by the state are obviously meant to prevent such outcomes, but factors such as low staffing numbers, overwork, and empowering companies to self-regulate impede the effectiveness of regulatory agencies (Sharma et al., 2010). Further, scholars such as Silbergeld (2016) show that efforts to stop the spread of pathogens, such as HAACP, are controlled by industry and not at all effective due to serious gaps, as noted earlier.

Keep in mind that an enormous amount of food is imported into the United States each year, and from about 200 countries, produced by hundreds of thousands of growers and manufacturers. Walker (2019, p. 164) notes: "Around 70 percent are processed food products. Only about 1 percent of imported food is examined, which can range from a closers inspection of the manifest to a quick visual check of the product to dispatching samples of laboratory analysis." An Inspector General report of the FDA sampled "thirty recalls from 1,557 cases over thirty-two months," and it "revealed systemic deficiencies in evaluating health hazards, carrying out audits, ensuring compliance where recalls were initiated, tracking, and maintaining accurate recall data." Add on to that fact that pork inspections are voluntary, including which food-borne pathogens to monitor. Pork line speeds are also up to the manufacturers. Faster line speeds raise probabilities not only for more injuries to workers but also for greater contamination of food products.

One major problem with the conventional food system is that there is no monitoring of secondary processing. That is, companies buy carcasses from major corporations and then turn them into things like chicken nuggets or chicken breasts that are then bought by consumers. Silbergeld (2016, p. 198) says:

> We do not know much about these secondary processors, including how many times packages are opened and reopened, how products are manipulated, and how the packages are resealed. Each of these events represents another opportunity for contamination. We don't know how often these plants are inspected, and we have little data on contamination issues in these plants.

The USDA does require that all meat packages come with instructions about how to store and prepare the products. According to Silbergeld, a review by CDC and FDA found that they do not work. Less than half of the people surveyed even recalled seeing the information on products they bought.

When foods are known to be unsafe, the correct thing to do is recall them. Silbergeld (2016, p. 194) says that "only in 2012 did FDA receive legislative authority to enforce recalls; before that time, they could only announce and problem and advise corporations to take their products out of the food supply." Yet, even with this new power, FDA rarely acts and when it does, it is not always done quickly.

Leighton (2019) outlines the case of Salmonella poisoning in peanuts sold by the Peanut Corporation of America, an outbreak that killed nine people, sickened 714 others, and led to 166 hospitalizations; total cases were estimated to be between 11,000 and 20,700. Leighton shows that the peanuts were knowingly shipped after being contaminated, and thus the corporation was culpable for the harms caused. According to Nestle (2013), the company knew the peanuts contained salmonella because its tests for salmonella came back positive, but the company retested samples until they came back negative. She notes: "Despite hazardous production standards, the company had passed inspections" (p. 119). The CEO of the company was convicted and sentenced to 28 years in prison, though the crime he was convicted of was not for killing or sickening consumers, as none were even mentioned in the indictment; instead, it names corporate victims including Kellogg's, "who were the immediate purchasers" of the peanut products (p. 177).

Asomah and Cheng (2019, p. 193) claim that about 350,000 around the world die from food poisoning. While many of these cases are likely accidental, it is still a certainty that some of them result from culpable acts of manufacturers and sellers of food products contaminated with deadly pathogens. In their chapter, Asomah and Cheng (2019, pp. 194, 196–199) discuss the following real-life cases, some of which pertain to food poisoning and others to mislabeled or fraudulent products:

- Sanlu Group "consciously sold its melamine-contaminated infant formula to consumers, which caused the deaths of at least six babies, and over 300,000 illness."
- Caraga sold poisoned candy in 2015 that sickened 2,000 people.
- Punjab sweet poisoning killed 33 people including five children.
- XL Foods sold food products contaminated with *E. coli* that killed at least 15 people.
- Cooking oil contaminated with recycled waste, animal feed, and gutter oil was sold in Taiwan.
- Pork Pie products were recalled in Australia due to possible Salmonella contamination.
- Pinnacle Foods in the United States recalled pancakes, waffles, and French toast due to possible listeria contamination.
- Frito-Lay recalled potato chips due to possible Salmonella contamination.
- Lord Organic recalled a ginger powder product due to Salmonella contamination.
- Maple Leaf Foods in Canada had a listeriosis outbreak that killed 22 people in 2008.
- Chocolate bars from Mars containing plastics were recalled in 55 countries, including France, Germany, and the Netherlands.
- Lamb-containing beef was sold to consumers in the United Kingdom.
- Rat meat was sold to consumers in China in 2013.
- Misleading food labeling scandals have hit Frito-Lay, Heinz, Kraft, and Unilever.

Barbarosa (2019, p. 385) adds other cases including Perrier's benzene-contaminated mineral water, Coca-Cola's fungicide-contaminated soft drinks, an E Coli outbreak in Taco Bell's lettuce, and KFC and Pizza Hut's sale of expired meat.

## Hazardous Working Conditions

Food production is also often inherently dangerous. One of the most dangerous jobs in America is agricultural production, and in particular, one of the most hazardous occupations is farming: "From 2006–2009, the occupational fatality rate for workers in agriculture, forestry, and fishing was significantly higher than for all other industries" (Nesheim et al., 2015, p. 180). Perhaps the highest of all fatality rates is found in the commercial fishing industry, where the rate of death is 124 per 100,000 workers, versus the rate of only 4 per 100,000 among all workers (Nesheim et al., 2015, p. 181).

Approximately 170,000 agricultural workers and farmers die every year, "representing twice the fatality rate in other sectors of the labour economy" (Del Prado-Lu, 2019, p. 93). According to Del Prado-Lu (2019, p. 94): "Farmers are among the occupational groups that suffer the most from health risks as they are usually unprotected, poorly covered for health insurance, and experience more barriers in accessing healthcare services and programmes." One major hazard to farmers and their families and communities is the heavy use of pesticides: "Studies have correlated the extent of direct and indirect pesticide exposure to health hazards such as increased mortality, dermal contamination, a depression in cholinesterase levels, and bother foetal abnormalities and spontaneous abortion among pregnant women."

Major outcomes of pesticide exposure include prostate and bladder cancer and lymphoma, reproductive disorders, disruptions to neurological functioning, neuromuscular dysfunction, cognitive memory deficits, attention deficits, dermatoses, respiratory distress and illness, organic dust toxic syndrome and Farmer's lung, chemical poisoning (with 3.6 million calls per year, with children making up more than half of the victims), and even suicidal ideation (Del Prado-Lu, 2019, pp. 96–98). Both acute effects (such as "vomiting, diarrhea, cough, seizures, changes in the sensorium, headache, and skin irritation or itchiness") and chronic effects (such as "endocrine disruption, hypertension, neuropathies, cancer, bone marrow effects, skin lesions, and cytogenic or immunotoxic effects") are common (Del Prado-Lu, 2019, p. 98). Del Prado-Lu (2019, p. 99) notes the use of extremely hazardous chemicals in Asia that are found to produce outcomes such as Parkinson's disease, chromosomal disorders, pregnancy abnormalities, and hearing loss. Del Prado-Lu (2019, p. 103) concludes that "the global food system . . . corroborated by the lack of regulation and social safety nets, is producing significant health and socioeconomic harms against farmers and farming communities."

Farm workers are more likely to be exposed to chemical poisoning with a rate of 53.6 per 100,000 workers versus 1.38 per 100,000 for non-farm workers. This risk is highest for those working in food processing and plant packing workers (Nesheim et al., 2015, p. 181). This work also comes with higher risks of respiratory and skin disorders, chronic pain from injuries, stress, anxiety, depression, as well as suicide.

Food processing is particularly risky work. In addition to chemical exposure, extreme heat and cold, loud noise, long periods of standing, heavy lifting, and working with sharp instruments and electrical equipment, the "meat and poultry slaughtering and processing industries have long been associated with a high rate of injuries, fatalities, and illnesses" (Nesheim et al., 2015, p. 188). According to Hauter (2012), the repetitive nature of some food workplaces (e.g., poultry processing) leads to both painful and debilitating injuries. She claims that 20% of poultry workers get injured on the job, for example. Hauter claims: "Processing plants are a miserable work environment—bone-chilling cold, wet, and slippery" (p. 205). Transportation and warehouse workers also face the risks of musculo-skeletal injuries associated with heavy lifting and repetitive motions.

Overall, the risks of death are about ten times higher in the food industry than in other industries. Deaths are caused by transportation incidents, fires and explosions, contact with objects and equipment, exposure to harmful substances, and falls (Bhushan, 2011). In 2017, 5,147 people were reportedly killed at work, including 258 farmers, ranchers, and agricultural managers, 89 food preparation workers, 60 food and beverage store workers, 41 fishers and fishing workers, and 987 drivers (some of whom were working in the conventional food system) (Bureau of Labor Statistics, 2018). Rates of death in the food industry are higher than rates of death in the non-food industry work in production, processing, storage, and retail. Further: "Food industries, especially those involved in processing, storage, and retail, have significantly higher rates of illnesses and injuries, as compared to non-food private industries in the U.S." (Newman et al., 2015).

According to the Bureau of Labor Statistics (2018): "Private industry employers reported 2.6 million nonfatal workplace injuries and illnesses in 2021. . . . In 2021, the incidence rate of total recordable cases (TRC) in private industry was 2.7 cases per 100 full-time equivalent (FTE) workers." A large majority of reported cases were injuries, as opposed to illnesses obtained at work. Only about 365,200 nonfatal illnesses were reported in 2021, including 269,600 respiratory illnesses. Thus, about 2.2 million injuries were reported at work in 2021. Of these, about 1,062,700 nonfatal injuries and illnesses caused a worker to miss at least one day at work, for a rate of 1.1 per 100 workers. The rate of days away from work in private industry food manufacturing was reported as 2.1 cases per 100 full-time employees.

Figure 7.3 illustrates the total number of illnesses and injuries in selected industry sectors in 2021. As you can see, the highest incidence of injury and illness was found in health care and social assistance, followed by retail trade, manufacturing, transportation and

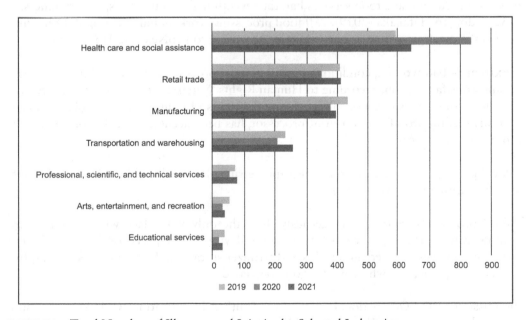

**FIGURE 7.3**    Total Number of Illnesses and Injuries by Selected Industries

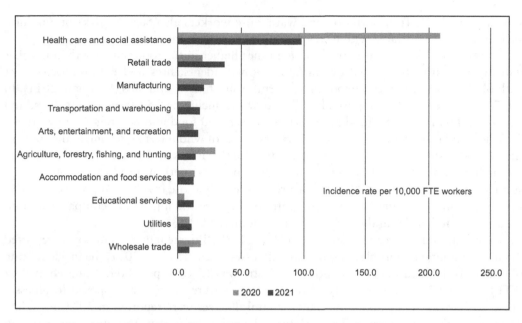

**FIGURE 7.4**  Total Number of Respiratory Illnesses by Selected Industries

*Source:* Bureau of Labor Statistics (2018). Employer-reported workplace injuries and illnesses-2021. Downloaded from: https://www.bls.gov/news.release/pdf/osh.pdf

warehousing, and then other areas. Three of these four areas would include elements of the conventional food system.

Figure 7.4 shows the rate of respiratory illness in selected industries in 2021. As you can see, accommodation and food services had the seventh highest rate of respiratory illness.

According to Hatanaka (2019), 750 food processing workers suffered amputations since 2010. Take meat processing as one example. According to EcoNexus (2015, p. 15):

Extremely bad working conditions prevail in meat processing. In the U.S. it is the most dangerous factory job, according to Human Rights Watch. Furthermore, wages are low, trade unions are usually not tolerated, and the rights of immigrants are violated. Tyson Foods Inc. the world's largest meat processor, has been accused of such working conditions for many years.

Workplace injuries are common in meat-processing plants. Silbergeld (2016, p. 162) outlines the nature of this work, saying:

food production starts off dangerously. It is the only workplace where children are allowed to work, where few—if any—safety devices are required on large machinery with cutting edges, where grain silos and fertilizer storage tanks go uninspected and frequently explode, and where wastes go unregulated.

The mission of the Occupational Safety and Health Administration (OSHA) is to assure that workplaces are safe. Yet, one OSHA proposal on injury prevention was repealed by

Congress in 2001. Since then, OSHA issues "nonbinding guidelines for workers in meat-packing and poultry processing. In OSHA's own words, these are 'advisory in nature and informational in content'; that is, they do not include legal means of enforcement." Incredibly, exposure to pathogens at work is not even considered work-related by OSHA or any "other occupational health and safety agency in the world" (Silbergeld, 2016, p. 172). Silbergeld shows how OSHA has been interfered with by other agencies as well as by the courts on issues such as line speeds for processing plants.

In 2014, a proposal from USDA on eliminating limits on line speeds became law. Silbergeld (2016, p. 166) notes

> While it steps back from the original proposal that would have taken all controls off line speed in processing plant, at the same time it completely removes the requirement for on-site inspection by USDA personnel. This proposal was supported by USDA with the astounding assertion that these changes would actually improve food safety with no adverse effects on worker safety.

A review of the evidence by the GAO found that the research into the issue was too poorly conducted to provide any usable data for analysis purposes.

According to Silbergeld (2016, p. 180): "The USDA has claimed repeatedly that there is no proven association between line speed and worker health and safety. They have no data on this topic, and, in fact, studies by the poultry industry refute this assertion." She asserts that faster speed processing of poultry and similar products is associated with high rates of chronic and acute injury, but these injuries are underreported by the poultry slaughter and processing industry.

So, Silbergeld (2016, p. 171) claims:

> As of 2015, the safety net of HAACP had been almost completely shredded. In its final rule about line speeds and food inspection in poultry production, the USDA maintained an upper limit for line speed by acquiesced entirely to industry, and relegated government inspectors to inspecting industry records.

She says:

> The responsibility—and opportunity—for ensuring HAACP is placed almost entirely on the industry, with government engagement focused on ensuring compliance through inspection of industry records and programs rather than any actual data. This approach is referred to as *management-based regulation* and is praised by policy analysts for shifting the burden of policy implementation from government and for permitting some degree of flexibility by the industry in meeting policy goals. Yet, a needed reform is clearly to put safety regulation back in the hands of governments, who are ultimately responsible to the citizens they represent.

Studies of actual workers from states such as North Carolina show higher rates of carpal syndrome than in other workplaces, as well of other musculoskeletal disorders. A study in Sweden found those workers involved in deboning chickens had the highest rates of carpal tunnel. Silbergeld (2016, p. 181) notes that OSHA conducted its own review of injuries in infections in hog and poultry slaughter plants, using data from reports filed by the industry itself, meaning they are likely vastly underreported. A review of the OSHA reports found "the injury rates reported by the eighteen hog and poultry slaughter/processing plants from 2004 to 2009 were the highest in the private sector industry workforce" (p. 183). The data also showed a strong link between lacerations suffered at work and subsequent infection with pathogens.

Of course, most cases of injuries are not reported and most causes are unknown. And these numbers do not also include cases of antibiotic-resistant infections caused by antibiotics used in foods. The National Antibiotic Resistance Monitoring System, meant to capture the incidence and prevalence of this problem, is actually not even national but rather represents ten states and about 15% of the population. Silbergeld (2016, p. 189) claims: "Surveillance of resistance in bacteria isolated from humans is also a voluntary program, and less than half of the states are currently participants." Here is another failure of the government to develop and utilize such programs.

Silbergeld (2016, p. 164) asks:

Where is [Occupation Safety and Health Administration]? Largely absent. It is easier to blame the victim, to assert and assume that the workers are part of the problem, rather than acknowledging that this workforce faces some of the highest risks to health and safety in American industry.

Meanwhile, the industry, motivated by profit at all costs, resists "intrusions related to environmental protection, animal welfare and health or even sanitation for workers solely in the name of profit without contesting the allegations of injury and illness." Silbergeld continues, writing "OSHA has never been responsive to its real constituency of American workers; rather, it has been almost completely captured by the industry." She claims, incredibly: "Killing a worker is literally cheaper by the dozen—the pathetic limit on OSHA fines for a death at work is only $20,000 for the first incident and $10,000 per death thereafter."

So, Silbergeld concludes that OSHA is not really any help to workers because

it has been limited from intervening too much in regulating work in agriculture . . . and it has been extraordinarily passive in dealing with growing operations, which are clearly not farms and thus should never have received the traditional immunity of agriculture to regulation related to occupational health and safety.

She continues: "OSHA has delegated responsibility for inspection and enforcement to state programs, which have largely distinguished themselves by inactivity and, in at least two cases, catastrophic failure" (Silbergeld, 2016, p. 171). She sees OSHA as more a protector of industry profit than workers. If so, this is a failure of government regulation at the expense of worker safety.

Silbergeld (2016, p. 49) argues that benefits to industrial agriculture come at the cost to workers, such as

the exemptions granted from regulations in occupational health and safety as well as environmental regulations that increase risks to workers and communities. Environmental practices that are illegal in other businesses, such as employing workers younger than age eighteen or undocumented persons, are exempted in law or practice.

Walker (2019, p. 30) lays out the realities of the modern meat industry:

The meat industry follows a simple strategy—high volume from largely three kinds of meat controlled through a handful of multinational companies that own or take control of most meat animals. To increase profitability, low-wage workers now slaughter and fabricate the meat; feed is subsidized via government agricultural policies; environmental regulations meet strong resistance; and opposing food safety regulations is a routine part of doing business.

Walker (2019, p. 31) even offers up the reality of contemporary meat:

Separate two-thousand-pound 'combo bins' are filled with fat from feedlot steers, lean meat from culled dairy cows, and imported scrap trimmings. The bins are mixed, ground, shipped, remixed, reground based on market opportunities. The typical meat patty can contain buts of multiple cows from different countries.

Even meat embedded in fat can be removed with technology—heating and centrifuging meat, plus adding ammonia, called by the industry "beef slurry." This "pink slime" made the national news, leading to a revolt among consumers and embarrassment to the industry.

Walker (2019, pp. 157–158) lays out what he saw when visiting a slaughterhouse, writing:

we entered a massive, windowless concrete building and walked long cement corridors alongside workers coming on shit. In the changing room we were outfitted with ear plugs, hairnets, hard hats, safety glasses, white smocks, and knee-high rubber boots, then escorted to where the line began. As expected, blood was being splattered about, and the noise from motors, chains, and saws made it difficult to be heard. . .. The rear legs of warm, freshly killed cattle were shackled to the overhead line that serpentined through the plant before exiting near the cold-storage lockers. Workers standing next to each other were draped in eight pounds of chain mail as protection against errant knife or hook. Underneath the layers of safeguards were people who repeated the same repetitive task thousands of times for at least eight hours each day. Behind them were supervisors who walked back and forth, making sure their section of the line kept pace. Profitability was measured in volume, volume was measured in efficiency, efficiency was measured by ensuring no worker slowed down the line. High rates of injury made meat-packing one of the most dangerous jobs. Whenever a worker was injured, a "floater" was brought in as a replacement.

Is it consumer demand that is to blame for these conditions? Or is demand so high due to the actions of the industry? Either way, companies are culpable for the safety of their workers, and current conditions in much of the food industry are unsafe. And the government has some culpability, as well, or failing to assure healthy workplaces for workers.

Then there is the issue of slavery and forced labor in food! Hinch (2019) discusses the history of slavery and forced labor in the chocolate industry, and he shows that the issue is still a problem for hundreds of thousands of people. He claims it survives for the following reasons:

- There is a high demand for cheap labor.
- People with little power or resources have few other opportunities for work in certain communities and countries.

*(p. 80)*

There are about 450,000 cocoa farms in the Ivory Coast alone, and half of the country's population depends on cocoa for income. Hinch claims that about 40% of kids in the country work in agriculture, while about 23% work in the cocoa sector. Further: "There are an estimated 144,000 slaves" in the country, "a significant amount of whom are children" (p. 83). Then, there are more than 100,000 more in countries such as Cameroon, Democratic Republic of the Congo, Ghana, Mali, and Nigeria. Laws in the United States, United Kingdom, and European Union forbid the importation of goods produced by slave or forced labor, yet the problem nevertheless persists.

## Environmental Damage

Agricultural activities (such as farming) and processing activities (such as slaughtering animals) obviously take an enormous toll on the physical environment. A review of 148 papers on the issue of the impact of animal-sourced food found the food system impacts the environment through electricity use, greenhouse gas emissions, nutrient runoff into the soil and water, and air pollution (Hilborn et al., 2018). Hauter (2012, p. 12) explains:

> The long list of the consequences of industrialized agriculture includes the polluting of lakes, rivers, streams, and marine ecosystems with agrochemicals, excess fertilizer, and animal waste. Nutrient runoff (nitrogen and phosphorous) from row crops and animal factory farms, one of the foremost causes of the conditions that starve waterways and the ocean of oxygen, is creating massive dead areas of the ocean, such as one at the mouth of the Mississippi River the size of the state of New Jersey.

Further: "Planting and irrigating row crops has caused serious erosion, as irrigation and rainwater wash the topsoil away at the rate of 1.3 billion tons per year."

White and Yeates (2018, p. 315) claim that "farming of various kinds . . . and food production specifically . . . are inextricably intertwined with climate change issues." One example is the link between livestock production and greenhouse gas emissions: it "contributes to an estimated 14.5 per cent of global anthropogenic greenhouse gas emissions." Another example is the link between deforestation and agricultural production. And still another example is the threat of agriculture to biodiversity.

Nesheim and colleagues (2015, p.42) note these negative effects occur because of "consuming scarce resources (like energy or nutrients), contaminating resources, or generating outputs (often waste products) that diminish the quality of the biophysical environment." Numerous contaminants are created by the conventional food system, including "nutrients (i.e., nitrogen and phosphorus), pesticides, pharmaceuticals, pathogens, gases and inhalants (i.e., ammonia, nitrogen oxide, methane, odors, and fine particulate matter, or PM), and soil sediment (including the chemicals and organisms it may contain)" (Nesheim et al., 2015, p. 131).

According to Walker (2019, p. 228): "Humans have appropriated up to 40 percent of the Earth's biological production potential. Through our efforts alone we have modified an estimated 50 percent of the Earth's surface." Much of this is attributable to agriculture. And all of this owes to our belief in the infinity of natural resources. Walker outlines two parts of this illusion: "The first was humankind's long passage to farming. The second, much more recent, was the transition to becoming a nation of consumers. As farmers, our lives were still subservient to food. As consumers, food became subservient to us." Whereas, throughout much of human history, food was used for survival, today it is largely a matter of pleasure and freedom: "As food became ever more readily available, the easier it was to believe that we deserved more while being free of responsibilities—and the simpler it became to ignore the forces that made abundance possible." An anticipated outcome of this reality is that we will continue to use land as we see fit, without much regard to the sustainability of doing so. Efforts of sustainability are underway in various aspects of the conventional food system, but our current overuse of land for the production of food is likely not sustainable.

According to Gillon (2019, p. 208), "over 38 percent of the world's land is agricultural, with pasture accounting for 26 percent of the total and cropland accounting for the remainder." There are about five billion hectares of land being used for these purposes. It should be obvious that every single "industrial, productive, or extractive process" depend on "environmental resources to create products; and they typically depend on the environment as a place to lodge waste" (p. 205).

Walker (2019, pp. 52–53) points out:

Seven out of every ten acres planted are devoted to just three crops—corn, soybeans, and wheat. Cattle, hogs, and poultry provide 92 percent of all meat consumed. Some fifty thousand pesticides have been used on American farms. Eighty percent of all antibiotics consumed are in agriculture. The majority of food sold comes from a small proportion of farms that specialize in one or two enterprises such as raising hogs or growing wheat.

These farms, especially those that grow meat, dominate land usage, and the rest use vast natural resources that are the property of the people (Walker, 2019). The conventional food system does not take into account these facts. One example is that about 70–80% of ground and surface water in the United States is used by agriculture. Research also shows that agricultural companies across the United States are using more water than is being replenished. Not only is this a threat to the continued production of crops but also ultimately to life itself. This is particularly important given that agricultural activities occupy 54% of all US land (Nesheim et al., 2015, p. 142).

Silbergeld (2016, p. 145) notes: "The scale of agriculture—and its industrial production of catfish, pultry, swine, and salmon—is now so large that it indents a new, artificial, and visible patchwork on the landscape as well as an invisible footprint on the environment." Scholars claim that one of the most significant, far-reaching consequences of the contemporary food system includes the impacts on "the environment (e.g., effects on biodiversity, water, soil, air, and climate)"—many of which are negative, as well as "human health (e.g., direct effects on diet-related chronic disease risk, and indirect effects associated with soil, air, and water pollution)" (Nesheim et al., 2015).

Consider Concentrated Animal Feeding Operations (CAFOs), discussed earlier. CAFOs are linked to extensive environmental problems. For example, James (2019, p. 70) claims that they are linked to "increased greenhouse gas emissions, land degradation, water pollution, and increased health problems because of the concentration of animal waste such farm operations produce."

Fundamentally, it is our demand that is to blame:

> To satisfy our appetites for meat, the enormous appetites of billions of chickens and millions of swine require the production of enormous amounts of feed crops and water. . . . Accompanying them are major increases in the basic feedstocks of animal feeds—corn and soybeans—which, in turn, have exerted demands for land and water as well as the ancillary inputs of synthetic fertilizers and pesticides.
>
> *(Silbergeld, 2016, p. 147)*

According to Silbergeld, the real and major cause of deforestation in the Amazon is the introduction of soybeans into the agricultural reality of the area.

Whereas the introduction of mineral fertilizers has greatly improved the productivity of the farming industry, it has also "led to negative impacts on the environment, such as greater greenhouse gas (GHG) emissions and deterioration of water quality. GHG emissions also can result from the burning of fossil fuels in the food manufacturing process and during food distribution" (Nesheim et al., 2015, p. 127). Research from the EPA shows that agricultural practices, "principally fertilizer use and manure management, are responsible for about 74 percent of U.S. emissions of the greenhouse gas nitrous oxide and 84 percent of the nation's emissions of ammonia and other NHx-nitrogen compounds" (Nesheim et al., 2015, p. 143).

Incredibly, when samples are taken in agricultural areas of the United States,

> pesticides were detected in 97 percent of samples in streams and 61 percent of samples in shallow groundwater areas. Additionally, organochlorine compounds, the majority of which are no longer used and which are considered "legacy" pesticides, were detected in 92 percent of fish tissue samples and 57 percent of aquatic bed sediment samples.
>
> *(Nesheim et al., 2015, p. 134)*

According to Walker (2019, pp. 195–196):

> Most pesticides are the by-product of research into chemical warfare. After World War II ended, Congress passed legislation governing their sale and intended use—but not their long-term effects. In just five years, ten thousand new pesticides were registered. Over the next forty years, fifty thousand different pesticides came into use across America's farms.

Logically, no one really knows how safe these chemicals are for human consumption, or fully understands the implications of using them in the physical environment.

Chemicals that pose a risk to humans and/or the environment are listed as "restricted use." According to Walker (2019), the EPA has approved some 700 of these pesticides. There is no limit on how much these chemicals can be used, and some of them are widely used to "produce foods like soybeans, nuts, citrus, and fruit trees, as well as specialty crops like brussels sprouts, cranberries, broccoli, and cauliflower." According to Walker (2019, p. 199): "Research from animal models and young children show that even low exposure levels are associated with irreversible brain development and impaired cognitive functioning." Even an assessment by the EPA found that "1,800 animals and plants were likely adversely affected" by the chemical chlorpyrifos. Three CEOs of chemical companies argue the studies are flawed, and they offered up their own studies as proof; the EPA backed off (pp. 199–200).

EcoNexus (2015, p. 11) reports:

> Millions of farmers and agricultural workers are poisoned by pesticides every year— around 40,000 of them fatally. The number of unreported cases is high. And medical care is often missing. Poisoning mainly occurs in developing countries where pesticide users cannot protect themselves appropriately, and where products are sold that have been banned in the North for many years.

Like other areas of industry, pesticide production is highly concentrated, and market leaders such as Bayer, BASF, and Syngenta "each distribute more than 50 active ingredients that are classified as highly hazardous."

Silbergeld (2016, p. 157) shows that there is no controlled density in the United States when it comes to intensive agricultural operations:

> The World Bank and the FAO have developed guidelines and methods to control density. No such steps have been implemented in the United States, where industrialized agriculture is still immunized from public criticism through the persistence of idealized, false images as well as legislation that prohibits public access to and criticism of industrial operations, including farms and slaughterhouses. In several states, the industry has acted to influence legislatures to pass laws protecting farms from political intrusions (all fifty states have so-called right-to-farm laws), limiting investigations of arming and food-processing practices (six states have so-called gag laws) and even criminalizing statements about farming and food that are considered defamatory (thirteen states have so-called food libel laws).

Alkon (2019, p. 352) calls industrial agriculture "environmentally and socially destructive" and claims that it

> increases erosion, through which chemical pesticides and fertilizers can cause ecological harm to nearby waterways. Additionally, fertilizers and pesticides are largely made from fossil fuels, making agriculture responsible for approximately 12 percent of climate change-causing greenhouse gas emissions globally.

Red meat production is another major contributor to climate change. Interestingly, a study testing the effects of adding climate impact labels to foods found that labels indicating high climate impact made consumers tend to choose lower-impact meals (Paddison, 2022). This suggests educational campaigns about food and climate change might help make a difference in this area.

When farmers are incentivized to grow the same crops over and over again— monocropping—this "depletes soil of nutrients and can lead to significant erosion. The practice requires synthetic fertilizers to compensate for the lost nutrients, and pesticides to combat fungi and insect predators that thrive in these conditions" (Lakhani et al., 2021). Then, there is agricultural runoff from crop production, which "is now responsible for 80% of excessive nutrients in our freshwater and oceans, which cause dense growth of plant life like algae that block oxygen from reaching fish and other animals." And finally, there is the issue of threats to species: "In 2019, agriculture and aquaculture were identified as a threat to 24,000 of the almost 28,000 species threatened with extinction."

Gillon (2019) lists some of the environmental impacts of the conventional food system, including climate change, land degradation and stewardship, soil loss caused by agriculture, threats to biodiversity, toxic environments, and water quality and availability. With regard to biodiversity loss, a report by the Ellen MacArthur Foundation holds that "90% of biodiversity loss to date is attributable to how we make and use products and food, with agriculture being a primary deriver" (Edie, 2022). With regard to climate change, it is important to note that agriculture uses about 16% of all energy used in the United States. Energy use per capita has been growing, and one study found that

half of the growth in energy use is attributed to energy-intensive technologies in food processing (e.g., freezing, baking, and cleaning food) and the conversion of food preparation labor from human to automated processes (e.g., the trend toward processed, packaged convenience foods that require less household preparation time).

And then, the majority of greenhouse gases in the conventional food system are produced by agriculture. Amazingly, producing biofuels such as corn ethanol and soy-based biodiesel appears to release more greenhouse gas emissions than they end up saving relative to fossil fuels.

It is now well-known and widely reported that agriculture is associated with global GHG emissions; as much as 10–15% of all global GHG emissions come from animal agriculture (Barnhill et al., 2018). Wolf (2018, p. 45) reports on a 2011 Environmental Working Group (EWG) study which found that "beef production results in about 27 units of $CO_2$, for every unit of beef produced. This compares with 2 units of $CO_2$ produced for every unit of beans produced." Interestingly, lamb is the least efficient meat, with 39.2 units of $CO_2$ per unit of lamb, while lentils are the most efficient, with only 0.9 units of $CO_2$ per unit of lentils.

With regard to water, nitrogen fertilizers have been linked to "dead zones"—"areas where aquatic life is impossible because of nutrient pollution and the dissolved oxygen depletion that results." According to Gillon (2019, p. 209): "The FAO indicates that the number of dead zones increased from less than fifty in 1950 to over five hundred by 2010, covering well over 245,000 square kilometers." There is, stated simply, massive pesticide use in agriculture, as evidenced by the growth of pesticide sales globally from about $8 billion to $35 billion from 199 to 2014.

A 2008 report in the journal, *Science*, found about 400 coastal regions around the world now house dead zones, "dead because of fertilizer run off and oxygen deficiency" (EcoNexus, 2015, p. 11). Silbergeld (2016, p. 147) describes the Chesapeake Bay region as "an ecosystem heavily impacted by broiler poultry production . . . that has created a larger dead zone in which the water is depleted of oxygen to the extent that the normal ecosystem of plants and animals cannot survive." He notes that

excessive land disposal of animal waste—from chickens and hogs—resulted in overloads of nutrients, termed *eutrophication* by ecologists, sufficient to stimulate the overgrowth of these toxic microorganisms in the estuarine streams and rivers of the bay as well as Pamlico Sound in North Carolina.

*(p. xi)*

A few examples illustrate how significant this problem can be. One study suggests that "nearly 1 million metric tons of nitrogen are delivered annually into the Gulf of Mexico from agricultural lands lying upstream in the Mississippi River Basin, leading to formation of a coastal hypoxic zone." Another study found that about 1% of the herbicide atrazine applied to crops "moves into associated streams, creating conditions that can exceed thresholds for safeguarding aquatic organisms and human health." Still another study suggests that "agricultural intensification over the past 50 years had led to accelerating increases in soil sediment deposition in . . . lakes due to erosion, despite soil conservation efforts" (Nesheim et al., 2015, p. 143).

A non-profit organization focused on issues of sustainability—Ceres—ranked the 42 largest global food and beverage companies (most based in the United States), on how well they were responding to issues such as water dependence, water security, and operational water use efficiency. Its analysis divided companies into four categories: packaged food, beverage, agricultural products, and meat. The analysis showed these as the top scoring companies: Nestle (packaged food) 82, up from 64 in 2015; Coca-Cola (beverage) 72, up from 67 in 2015; Smithfield Foods (meat) 33, no change from 2015; and Olam (agricultural products) 49, not part of the 2015 analysis. The study noted a 10% improvement "in the average score of the food sector's management of water risk since 2015. The packaged food and meat industries made the biggest gains in improvement at 16 and 20 percent, respectively." However, a major caveat was also noted, and that is "the average score for the 41 companies benchmarked was still only 31 point and despite big gains, the meat and agricultural products industries continue to lag far behind the packaged food and beverage industries" (Source Magazine, 2017).

Global investors have taken actions to urge food and beverage companies to better manage water risks, given major threats such a fertilizer runoff from farms, "the most significant source of water pollution in the U.S," which "can create toxic algae blooms that trigger oxygen deprived 'dead zones'" (Sustainable Brands, 2015). Fifteen companies were targeted in a letter-writing campaign by these investors, largely because of failing to have major business planning activities and/or investment decision-making plans addressing water issues.

The concept of a "harm footprint" is used for capturing and depicting the deleterious impacts on the environment of various forms of agriculture (Budolfson, 2016). A chart by Budolfson illustrates that lamb, potatoes, and apples are particularly problematic when it comes to GHGs; apples also strain water resources. Beef poses the heaviest impact on land resources, while strawberries posed the heaviest impact on workers (p. 90). Bananas and other fruits and vegetables are also difficult to farm. Budolfson (2016, p. 171) thus claims that "many of the fruits and vegetables that we consume require intensive backbreaking work by humans to harvest, where that work is ultimately debilitating to many of those farm workers before they reach middle age." Budolfson (2016) also shows that potatoes, rice, tomatoes, and beef also produce high rates of GHGs, while eggs, milk, and other dairy products produce high rates of other pollutants. He also claims that "many vegan staples do *worse* than animal product alternatives." Sebo (2018, p. 420) notes that "our food system is arguably responsible for more harm than any other industry."

We can end this section with a question and answer. Singer asks:

Would you ever open your refrigerator, pull out 16 plates of pasta, toss 15 in the trash, and then eat just one plate of food? How about leveling 55 square feet of rain forest for a single meal or dumping 2,400 gallons of water down the drain? Of course you wouldn't. But if you're eating chicken, fish, turkeys, pigs, cows, milk, or eggs, that's what you're doing—wasting resources and destroying our environment.

*(cited in Budolfson, 2016, p. 163)*

The total economic costs of all these environmental challenges are unknown, but the estimated costs of climate change alone range in the hundreds of billions of dollars per year, a large portion of which is attributable to food production. A reasonable estimate of the economic impact on the environment of the conventional food system would be in the hundreds of billions of dollars.

### Food Insecurity

*Food insecurity* essentially means not having enough reliable access to food. Silbergeld (2016, p. 214) estimates that about two billion people worldwide, and about 10% of the US population, are at risk of food insecurity. Data from 2012 suggest that 15% of US households faced food insecurity (Nesheim et al., 2015, p. 105). In that year, nearly 9% of households (10.6 million) were characterized as facing "low food security (reports of 'reduced quality, variety, or desirability of diet, with little or no indication of reduced food intake')," and about 6% of households (seven million) were characterized as having "very low food security" where food intake "was reduced and normal eating patterns were disrupted at times during the year due to limited resources" (Nesheim et al., 2015, p. 197). About 42 million Americans are food insecure (Konefal & Hatanaka, 2019). The estimated costs of this problem are at least $90 billion (Brown et al., 2009).

Then, according to Gillon (2019), nearly 800 million people in the world, or 13% of the population, are malnourished. So, hundreds of millions of people around the world suffer from either poor-nutrient foods and/or a lack of food (Mardirossian & Muiuri, 2021). Walters (2019, p. 265) claims there are 795 million people who are malnourished in the world. Of those, about one in nine is chronically malnourished (Myers, 2019). Figure 7.5 illustrates the World Hunger Map from the United Nations. Incredibly, "the world produces re than one and a half times the quantity of food needed to feed its seven billion inhabitants" (Gillon, 2019, p. 215). But 1.3 billion tons of food, about one-third of the food intended for humans, is wasted. In the United States alone, we waste enough food to provide for the entire world's food deficit. Walters (2019, p. 267) notes: "The Global Sustainability Institute has predicted that catastrophic food shortages by 2040 will cause worldwide riots and death, and society as it is known 'will collapse.' "

The problem of food insecurity is structural in nature. For example, consider the issue of "food deserts." The US FDA defines a food desert as "a low-income census tract where wither a substantial number or share of residents has low access to a supermarket of large grocery store." Del Canto and Engler-Stringer (2019, p. 145) define a food desert as "a geographic area, or neighbourhood, where affordable and nutritious foods are unavailable, particularly in low socioeconomic status neighbourhoods and communities."

**FIGURE 7.5**  World Hunger Map

*Source:* https://reliefweb.int/map/world/hunger-map-2020

Colas e al. (2018, p. 197) add that food deserts are defined based on the number of food stores in an area and one's distance from them. They claim that

> food deserts generally arise as a result of planning failure, market failure, and a consumer-ist approach to health policy that automatically disadvantages those with lower incomes. People with limited access to, in particular, fresh produce are made more dependent on processes, packaged food.

Increased convenience of access to these foods would logically be expected to be associated with increased weight as it "allows people to eat more of them and more often, whether as part of a meal or in the form of snacks."

According to Myers (2019), about 30 million people in the United States, or about 10% of the US population, lives in food deserts. He notes that "a majority of these people were located in urban areas and were disproportionately black and Latno/a" (pp. 226–227). An example of a food desert is West Oakland, California, which has 40,000 residents and not a single full-scale grocery store (Alkon, 2019). Meanwhile, food swamps are also very common in the United States, "areas in which large relative amounts of energy-dense snack foods inundate healthy food options" (p. 227). Research shows that people living in low-income neighborhoods and with high minority presence tend to have less stores to choose from and less products within those stores to choose (Hatanaka, 2019).

Del Canto and Engler-Stringer (2019, p. 145) define a food swamp as "an area of low socioeconomic status with high geographic access to non-nutritious food sources." Living in either condition can lead to food poverty; Long and Lynch (2018, p. 334) define food poverty as "the inability of individuals and households to obtain an adequate and nutritious diet, often because they cannot afford healthy food or there is a lack of shops in their area that are easy to reach." Fuhrman (2017, p. 154) notes:

> Food deserts are predominantly located in low-income areas where people typically don't have easy access to transportation. People who can't afford to drive the mile or more to a grocery store are forced to rely on corner stores, bodegas, and fast food joints that sell commercial foods that create health problems.

In these areas, people often turn mostly to ultra-processed foods, which Simon (2006, p. 26) describes as "the only game in town."

Del Canto and Engler-Stringer (2019, p. 142) explain the logic of how food environments impact consumer behavior, writing "the structure and organisation of the neighbourhood or community food environment might influence food purchasing patterns, and subsequently, diet-related health outcomes." They note that "there is a marked difference in the price, quality, and availability or nutritious foods in neighbourhoods of different socioeconomic status, with lower access and affordability documented in low socioeconomic status communities" (pp. 142–143). Del Canto and Engler-Stringer (2019, p. 143), for example, mention a study from 2007 which found the ratio of grocery stores to residents was 1: 3,816 in wealthier, white neighborhoods but 1: 23,582 in lower SES neighborhoods with larger numbers of African Americans and Latinos.

Food access is determined by "the spatial accessibility and affordability of food retailers" (Integrate, 2018). It is also impacted by access to transportation, availability of produce/healthy foods via food retailer type, and of course, food price itself. Retail food establishments are those that manufacture, process, or hold food with the intent to sell the product for profit. The locations and types of retail food establishments determine a consumer's benefit from the conventional food system based on the location of their residency as well as their socioeconomic status. Prices determine the quantity and quality of food consumed.

Factors that influence the price of food often depend on the presence of other retailers in the area as well as the type of retailer. Convenience stores and gas stations offer increased prices for the convenience factor and a limited selection of primarily unhealthy food not recommended to sustain a healthy lifestyle (Blanchard & Matthews, 2009). Globalization in the food market has caused an increase in super stores like Walmart taking up large geographic areas, meaning they take the majority of consumers in the area away from other smaller, independently owned businesses that are typically in town and closer to residential areas. Increasing trends in globalization also reflect increase monopolization in the food market industry. This transition puts citizens at risk of less local retail job opportunities in addition to decreased access to healthy foods.

Many areas across the United States have limited access to food; in some areas, the prevalence of fast-food and convenience store establishments is significantly high. In these places, there are also fewer supermarkets—which tend to offer healthier and more affordable foods (Morland et al., 2002). There are also more gas stations and convenience stores in rural and low-income areas compared to larger metropolitan areas.

A discrepancy in the type of available food sources between areas depicts a larger problem. While there are large superstores such as Walmart available primarily on the outskirts of inner-cities, these locations require transportation or finances to pay for temporary transportation (taxi, bus fare) for access. These bigger and more affordable establishments push out previously available closer, smaller, independent stores. This effect causes increased unemployment for local residents and hardship to access necessities and food resources for those without easy access to transportation or finances.

Keep in mind the context of all of this. Myers (2019, p. 230), for example, notes: "Compared to other wealthy nations, the United States have the highest proportion of workers in low-wage jobs, defined as those where employees earn less than two-thirds of the media wage." The working poor are those who work but still cannot afford to fully provide for themselves or their families. One must ask, why do these types of jobs exist in the first place? Further, doesn't there existence prove clearly and unequivocally, that the economic system is designed to force people to work in such jobs, because, after all, someone must work them for the system to function? That is, capitalism produces and even mandates poverty. Myers (2019, p. 230) writes that "it is difficult to address the roots of hunger in the United States today without focusing on the inability of minimum wage employment and near minimum wage employment to feed a person, let alone a family."

Of course, the federal government has programs to assist those in need. One program, the SNAP, allots at least $74 billion, going to about 47 million people living in about 23 million households. Myers (2019, p. 229) explains: "This represents about 14 percent of the US population and means that more than one in seven US residents obtains food stamps, with the average SNAP participant receiving about $126 per month, or $1.40 per meal." About 44% of recipients were children, 10% were elderly, and 10% were disabled. Interestingly, 43% of people receiving benefits live in a household where someone works, but most "SNAP participants are those who are unable to work, those marginalized from work, and those who are the working poor." Myers (2019, p. 233) concludes that "the problem of hunger is not generally caused by the lack of a work ethic, but by the lack of well-payed work and the lack of access to income when one cannot work."

Additionally, companies are now pledging hundreds of millions of dollars in dozens of countries to help reduce food insecurity. CEOs at companies including Nestle, PepsiCo, and Google have signed on to help in new initiatives aimed at this goal. But, there is obviously far more that needs to be done in this regard.

The Global Index, released by the Access to Nutrition Initiative, assesses 25 leading food and beverage manufacturers "on their commitments, practices, and disclosure—with regards to governance and management; the production and distribution of healthy, affordable, accessible products; and how they influence consumer choices and behavior" (Access to Nutrition Initiative, 2022). The companies are ranked into seven categories including governance, products, accessibility, marketing, lifestyles, labeling, and engagement (Access to Nutrition Initiative, 2022). Nestle was ranked the highest, with a score of 6.7, Unilever was second with a score of 6.3, and FrieslandCampina was third with a score of 5.9 points. The average score was only 3.3. Amazingly, only 29% of Nestle's products are assessed as healthy (Chandrasekhar, 2021). According to Chandrasekhar (2021): "Nestle made headlines in May when the media got hold of a presentation circulated to executives that acknowledged that 60% of its mainstream products could not be called healthy and will never become so."

Documented outcomes of food insecurity include heart disease, diabetes, respiratory illnesses such as asthma, problems with cognitive functioning, behavioral problems, educational impairments, and mental health problems such as depression and suicide. Clearly, much more needs to be done to end food insecurity.

## Food Waste

Walker (2019, p. 215) calls food waste "chronic," saying that

> from one-third to one-half of food grown is never consumed at all. In developing countries, more than 40 percent of loss happens near the farm in storage, transportation, and processing. In countries like the United States, less food is lost on the front end, while more is wasted in stores, restaurant tables, and home refrigerators, where up to 40 percent is never eaten.

Long and Lynch (2018, p. 331) claim that between 30% and 50% of food is wasted: "To put the level of food loss into perspective, roughly 100 million tons of food is wasted annually in the European Union (EU), which is enough food to feed the hungry people throughout the world—twice." Further, "the volume of food waste in the US (34 million tons) is sufficient to cover Manhattan, one of the boroughs of New York City, in six feet of food waste annually" (p. 337). The nature of food waste varies in low-income countries, where it tends to happen earlier in the food chain, and high-income countries, where it tends to happen later. Long and Lynch (2018, p. 334) add that food waste "accounts for over 25 over cent of the country's total freshwater consumption" and also accounts for about 2% of the energy usage of the country. In the United States, most food waste ends up in landfills, where it rots away, producing methane, and thus contributing to climate change. But, perhaps the greatest shame of food waste is that much of the wasted food is edible "yet there are hundreds of millions of people throughout the world who are currently being denied their basic human right of access to sufficient food" (p. 340).

Matthes and Matthes (2018) outline four categories of food waste, including:

1. Inefficient resource use (e.g., fossil fuels and water)
2. Production waste (i.e., GHG emissions and pollution)
3. Byproducts
4. Wasted food (e.g., nearly 15% of total solid waste in city landfills was food, in 2013).

According to Linnekin (2016, p. 109), "the problem of food waste is immense. It stretches across the country and . . . is a problem in many other countries." He reports that we waste about $165 billion in food every year and that 10% of the money we spend on food goes to waste. Wasted food is a major source of methane, which, in landfills, contributes to climate change (and is about 20 times as potent as carbon dioxide). Linnekin reports on an estimate that every ton of food waste produces about 3.8 tons of GHG emissions and shows that a 2013 United Nations report called food waste the world's third-leading contributor to GHGs, producing 3.3 billion tons of GHGs into the atmosphere every year.

Another study found that restaurants, grocery stores, and convenience stores generate about 150 million tons of food waste every year (Linnekin, 2016). Most of that comes from fast food, but 40% comes from full-service restaurants and grocers. Even the USDA's National School Lunch Program is known to be a massive contributor to food waste. A recent change to the law, spearheads by First Lady Michelle Obama—Healthy, Hunger-Free Kids Act of 2010—is actually thought to have increased food waste in four out of five schools by requiring more servings of fruits and vegetables to children, which, when then not eaten, end up in the trash can. One study found that 96% of salad and 94% of unflavored milk ended up in the garbage. Another study found that 89% of fruits and vegetables ended up as food waste.

According to Bloom (2019), about 40% of food is wasted in the United States, and this represents about $166 billion. Barnhill et al. (2018, p. 11) agree, writing, "40% of the food produced" in the United States "is wasted—for example, it is left in the fields because it looks funny, it spoils in transport, or it is thrown out by processors, retailers, or consumers, much of it ending up as methane-producing waste in landfills." Part of this owes itself to the fact that "consumers increasingly expect that the fruits and vegetables they purchase in supermarkets will be uniform in size, shape, and color" (Bloom, 2019, p. 140). Further, in the United States, "retailers have historically refused to purchase so-called ugly fruit from farmers or wholesalers (who buy from the farmers)," whereas in the European Union, there are "regulations that prohibit oddly shaped fruits and vegetables from being sold" (Ransom, 2019, p. 147).

Bananas are the most popular fruit in America but are imported. Grocery stores do not profit off of them nor do they plan to. They are a " 'loss leader,' price to entice customers into the store so that they would buy other groceries" (Walker, 2019, p. 19). These fruits are shipped green and then ripened to turn yellow with ethylene gas in boxes in sealed chambers in giant warehouses, part of the conventional food system:

> By the time a consumer grabs a hand or two and scurries on to their next purchase, each box has been moved dozens of times, traveled thousands of miles, possibly undergone one or more inspections, and been the responsibility of thousands of people. For the average American, such breadth and complexity is invisible. Each person will consume ten thousand bananas by the age of forty, rarely, if ever, considering the workers' hands that touched them.
>
> *(Walker, 2019, p. 22)*

They also have no idea about the amount of food waste associated with bananas that occurs—bananas being thrown away for being too short, too long, too curved, or too damaged, so over time, "consumers had come to believe that every banana should be identical and perfect" (p. 23).

I am not sure who is responsible for the rejection of so-called ugly food—consumers or manufacturers—but I am confident that an industry-led educational campaign about the importance of not wasting food would go a long way to preventing such unnecessary waste. A consumer-led campaign could also work, but consumers cannot force food companies to make foods with cosmetic damages available for purchase; that is up to the companies.

## Focus on Fishing

Fishing is a major source of jobs, as about 12% of the world's population relies on fishing and fisheries for their livelihoods (Clausen et al., 2019). Much of the labor in fishing is very difficult as well as dangerous. Consider, for example, the deskilled labor working on the "slime line," a term used to capture assembly line fish processing plants. Further:

> In the contemporary period, there is growing concern for a sector of workers who are being forced onto commercial fishing boats against their will. These workers are called *sea slaves*. They remain on fishing ships year round, receive meager meals, and experience extreme hardships. The actual number of sea slaves is unknown; however, it has been estimated that 145,000 to 200,000 people are enslaved under such conditions in Thailand alone.

Fish are also a significant source of protein for people, accounting for about 17% of the world's consumption of animal proteins. Fish meal is also used as a protein to be incorporated into swine and poultry production. And: "Twenty-one million tons of world fish production was destined for nonfood products, of which 76 percent was reduced to fish meal and fish oil" (Clausen et al., 2019, p. 175). Taken together, these facts show the importance of fish in animal food production. Unfortunately, fishing comes with potential costs. For example, Clausen et al. (2019, p. 166) write: "Approximately 150 years ago, there was a major increase in fisheries captures. By the mid- to late-twentieth century, fisheries collapse, or the overexploitation and decline of a fishery, began to occur at an extraordinary rate." Then, after World War II, "the scale of fish production expanded to such a degree that it began to have global impacts and threatened the biodiversity of marine systems" (p. 167). Two examples of fishery collapses are the Peruvian anchoveta and Atlantic cod fisheries. When big fish start to disappear, Clausen et al. (2019) claim that fisherpersons just move down the chain to smaller fish in order to stay in business.

With the invention of steam engines and refrigeration, more capture and better storing were possible, leading to the great possibility of overfishing. With trawlers, longlines, and purse seines, *by-catch*—the capture of unwanted or unintended fish and sea animals—becomes a huge problem. About one-third of all fish caught are dumped overboard, a wasteful killing practice (Clausen et al., 2019). There is also illegal, unregulated, and unreported fishing, which can account for up to 15% of all fish caught in a given year, meaning the problem of overfishing is likely worse than we even know.

Unrelated to nutrition, some fish are of course viewed as worth more than others, and these fish are the ones most targeted by fisherpersons. For example: "The significant profit potential for Atlantic Bluefin tuna influences the fishery, rather than its potential to feed people who are hungry or malnourished" (Clausen et al., 2019, p. 171). The Maguson-Stevens Fishery Conservation and Management Act of 1976 created eight regional councils to support and protect fish stocks thorough quotas and other policies.

One way to address overfishing and the collapse of fisheries is farmed fish. Clausen et al. (2019, p. 172) report that farmed fish's contribution to human consumption now surpasses that of wild fish:

> Hailed as the blue revolution, fish farming is frequently compared to agriculture's green revolution as a way to achieve food security and economic growth to help the world's poor. Yet, a large amount of such farmed fish is reportedly going to feed wealthy consumers at home and abroad.

And the market is highly McDonaldized, as witnessed by this claim by Clausen et al. (2019, p. 172): "Salmon aquaculture practices emphasize conformity, control, and predictability of the fish that are being produced. In short, it is organized to obtain the most profit for the least amount of investment."

Two major drawbacks to farmed fish are the incredible amount of waste produced by the fish that obviously impacts water quality through increased pollution and the use of antibiotics to help fish fight off infections in such environments. Incredibly, pesticides are often used, as well. But I suspect that farmed fish likely produces high-quality products with less environmental destruction relative to open ocean fishing. If so, farmed fish will remain an important source of protein for people around the world.

The term "bycatch," as noted earlier, refers to the capture of nontarget fish and other marine life: it is

> what happens when fishermen literally cast a wide net, trawl the bottom of the ocean indiscriminately, or use other unsustainable fishing practices in the quest to catch a certain species . . . and end up also catching . . . other species the fishermen did not intend to catch.
>
> *(Linnekin, 2016, p. 135)*

One estimate suggests that about 40% of all fish caught each year is bycatch. Shark fishing is particularly wasteful when they are caught and killed for their fins and thus 95% of the shark is wasted. The Shark Conservation Act of 2010 banned shark fin fishing and required all sharks to arrive on land with their fins intact.

In addition to by-catch, many species of ocean life are now faced with the threat of extinction. Ransom (2019) claims that blue crabs, oysters, and rockfish are under threat of extinction due to pollution in the water as well as overharvesting. Agriculture is reportedly the major source of runoff into streams and waterways, leading to oxygen deprivation.

## Conclusion

In this chapter, the focus was on harms associated with the conventional food system. You saw that, first and foremost, billions of animals lose their lives as part of industrialized animal food production. Then there is the issue of the welfare of animals prior to their deaths, as well as of animals who are not killed but who are instead farmed for food products such as milk and eggs. One word that captures the way they are treated in factory farms and CAFOs is cruel. While it is not expected that some massive social movement will change the way Americans eat, much less people all around the globe, more attention must be focused on issues of animal welfare, and both food corporations and the government are responsible for assuring that animal welfare is improved as long as industrialized animal food production will continue.

Attention then turned to dangerous pathogens in our food, leading to serious illnesses and deaths associated with food poisoning. Both food companies and the government share culpability for incidences of food poisoning when they occur. Next, hazardous working conditions were examined, and you saw that a large portion of employees in the conventional food system work in hazardous conditions. The government must act to assure better working conditions for everyone involved in all aspects of the conventional food system.

Clearly, corporations will not assure (or at least are not assuring) safe working conditions for employees. This is a social justice issue that demands immediate action by the government.

Probably most notably, since it impacts everyone living on the planet, industrialized agriculture poses enormous environmental threats to the planet, including being a serious contributor to climate change and global warming. Here, though global food corporations are taking actions to reduce their environmental footprints, it is really up to consumers and the governments who represent them, to demand more change in this area.

Other issues addressed include the seemingly contradictory issues of food insecurity and food waste. I at least partially attribute both problems to the conventional food system itself, for the system seems to be built on the dual realities of large portions of people in the United States and around the globe who do not have full access to food in spite of enough of it being produced and a lot of it being wasted. More programs must be developed and carefully implemented to assure that no one goes hungry and that far less food is wasted. Surely we are smart enough to figure these issues out.

# 8
# CULPABILITY FOR FOOD CRIMES

## Introduction

In this chapter, the major arguments about what specifically, food companies are culpable for, are made. In addition to topics discussed in other chapters of the book, such as food addiction (Chapter 5) harms produced directly by the foods we eat (e.g., obesity and diabetes) (Chapter 6) and those caused by the system itself (e.g., harms to animals and workers and environmental degradation) (Chapter 7), this chapter identifies very specific actions of food companies for which they alone are culpable. These include producing excess calories, putting food in non-food environments, funding research to create one-sided studies, using front groups to confuse consumers, advertising unhealthy products, engaging in deceptive advertising, selling fraudulent foods, and shrinking products over time (i.e., selling less products over time in same size containers).

I make the argument that food companies are morally if not legally responsible for these behaviors. And from the perspective of a criminologist, some of these things could very well be made illegal and thus would actually be "food crimes." I do not argue to criminalize any behavior in this chapter (except food and possibly product shrinkage), but given that anything can be a crime (all is needed is a law to criminalize it), one day could bring about change in this area so that society cracks down on some of the behaviors discussed in this chapter.

None of this excuses the consumer, and so a section on consumer responsibility for these actions is also included. The chapter ends with a discussion of the responsible actions of companies, efforts to make food healthier and reduce the environmental harms associated with food production.

## Food Companies Are Culpable for Producing Excess Calories

It is food companies that are responsible for the fact that 4,000 calories per capita are produced per day (Nestle, 2018, p. 6) even though the average person only needs about 2,000 calories a day. Lorr (2020) notes that one of the main problems with the conventional food

DOI: 10.4324/9781003296454-8

system today is that we now have too much food, and he calls this a problem "entirely unprecedented in the history of humanity" (p. 6). And it is "Wall Street [that] expects publicly traded corporations to do more than make profits; it expects them to increase shareholder value every quarter." Pollan (2008, p. 122) notes: "A diet based on quantity rather than quality has ushered a new creature onto the world stage: the human being who manages to be both overfed and undernourished, two characteristics found in the same body in the long natural history of our species."

Given that "the most profitable products [by far] are highly processed 'junk foods,' and beverages, high in calories but low in nutritional value," what do you expect the companies will continue to produce, and in ever higher and higher quantities? It is important to note here that the problem is thus systemic in nature, for it is deeply rooted in capitalistic structures, suggesting that only systemic changes are likely to ever bring about a meaningful change to this reality. What this change looks like is hard to tell.

According to Herz (2018, p. 186),

if food companies would please put fewer pictures of tantalizing chips and biscuits on their package designs we would go through their wares less rapidly. However, this would translate into a slower rate of replenishing our stash and less money spent over time, which is not what food companies are in business for.

This example, though, shows you what we're up against.

Walker (2019, p. 94) notes that "businesses have grown fond of saying that their duty is to shareholders, which is a clever way of rejecting responsibility for the consequences of single-minded self-interest." A danger of the profit-motive when it comes to food is identified by Walker (2019), who claims that money has replaced memory, or lessons learned from the past, so that past mistakes of environmental damage caused by the food industry are likely to be repeated going forward. This is to say, as food companies insist on growing more and more over time, we can expect the harmful outcomes of their activities to grow as well.

Walker (2019, p. 29) claims: "Our food system trains us to want more in the way of *volume*, while offering less in terms of *nutrition*." Walker then says:

When always wanting more dominates our outlook, always *having* more becomes the solution for never seeing the value in living with less. The more that food is valued based on abundance, the less food can remind us of our reliance on other species, nature, and the environment.

*(p. 59)*

So, the overproduction of food leads us to lose touch with where food comes from and to understand its implications for where it comes from. Yet: "We have built an infrastructure and a lifestyle that presume that finite resources are always infinitely available" (p. 81), even though this is not the case.

All of this has happened over just the past several decades. Moss (2021, p. xxv) points out that, it has only been in the past 40 years, that we have such dramatic impacts on human health. He asks what happened and answers "The food is what happened." Whereas we are and have always been biologically driven to eat, "the companies changed the food."

Specifically, "the processed food companies have wielded salt, sugar, and fat not just in pursuing profits through the cheapest means of production" but also "to reach the primeval zones of our brain where we act by instinct rather than rationalization." Material presented in Chapters 4 and 5 of this book suggests this is true. Indeed, Americans now get about 75% of their calories from groceries that they do not need to cook but rather are ready to eat upon purchase or can merely be heated (Moss, 2021). These products tend not to be particularly healthy.

Consider grocery stores as one example of what the problem is when it comes to excess calorie production. According to Lorr (2020), Americans spent more than $700 billion in 38,000 different grocery stores in 2018. He estimates we spend about 2% of our lives inside these stores. Nestle (2013, p. 135) notes what we will inevitably see in these stores:

> aisles filled with food products laden with sugars, saturated fats, salt, and excess calories, all heavily promoted and highly profitable for makers and sellers . . . entire aisles are devoted to sodas, snack foods, cookies, candies, an sugary breakfast cereals . . . made with inexpensive ingredients, advertised with enormous budgets, and manufactured by some of the largest food corporations in the world.

So, it is not just the amount of calories being produced, it is also the nature of it.

Interestingly, the USDA has a category of foods called "foods of minimal nutritional value." These are shown in Table 8.1. This list needs to be greatly expanded based on what is being produced by the conventional food system now!

Fuhrman (2017, p. 1) notes: "Processed, fake, and fast foods have become the primary source of calories in this country, and they're on track to become the same in other countries." He argues these "Frankenfoods"—"unnatural, human-made, processed fast foods"—are addictive, marketed to us with the intent of getting us hooked, and the result is a genocide, "as these foods destroy life with frightening efficiency and this damage is worsening." The issue of food addiction is addressed in Chapter 5.

The defense of processed foods goes something like this: They give us what we want: "easy-to-eat foods that require little preparation and taste better than anything you could make yourself" (Nestle, 2013, p. 136). Further, the products are filled with ingredients to keep the foods fresher for longer so that the manufacturers can buy low-priced ingredients and sell products later at whatever prices they want. The reality of processed foods means that fruits and vegetables will never be part of the realm of processed products because they do not work within the constraints of the system (Moss, 2013, p. 201). In this way, unhealthy food can be seen as structural in nature—the system is not designed to provide it.

We must insist that food companies stop producing so much in terms of ultra-processed junk foods and insist that they produce more quality food. Unless and until they do, food companies remain culpable for the harms that result. Here, companies are acting reckless or with implied malice, making them morally if not legally responsible for harms associated with the production of so many especially high calorie foods.

### Food Companies Are Culpable for Putting Food in Non-food Environments

Surely you have noticed that when you go to hardware stores, bookstores, and other non-food places of business, you are often surrounded by food, at least at the cash registers

**TABLE 8.1**  Foods of Minimal Nutritional Value (USDA)

The following is taken from Appendix B of 7 CFR Part 210.

Appendix B to Part 210—Categories of Foods of Minimal Nutritional Value

(a) Foods of minimal nutritional value—Foods of minimal nutritional value are:

Soda Water—A class of beverages made by absorbing carbon dioxide in potable water. The amount of carbon dioxide used is not less than that which will be absorbed by the beverage at a pressure of one atmosphere and at a temperature of 60°F. It either contains no alcohol or only such alcohol, not in excess of 0.5% by weight of the finished beverage, as is contributed by the flavoring ingredient used. No product shall be excluded from this definition because it contains artificial sweeteners or discrete nutrients added to the food such as vitamins, minerals, and proteins.

Water Ices—As defined by 21 CFR 135.160 Food and Drug Administration Regulations except that water ices which contain fruit or fruit juices are not included in this definition.

Chewing Gum—Flavored products from natural or synthetic gums and other ingredients which form an insoluble mass for chewing.

Certain Candies—Processed foods made predominantly from sweeteners or artificial sweeteners with a variety of minor ingredients which characterize the following types:

Hard Candy—A product made predominantly from sugar (sucrose) and corn syrup which may be flavored and colored, is characterized by a hard, brittle texture, and includes such items as sour balls, fruit balls, candy sticks, lollipops, starlight mints, after dinner mints, sugar wafers, rock candy, cinnamon candies, breath mints, jaw breakers, and cough drops.

Jellies and Gums—A mixture of carbohydrates which are combined to form a stable gelatinous system of jelly-like character, and are generally flavored and colored, and include gum drops, jelly beans, and jellied and fruit-flavored slices.

Marshmallow Candies—An aerated confection composed as sugar, corn syrup, invert sugar, 20% water, and gelatin or egg white to which flavors and colors may be added.

Fondant—A product consisting of microscopic-sized sugar crystals which are separated by thin film of sugar and/or invert sugar in solution such as candy corn, soft mints.

Licorice—A product made predominantly from sugar and corn syrup which is flavored with an extract made from the licorice root.

Spun Candy—A product that is made from sugar that has been boiled at high temperature and spun at a high speed in a special machine.

Candy Coated Popcorn—Popcorn which is coated with a mixture made predominantly from sugar and corn syrup.

*Source:* https://www.fns.usda.gov/cn/foods-minimal-nutritional-value

when you check out. You'll notice, for example, that "grocery stores strategically place their high-profit low-nutrient food near the entrance and checkout line, thereby enticing shoppers" who might not have even come to the store for such products (Packer & Guthman, 2019, p. 239). Surely, this raises the issue of culpability in grocery stores.

This reality was discussed in Chapter 7 of this book. We can refer to the insertion of food products into non-food environments as "penetration." Figure 8.1 illustrates an example of this penetration. The image is of a local Walmart at the cash register. Notice the wide variety of "food" products as you check out, plus the Subway sandwich shop behind it.

Winson and Choi (2019, p. 48) note "the intensive penetration of industrial edible products and fast foods into *new* institutional domains, including public-sector institutions such as schools, airports, and hospitals, as well as nontraditional private institutions, such as gas stations and even banks." Winson and Choi (2019) explain that this penetration has occurred via three principal vectors: (10 Multinational snack food corporations; (2)

**FIGURE 8.1**  Penetration of Food Into Non-food Environments

*Source:* Photo by author

American fast-food corporations; and (3) supermarket and convenience chain stores. They show that each has invaded much of the world to grow in size and stature; that each offers unhealthy foods is evidence of the negative impact of industrialized food not only in the United States but also around the world.

Winson and Choi (2019, p. 54) discuss penetration under the concept of "spatial colonization," defined as "a way to ensure that a processed product is *available and visible* in as wide a variety of food environments (e.g., supermarkets and fast-food outlets) as possible, in addition to making products widely recognized through advertising." Penetration also includes the placement of fast-food outlets in spatial areas:

> The leading fast-food companies have come to dominate food environments by securing the most desirable high-traffic locations in new and existing suburban and urban areas, to the point where they have essentially saturated the market with their outlets in many developed countries.

They note that McDonald's has over 36,000 outlets in 199 countries! Wilson and Choi claim that spatial colonization is "a mechanism by which nutrient-poor edible products—pseudofoods—are coming to dominate food environments."

Penetration is also seen in the realm of convenience stores. Moss (2013, pp. 114–115) claims that the "up and down the street" marketing strategy of convenience stores—"as in driving the delivery truck up and down the streets of a neighborhood, from one C-store to the next"—was a "direct result of the marketing strategies developed by Coke and Pepsi, along with the snack food manufacturers, like Frito-Lay and Hostess." If true, the large number of convenience stores in certain neighborhoods (including disproportionately poor neighborhoods) may possibly be the fault of soda companies!

Gottlieb and Joshi (2010, p. 51) quote former FDA commissioner David Kessler who charged the food industry with placing fat, sugar, salt, and such "on every corner" and "available 24/7." And he adds: "They've added the emotional gloss of advertising. Look at an ad; you'll love it, you'll want it. They've made food into entertainment. We're living, in fact, in a food carnival."

Stated simply, and hopefully it is obvious, it is food companies that are responsible for surrounding us not only with food, but junk food, nearly everywhere we go. At the very least, we can assert this reality assures the creation of urges to buy and consume that make maintaining our weight and health more difficult. Here, companies are operating with the intent to get us to buy more of their products, making them morally if not legally responsible for harms associated with the consumption of their especially unhealthy products.

### Food Companies Are Culpable for Funding Research to Create One-Sided Studies

Food companies fund studies to prove points about food quality and food safety that tend to tell only one side of the story. Before turning to that issue, it is important to first acknowledge this is but one means of achieving outcomes consistent with their objectives. The food industry, in fact, has a whole "playbook" of strategies it uses to get what it wants. The playbook involves the following approaches:

1. Cast doubt on the science
2. Fund research to produce desired results
3. Offer gifts and consulting arrangements
4. Use front groups
5. Promote self-regulation
6. Promote personal responsibility as the fundamental issue
7. Use the courts to challenge critics and unfavorable regulations.

*(Nestle, 2018, p. 14)*

Those with a vested interest in the status quo and with profit in mind related to selling basic foodstuffs use well-known tricks to deny reality and promote pseudoscience. Walker (2019, p. 117) offers the following examples, consistent with the aforementioned playbook:

1. Using the media to promote two viewpoints to create the illusion that there are two evenly divided camps on an issue
2. Creating the illusion that there is uncertainty about an issue
3. Dismissing findings as false without offering any contrary evidence
4. Creating faulty science through industry-funded institutions and organizations

5. Attacking scientists who disagree with industry-generated research
6. Making donations to public institutions.

Kimura (2019, p. 63) also outlines how food companies influence nutrition science by funding research, creating front groups, and using lobbyists to defeat or modify legislation. Nestle (2002) outlines the various ways food corporations "work the system," including by lobbying Congress and state legislatures, co-opting nutrition professionals, cozying up to powerful friends, disarming critics, filing lawsuits, and even engaging in illegal price fixing.

Now, the industry-funded studies are conducted to generate support for greater sales of food products. So, we can be confident that companies are acting with intent, making them morally if not legally responsible. Marion Nestle, beginning in 2015,

> found 166 (laboratory) studies that had been supported by the (food) industry, and discovered that they almost always worked out great for the companies. Whether the investigation was on soda, breakfast cereal, pork, or nuts, the findings were consistent: Only twelve—less than 10 percent—could be interpreted as being contrary to the funder's interests.
>
> *(Moss, 2021, p. 150)*

Moss gives us the following examples of some of their findings:

> To wit, kids who eat more candy are skinnier, said the research by a candy association. Mars, the maker of M&M's, identified compounds in chocolate that are good for our hearts. From Kellog's: Sugar cereal makes you smarter. And not to be outdone, the company Nestle funded a study that found that skipping lunch will rob you of vital nutrients, and it just happened to have a ready solution: Hot Pockets.

Well, all of the aforementioned corporations are, of course, corporations, and: "Corporations, devoted as they must be to maximizing shareholder value, tend to push the research agenda in the direction of studies likely to contribute to profits" (Nestle, 2018, p. 165). Nestle claims that "'capturing' nutrition scientists and practitioners is a well-established strategy for influencing dietary advice and public policy" (p. 6). Her argument is that food companies intentionally "distort research to focus on topics useful for product development or marketing, influence investigators to put favorable spins on equivocal results, and encourage nutrition professional to offer favorable opinion about sponsors' products or to remain silent about unfavorable effects" (pp. 7–8).

According to Glenna and Tobin (2019), more than half of all research funding in the United States in food and agriculture comes from private sources. In other countries like the United Kingdom, the figure is even higher. And much of the research into crop seeds comes from just three companies—Dupont Pioneer, Monsanto, and Syngenta. This research is often done in conjunction with agricultural universities, and according to Glenna and Tobin (2019, p. 106), "the goals of the research often shift because of the funding sources and collaborations." The primary sponsor of public research in the United States is the federal government, and the primary sponsor of private research is agricultural companies.

Walker (2019, p. 116) describes a recent increase in agricultural production and discusses the reasons for it:

> From 1948 to 2013, agricultural production increased 169 percent. The basic research that set up major agricultural innovations in breeding, nutrition, pest management, machinery, etc., was publicly funded. The low-hanging research fruit that followed—new hybrid seeds chemicals, vaccines, nutrition, farm equipment, etc.—was commandeered by the private sector. Why? Because it was profitable. In 2013, more than three-quarters of the $16.2 billion spent on food and agricultural research came from private sources.

The investment in research by companies such as General Foods, Kraft, and Quaker Oats started after World War II when these companies built research facilities and hired scientists as staffers: "Today, Nestle (in Switzerland) and Unilever (in the Netherlands) are unique in maintaining large research operations that still do basic research, although they too partner with researchers at universities" (Nestle, 2018, p. 31).

An example of the type of research that food companies want to promote is noted by Nestle (2018, pp. 6–7): "Late in 2017, the *Journal of the American Heart Association* published the results of a clinical trial concluding that incorporating dark chocolate and almonds in your diet may reduce your risk of coronary heart disease?" She asks, guess who paid for the study? And she reveals, it was the Hershey Company and the Almond Board of California! Seven of the authors were paid for their work, and two others were employees of the funders. Presumably, this is an example of marketing research, which

> has a decidedly different purpose: to create and sell products. Food companies have always funded research aimed at product development but are now even more interested in research to demonstrate the health benefits of their products or to discredit evidence to the contrary.
>
> *(pp. 34–35)*

A review of 168 studies by Nestle found that 156 produced results in the interests of their sponsors, and only 12 did not. The studies included those for products including artificial sweeteners, canned foods, cereals, chocolate, coffee, corn, dairy foods, dietary supplements, formula for infants, garlic, gum, lentils, nuts, orange juice, potatoes, sodas, soy, and sugar. Another review of every article published in the top 15 most cited nutrition journals in 2014 found that 14% disclosed support by food companies. Of those that did, 60% reported results favorable to sponsors, versus only 3% that reported unfavorable results.

Another example is: "The supplement industry funds many studies that demonstrate health benefits from taking one supplement or another, but studies funded independently usually do not-and sometimes suggest that taking nutrients in pill form can be harmful." Nestle (2018, p. 11) adds: "Despite this evidence, half of American adults take supplements in the belief that they compensate for poor diets." This is troubling given reports that many supplement companies sell products that do not even contain the supplements they report and others that reportedly contain dangerous materials (Aschwanden, 2021).

As another example of the potential ramifications of corporate-sponsored research, the artificial sweetener NutraSweet (which makes aspartame) funded 74 studies, and all of them concluded that the sweetener was safe. Yet, in 92 studies that were independently funded, 90% questioned its safety! Currently, the US FDA considers aspartame safe, although at least one study links it to cancer in rodents (Safety of Aspartame, 2023). Similarly, with the substance Olestra, intended for use in snack foods, 80% of studies suggesting the substance had utility and was safe were sponsored by the company that produced it, P&G, while 89% of studies expressing doubts were sponsored by non-industry groups. Olestra was removed from products after it was found to cause significant digestive issues in consumers.

And with studies of the impact of soda and SSBs on obesity: "Studies funded by industry were eight times more likely to produce favorable conclusions than those funded by nonindustry sources" (Nestle, 2018, p. 39). Further, "studies funded by food companies reported smaller effects than those funded by nonindustry sources." Another study on the same issue concluded: "The industry seems to be manipulating contemporary scientific processes to create controversy and advance their business interests at the expense of the public's health" (pp. 39–40). And yet another study of the same issue concluded that industry-funded research "hindered the pursuit of scientific truth about the health effects of SSBs, and may have harmed public health" (p. 40). Here, food companies seem to be operating as tobacco companies once did, intentionally creating flawed science to counter valid science which points out harms associated with certain food products. This is both intentional and reckless behavior.

According to Nestle (2018, p. 84), the FDA did not use to allow supplement companies (or food companies) to claim on their products that "they could prevent, mitigate, or treat a disease; only drugs do these things." But, when Congress passed the Nutrition Labeling and Education Act of 1990, it allowed health claims to be stated if they were supported by "substantial scientific agreement." Nestle notes:

> All industries making products of questionable health benefit exert influence by diligent adherence to strategies—collectively referred to as "the playbook"—first established to great effect by tobacco companies and recently described in detail as a set of political and legal tactics to influence policy and shape public perceptions and to obtain research that helps with such efforts.
>
> *(p. 13)*

Campbell (2022) claims that major food companies, owned by major tobacco companies until the 2000s, used the tobacco "playbook" to get kids hooked on their products. Brownell and Warner (2009) claim that the playbook used by big tobacco is the same playbook used by big food. This playbook "emphasized personal responsibility" with regard to eating as well as weight management, paid scientists "who delivered research that instilled doubt," criticized so-called " 'junk' science that found harms associated with smoking," made "self-regulatory pledges," lobbied lawmakers "with massive resources to stifle government action," introduced " 'safer' products," and manipulated as well as denied "both the addictive nature of their products and their marketing to children."

The playbook of the food industry, according to Brownell and Warner (2009), includes the following:

- Focus on personal responsibility as the cause of the nation's unhealthy diet
- Raise fears that government action usurps personal freedom

- Vilify critics with totalitarian languages, characterizing them as food police, leaders of a nanny state, and even "food fascists," and accuse them of desiring to strip people of their civil liberties
- Criticize studies that hurt the industry as "junk science"
- Emphasize physical activity over diet
- State there are no good or bad foods; hence no food or food type (soft drinks, fast foods, etc.) should be targeted for change
- Plant doubt when concerns are raised about the industry.

According to Gewin (2018): "A close look at food companies that were purchased by tobacco companies found a 'systematic transfer of people, knowledge, information, and technology from tobacco to the sugar and beverage companies.'" Campbell (2022) clarifies: "Executives in the two largest U.S.-based tobacco companies had developed colors and flavors as additives for cigarettes and used them to build major children's beverage product lines, including Hawaiian Punch, Kool-Aid, Tang, and Capri Sun." Glantz (2019) agrees, noting: "Tobacco conglomerates that used colors, favors and marketing techniques to entice children as future smokers transferred these same strategies to sweetened beverages when they both food and drinks companies starting in 1963."

Imagine, big tobacco-selling SSBs! Young people, who already consume more than 140 calories every day in SSBs, do not need any extra encouragement by corporate powers to drink even more of substances that (1) have no nutritional value and (2) are filled with harmful sugars. According to Nguyen et al. (2019, p. 1), "Sugar-sweetened beverages are a risk factor for obesity and cardiometabolic disease." The companies, of course, generally deny this.

According to Nguyen et al. (2019, p. 1): "R.J. Reynolds and Philip Morris, the two largest US based tobacco conglomerates, began acquiring soft drink brands in the 1960s and were instrumental in developing leading children's drink brands, including Hawaiian Punch, Kool-Aid, Capri-Sun, and Tang." Glantz (2019) writes:

> Tobacco giant R.J. Reynolds led the transition to sweetened beverages in 1963 when it purchased Hawaiian Punch from Pacific Hawaiian Products Company. . . . . The beverage previously had been promoted to adults as a cocktail mixer, but R.J. Reynolds sought to beef up the drink's "Punchy" mascot—a counterpart to the "Joe Camel" cartoon character the company used to promote cigarettes—and featured it on toys, schoolbook covers, comics, tumblers, clothing and TV commercials. Punchy became the "best salesman the beverage ever had," according to tobacco industry documents.

So, tobacco companies used cartoon characters to promote their products to kids and young adults and continued this same strategy with some of its "food" products.

Further, "the company conducted taste tests with children and mothers to evaluate sweetness, colors and flavors for Hawaiian Punch line extensions. The children's preferences were prioritized" (Glantz, 2019). This suggests the company not only targeted kids with unhealthy products but also used children in their research to figure out how best to get them hooked. By the way, it was also R.J. Reynolds who created the first juice box, in 1983! Juice boxes are essentially just sugar water.

R.J. Reynolds, strangely enough, saw itself not as a tobacco company but instead as being in the flavor business. In this way, it made sense for it to transfer over lessons learned from the tobacco business to the SSB business. Both businesses, however, should be viewed as what they are—promoting addictive drugs.

Similarly, Phillip Morris owned Kool-Aid via General Foods (acquired in 1985) and used the "Kool-Aid Man" mascot "with branded toys, including Barbie and Hot Wheels." It also "developed a children's Kool-Aid loyalty program, described as 'our version of the Marlboro Country Store,' a cigarette incentive program" (Glantz, 2019). By the year 2004, Phillip Morris "had developed at least 36 child-tests flavors to its Kool-Aid line, of which some—like the 'Great Bluedini'—integrated colors with cartoon characters." The company also purchased Capri Sun and Tang, "and used similar child-focused integrated marketing strategies to drive those sales." Capri Sun would ultimately be teamed up with the Lunchables food product to be sold to kids and their parents (Nguyen et al., 2019).

Philip Morris would also rigorously test its products on kids. The goal was to discover which flavors, colors, cartoons, and product placements would have the biggest impact on purchasing decisions. In addition to the branded toys noted earlier, Philip Morris would get its famous Kool-Aid man in Marvel comics and on Nickelodeon TV, targeting 2- to 11-year-olds (Nguyen et al., 2019)!

Phillip Morris, which would go on to change its name to Altria after a lot of bad press associated with being part of Big Tobacco, "had substantial exposure to the food industry in the late 1980s and 1990s" (Caplinger, 2016). In 1985, it purchased General Foods. In 1988, it bought Kraft Foods, "combining the company a year later with its other food unit, General Foods Corp. to form Kraft General Foods" (Valenti, 2001). In 1985, Reynolds American bought Nabisco, but it would sell that to Altria in 2000. Ultimately, these companies would become public, independent companies.

According to Nguyen et al. (2019, p. 3):

> Both Reynolds and Philip Morris used cartoon mascots, child sized packaging technologies, and advertising messages found to appeal to children's desire for autonomy, play, and novelty. . . . Marketing campaigns used cartoon characters that appealed to children's aspirations, an approach also used to create brand loyalty to cigarettes.

Further, these tobacco companies "promoted their drinks using integrated marketing strategies that had been originally designed to sell cigarettes, surrounding children with consistent product messages in the home, store, school, sports stadium, and theme park." So, it was tobacco companies, engaged in tobacco company behavior, selling addictive sugary products to kids, just as they targeted kids with addictive cigarettes for decades before this.

The Center for Science in the Public Interest (2022) discusses the implications of these realities, noting:

> RJR Nabisco . . . once simultaneously contained the companies that made Camel cigarettes and Chips Ahoy! cookies. Until the mid-2000s, the companies that manufacture Marlboro and Virginia Slims cigarettes were part of the same conglomerate, Philip Morris (now Altria), which manufactured Kraft Macaroni & Cheese and Kool Aid. Those companies have since split their tobacco businesses from their food businesses, but heavy-handed product marketing may be ingrained in the companies' DNA.

In the context of these sugary beverages, we see efforts of the major food companies to self-regulate their behavior. Glantz (2019), for example, writes:

> Most sweetened beverage manufacturers claim to limit marketing to children of unhealthy foods and drinks. The industry launched both the Children's Advertising Review Unit, to promote responsible advertising to children through industry self-policing, and the Children's Food and Beverage Advertising initiative, which states that it devotes 100 percent of "child-directed advertising to better-for-you foods."

It should be clear that large food and beverage companies continue to target children, in spite of such initiatives.

We also see denials of any wrongdoing. The Center for Science in the Public Interest (CSPI) (2022) provides examples of claims made by industry front groups and compares them to those made by tobacco companies in the past. For example, in 2012, the ABA claimed that "sugar-sweetened beverages are not driving obesity." In 1954, the Tobacco Industry Research Committee stated "The products we make are not injurious to health." CSPI claims: "They were lying then. They are lying now. Soda and other sugary drinks are the top source of calories in the U.S., and the only food directly linked to obesity." CSPIA writes:

> When it comes to corporate responsibility, executives at some of the nation's largest food and beverage companies seem to have learned a lot from their counterparts at Big Tobacco in aggressively promoting consumption of unhealthy foods, and, in the same breath, blaming the consumer.

CSPI claims: "For years, the tobacco industry told parents it was their fault their kids smoked, all the while aggressively marketing cigarettes to youth. The food industry does the same thing."

CSPI provides the following examples:

- "We provide many choices that fit with the balanced, active lifestyle. It is up to [kids] to choose and their parents to choose, and it is their responsibility to do so" (Jim Skinner, McDonald's CEO, 2011).
- "It is the responsibility of every parent to encourage their children to make proper choices about lifestyle decisions. It is not the role of the federal government to mandate how children ought to behave."

*(R.J. Reynolds Tobacco Co., 1996)*

CSPI continues: "Just because parents share in the responsibility of keeping children healthy doesn't give food and beverage companies the green light to aggressively market products to kids that promote diabetes and obesity." Indeed, there is moral and even potential legal culpability for the companies that do this.

One reason all this matters is that some foods, like tobacco, are addictive, as shown in Chapter 5. Gearhardt (2021), for example, addresses the issue of ultra-processed foods—"industrial formulations that are typically high in added fat, refined carbohydrates or both"—and shows how they meet the same conception of addiction as that of tobacco.

Specifically, they "alter mood in a similar way by increasing pleasurable feelings and reducing negative ones." They "are highly reinforcing—they can shape your behavior to keep you coming back for more." They "can trigger strong, often irresistible urges to use despite a desire to quit." Addictive foods included are things like "chocolate, ice cream, French fries, pizza and cookies." Chapter 5 in this book verifies this conclusion.

Gearhardt (2021) adds that unhealthy diets are also dangerous, like tobacco, more dangerous, in fact, than alcohol, tobacco, and illicit drug use combined. And she notes that "Philip Morris and RJ Reynolds took their scientific, marketing and industrial knowledge in designing and selling addictive, highly profitable tobacco products and applied it to their ultra-processed food portfolios." Further, they deliberately worked "to enhance 'craveability' and create 'heavy users,' " just like big tobacco.

But the food industry, unlike big tobacco, has a wide variety of trade organizations working on its behalf. They include those that represent types of foods (e.g., Snack Food Association and ABA), a certain part of the industry (e.g., NRA), some constituent of food (e.g., Sugar Association and Corn Refiners Association), or even the entire industry (e.g., Grocery Manufacturers of America). Each of these groups lobbies lawmakers to protect the industry, advertises products directly to consumers, and generally uses the playbook to resist any needed changes to unhealthy food products. One could argue that these groups are at least partially culpable for the products they represent.

Perhaps not surprisingly, the National Cattlemen's Beef Board sought out research to spell out the health benefits of eating beef, even though population studies show that meat is a risk factor for cancer: "People who eat the most meat display about a 20 percent higher risk of colon and rectal cancers, but they also seem to be at higher risk for cancers of the esophagus, liver, lunch, and pancreas" (Nestle, 2018, p. 63). The WHO classifies red meat as "probably carcinogenic to humans" and processed meats as "carcinogenic to humans." Yet, incredibly, the North American Meat Institute claimed in a study that children who ate lunches with processed meats have healthier diets!

Nestle (2018, p. 64) states it this way: "Science funded by the meat industry argues that meat is nutritious, necessary, and safe. Independently funded scientists advise eating less meat. Take your pick." Ask yourself, which side is benefitting financially from its claims? Also consider the dairy industry checkoffs, the National Dairy Promotion and Research Board, and the Fluid Milk Processors Promotion Program. Remember the "Got Milk?" campaign? That's them. At the time of this writing, the official website for the campaign, www.gotmilk.com, says this about dairy milk, which they refer to as "real milk": "Milk does more than build strong bones. Milk's many vitamins and minerals benefit your overall health, inside and out. From disease prevention to muscle building, milk is an essential part of helping every body thrive."

By the way, USDA "checkoff programs" are used for beer, pork, poultry, and other foods, to raise funds for groups such as the National Cattleman's Beef Association. The money is used by the industry to encourage you to eat more meat. According to Linnekin (2016, p. 81): "USDA daily marketing orders set minimum dairy prices, which the checkoff program takes money from dairy farmers to promote milk and other dairy products." Tax money even goes directly to the industry to buy up surplus meat.

Critics claim that dairy checkoff funds are being used to promote junk foods such as flavored milk, frozen desserts, and cheeses. To the degree this is true, change is certainly needed. And of course, much research into dairy is being paid for by the industry. The National

Dairy Council is the marketing arm of the dairy checkoff programs and regularly promotes studies that supposedly demonstrate the health benefits of dairy products such as milk. They make claims such as dairy reduces the risk of conditions including coronary heart disease, stroke, type 2 diabetes, and much more. Further: "A consortium of dairy trade associations from Canada, the Netherlands, and Denmark sponsored a study concluding that dairy fats have not effect on 'a large array of cardiometabolic variables'" (Nestle, 2018, p. 69). Yet, Nestle points out that not all studies funded by the industry find the health benefits of their products. And independently funded studies sometimes find negative health outcomes associated with the consumption of dairy products.

According to Nestle (2018, p. 63):

The explicit purpose of checkoff programs is to increase demand for commodity agricultural products. Producers pay fees per weight of product: these fees go into a common fund distributed to national and state programs. The US Department of Agriculture (USDA) oversees and administers the programs, sets guidelines, approves board members, and monitors advertising, budgets, and contracts—as well as the research. The checkoff boards reimburse the USDA for the expense it occurs. In theory, the boards advertise, educate, and do research; they are not supposed to lobby. In practice, the lines sometimes blur.

Checkoffs include the American Lamb Board, Cattlemen's Beef Board, and National Pork Board. So, consider the reality that the organization (USDA) that is responsible for advising Americans on what to eat is basically in bed with the meat industry! Here is a serious problem begging for reform.

By the way, Nestle (2018, p. 80) also identifies the USDA marketing programs for fruits, vegetables, and nuts. They include the Hass Avocado Board, Mushroom Council, National Mango Board, National Peanut Board, National Potato Promotion Board, National Processed Raspberry Council, National Watermelon Promotion Board, Popcorn Board, United States Potato Board, and US Highbush Blueberry Council. And there are marketing orders for almonds, apricots, avocados, sweet cherries, tart cherries, citrus, cranberries, grapes, hazelnuts, kiwi, olives, onions, pears, pecans, pistachios, plums, prunes, potatoes, raisins, spearmint oil, tomatoes, and walnuts. Isn't it nice to know that some healthy foods are being promoted by the government? By the way, "marketing orders" are used by the USDA to "restrict the supply of a designated agricultural product in order to make the product more expensive" (Linnekin, 2016, p. 76). Linnekin provides an example of how this works with the USDA's Raisin Administrative Committee which determines how many raisins are available for sale each year, based in part on how many they seize in order to use for various purposes.

Nestle (2018) examines evidence that the USDA is very closely aligned with the food industry, and of course, they are given how closely they work together even in promoting certain food products. They also engage in cooperative studies, typically but not always, finding evidence in support of further sales of products. Examples include how avocadoes improve dietary intake and weight (sponsored by the Hass Avocado Board), blueberries improve cognition in older adults (sponsored by the US Highbush Blueberry Council), pears improve dietary quality and reduce obesity (sponsored by Pears Bureau Northwest), rice improves nutrient intake (sponsored by the USA Rice Federation), and farmed salmon reduces cardiovascular risk (sponsored by Cooke Aquaculture, Canada).

Nestle (2018, p. 88) provides the following many more examples of industry-funded studies that have found positive health claims: almonds reduce body fat and blood pressure (sponsored by the Almond Board of California); avocados improve cognitive health (sponsored by the Hass Avocado Board); bananas improve metabolic recovery from exercise (sponsored by Dole Foods); cashews decrease blood cholesterol (sponsored by Kraft Heinz); cranberries reduce urinary tract infections (sponsored by Ocean Spray Cranberries); aged garlic improves the immune system and reduces the severity of colds and the flu (sponsored by Wakunaga of America); concord grapes improve cognitive function and driving ability (sponsored by Welch Foods); mangos improve the microbiome and tolerance to high-fat diets (sponsored by the National Mango Board); peanuts improve metabolic and blood-vessel function (sponsored by The Peanut Institute); raisins improve blood sugar and blood pressure (sponsored by the California Marketing Raisin Board); raspberries reduce the risk of chronic disease (sponsored by the National Processed Raspberry Council); soy snacks improve satiety, diet quality, as well as mood and cognition (sponsored by DuPont Nutrition & Health); canned vegetables and fruits improve nutrient intake and diet quality (sponsored by the Canned Food Alliance); walnuts improve diet quality, blood cholesterol, and blood vessels (sponsored by the California Walnut Commission); and whole grains improve nutrient intakes and body weight (sponsored by General Mills). If nothing else, these examples show the incredible variety of food groups involved in conducting and promoting research to consumers. Nestle also notes that journals tend to be supportive of the industry, as well, if only for the reason that journals do not like to publish studies that produce negative results (Nestle, 2018, p. 161).

Who does not like candy, at least on occasion? According to Nestle (2018, p. 51), it is at least a $35 billion industry. Of course, there is literally no health benefit of eating candy, and eating too much will ultimately lead to bad outcomes; this is beyond dispute. Yet, studies funded by the NCA found that eating candy does not negatively affect the health or weight of children who eat it. In fact, they even claimed that eating candy was associated with weight loss! Such outlandish claims make candy companies at least partially culpable for negative outcomes associated with candy consumption.

Congress has taken action to try to counteract the impact of corporate-funded research of science, at least on paper. Nestle (2018, p. 25) explains:

In 1995, it established standards to ensure that research design, conduct, and reporting would not be biased by conflicting interests resulting from financial ties to corporations. Congress required recipients of research grants to disclose such ties, which it specified as salaries, consulting fees, honoria, stocks, patents, copyrights, or royalties.

It even specified that researchers must, if necessary, "modify their research plans, end their participation, of divest their conflicted holdings." All research was included, even nutrition research.

Yet, Nestle (2018, p. 43) claims that funded research by food companies "has an especially high probability of bias." First, she says that "food companies have no reason to look for unfavorable effects of their products and much prefer studies that allow them to adjust product dosages to increase the probability of finding benefits or of keeping adverse effects below statistical significance." Second, it is also possible that funding "can influence investigators to overlook unfavorable data, downplay negative results, or avoid publishing them out of reluctance to displease a sponsor."

Of course, there is a legitimate place for food research, or food science. Nestle (2018, p. 218–219) clarifies, writing

> when companies want research to make health claims for their products, they are distorting the research agenda. There is a big difference between "Let's fund a study to prove our product is healthy" (marketing) and "Let's fund a study to learn about whether or how diets might affect health" (science).

Elsewhere, Nestle (2018, p. 143) writes:

> The goal of nutrition education is to help the public choose foods and meals that promote health, but the goal of food (and restaurant) companies is to sell as much of the most profitable items as possible. Because these aims are often contradictory, they make nutrition education inherently incompatible with the goals of food companies.

According to Nestle, credit goes to Unilever, which "has an explicit policy on research integrity: its science must be driven by hypothesis and be rigorous, objective, and transparent." Nestle "insists that its partnerships adhere to principles of academic freedom, ethics, and integrity." And even Coca-Cola "now limits its funding to 50 percent of a study's cost" (p. 219).

## Food Companies Are Culpable for Using Front Groups to Confuse Consumers

Numerous "front groups" exist to represent, present research for, make claims for, and file lawsuits to protect food corporations. Companies form scientific front groups ("groups that do the dirty work that image-conscious companies shy away from, and that generally do not reveal their funding sources or are otherwise deceptive about their motives," buying off or paying for the work of health experts, and making backroom deals) (Simon, 2006, p. xiii). In the following, I provide several examples. The goals of these efforts, according to Simon (2006, pp. 191–192), are to

1. Cast doubt on findings that might threaten financial interests
2. Position food makers' own (biased) contributions to the scientific debate as legitimate and authoritative
3. Co-opt experts who would otherwise be critical of business practices
4. Ensure, ultimately, that people continue to consume food companies' unhealthy products.

### *Center for Consumer Freedom*

The Center for Consumer Freedom (CCF) is a "lobby group for the restaurant, alcohol, and tobacco industries" (Minger, 2013, p. 66). Its website says it

> is supported by restaurants, food companies and thousands of individual consumers. From farm to fork, from urban to rural, our friends and supporters include businesses, their employees, and their customers. The Center is a nonprofit 501(c)(3) organization. We file regular statements with the Internal Revenue Service, which are open to public inspection. Many of the companies and individuals who support the Center financially

have indicated that they want anonymity as contributors. They are reasonably apprehensive about privacy and safety in light of the violence and other forms of aggression some activists have adopted as a 'game plan' to impose their views, so we respect their wishes.

Nestle (2018, pp. 48–49) notes that the CCF is "infamous for [its] aggressive public relations campaigns on behalf of clients whose identities they keep deeply secret, never revealed." It also funded one researcher to the tune of $10 million over four years for studies showing no link between HFCS and illness. It also paid him $41,000 a month to write editorials to confront claims about risks associated with HFCS. Clearly, I am in the wrong business!

CCF is "the food industry's master front group and spin-maker," whose mission is to promote personal responsibility and protect consumer choices" rather than to, say, assure healthy food for its customers (Simon, 2006, p. 23). CCF says: "Consumer freedom is the right of adults and parents to choose what they eat, drink, and how they enjoy themselves. Defending enjoyment is what we're all about." Never mind if those choices and that enjoyment is ultimately unhealthy for the consumer. CCF's strategy includes "1) lobbying against nutrition legislation unfriendly to industry interests, 2) preparing well-timed press releases, 3) publishing op-ed articles and letters to the editor, and 4) advertising its views in print and electronic media" (p. 173).

CCF once mocked government efforts to restrict access to unhealthy foods, using pictured ads saying "You are too stupid" to make your own food choices. Consider what this would look like if we allowed drug dealers to advertise such slogans, asserting that the government should get out of the way of users making their own consumption choices!

The website for CCF says it "is a nonprofit organization devoted to promoting personal responsibility and protecting consumer choices. We believe that the consumer is King. And Queen." It goes on to essentially mock those aiming for healthier food options: "A growing cabal of activists has meddled in Americans' lives in recent years. They include self-anointed "food police," health campaigners, trial lawyers, personal-finance do-gooders, animal-rights misanthropes, and meddling bureaucrats."

> Their common denominator? They all claim to know "what's best for you." In reality, they're eroding our basic freedoms—the freedom to buy what we want, eat what we want, drink what we want, and raise our children as we see fit. When they push ordinary Americans around, we're here to push back.

They go on to complain what they mean by consumer choice: "Consumer freedom is the right of adults and parents to choose how they live their lives, what they eat and drink, how they manage their finances and insurance, and how they enjoy themselves."

Make no mistake: Those against consumer freedom have a broad agenda. Consider PETA, which wants to impose a vegan lifestyle on everyone. That means banning leather, fur, cashmere, down, silk, wool, meat, chicken, lobster, crabs, and fish—to name just a few products.

And they claim: "Activists are maneuvering to tax your favorite foods, throw the book at popular restaurants with tobacco-style lawsuits, limit the options you have for managing your money, and even criminally harass people with whom they disagree." The latter is a pretty startling claim, and they provide no evidence in support of it.

With regard to the issue of gluttony, they say bluntly that they do not support it: "Of course not! No one would endorse the 'Super Size Me' diet. Like all things in life, moderation is key. Eating a balanced diet and getting plenty of physical activity is crucial." Notice the allegiance to energy balance here! They go on to claim:

> Unfortunately, Americans have been force-fed a diet of bloated statistics hyping the problem of obesity. Those statistics have been used by Big Brother government bureaucrats and greedy trial lawyers to justify a host of noxious "solutions," like extra taxes on certain foods and lawsuits against anyone who grows, makes, or serves anything tasty.

There are and have been efforts to tax certain food or beverage choices, but then, there are also taxes on other harmful and addictive choices, such as alcohol and tobacco.

### Grocery Manufacturers Association

According to Hauter (2012, p. 46), the agenda of the Grocery Manufacturers Association (GMA) includes

> weakening federal pesticide and toxics laws; stopping the creation of a consumer protection agency; opposing liberal appointees to the FTC; promoting lax global trade and investment rules; allowing dangerous preservatives, additives, and colors to be used in food; weakening antitrust laws; advocating for food irradiation; stopping mandatory food labeling; warping nutrition standards; weakening food safety regulations . . . and promoting genetically engineered food.

GMA claims that it provides "a wide variety of nutritious foods and beverages" and helps "parents make the right choices for their families"—it's their "industry's top priority" (p. 25). GMA uses its enormous powers, with the backing of hundreds of companies, to encourage the US government to go easy on things like salt, sugar, and fat in its dietary guidelines. GMA is "on record as opposing virtually every state bill across the nation that would restrict the sale of junk food or soda in schools" (Simon, 2006, p. 223). Moss (2013, p. 221) notes, for example, this wording from the GMA about processed foods:

> We find that the Dietary Guidelines Advisory Committee report repeatedly suggests that Americans would benefit from consuming less processed foods. This supposition is not science-based, discounts the value of the U.S. food supply, and perpetuates a misguided belief that processed foods are inherently nutrient poor.

But, of course, that is exactly the case about processed foods! GMA promoted MyPyramid to the public after its release, not surprising given that it in no way attacks the processed food industry.

The group is now known as the Consumer Brands Association, and its website says its mission is: "We champion growth and innovation for the industry whose products consumers depend on every day." Among its programs, it created a SmartLabel program for smart devices that allow consumers access to information on thousands of consumer products, both food and beverage. It also runs "Facts Up Front (FUF), a program led by the Consumer Brands Association and the Food Marketing Institute, [which] is a simple and easy-to-use labeling system that displays key nutrition information on the front of food and beverage packages." Critics claim this program was an effort to shut down other efforts to provide more complete labeling information about products, including facts related to processing and ultra-processing of goods.

It also notes that it has its own PAC, which "provides the opportunity to advocate for the work of millions, champion the interests of consumers and shape the future of consumer packaged goods." This allows the organization to directly influence laws related to nutrition and regulation of it in issues such as food safety. PACs, in operation since the 1940s, are groups organized to raise and spend money on political campaigns. PACs typically represent corporate, labor, and ideological interests, and are allowed to give $5,000 to a candidate committee per election (primary, general, or special), up to $15,000 annually to any national party committee, and $5,000 annually to any other PAC. They may also receive up to $5,000 from any one individual, PAC, or party committee per calendar year (Center for Responsive Politics, 2022).

### National Restaurant Association

The NRA was created in 1919. According to its website, the NRA represents and is an advocate for more than 500,000 restaurants. It "strongly opposes providing nutrition information, and is determined to block access to the courtroom by consumers who might be harmed by eating its members' food" (Simon, 2006, p. xx). Moss (2021, p. 143) notes that the group "stepped up its game by creating a political action committee [PAC] to heighten the effectiveness of campaign donations by bundling money from its members."

You get a sense of the potential power of the NRA when you consider the numbers they bring to the table when meeting with government officials: 15 million jobs provided, $850 billion in annual sales, and roughly 170 million people served each day! The NRA made an attempt in 2003 to get Congress to act to protect restaurants from tort attorneys seeking to get damages for harms caused by processed foods. According to Moss, Congress did not see enough of a threat to take action. So the NRA took its case to the states, through lobbying at state legislatures to pass bills to achieve the same objective. A model bill—the Commonsense Consumption Act—"barred anyone from bringing a lawsuit that sought to win personal injury damages on the claim that the food they ate caused them to lose control of their eating" (the exact thing shown happens in a rich body of scientific evidence) (p. 146). At least 26 states passed versions of the bill, dubbed the "cheeseburger bill," and: "The entire food industry, including the manufacturer of groceries, was immunized from any litigation that sought to hold it accountable for our troubles with processed food" (p. 148). Here is a clear example of an industry using its power to shield itself from responsibility for its culpable acts. That government goes along with it is actually pretty stunning.

On its website, the organization has information on healthy food choices for families and children, as well as a lot of information for businesses about food safety. With regard to healthier food choices, NRA says, on its website:

The Association is working with leading restaurant brands and independent scientific experts to promote delicious, better-for-you meal options for their youngest guests.

KLW menu options meet added sugar, sodium, fat, and calorie thresholds established by the latest nutrition science.

The website continues:

> Participating restaurants offer a minimum of 2 meals and 2 sides that have undergone rigorous dietary analysis by nutritionists. This ensures the meals align with the current nutrition science, including the 2020–2025 Dietary Guidelines for Americans, published by the U.S. Departments of Agriculture (USDA) and Health and Human Services (HHS) every five years.

"Participating restaurants also commit to a default beverage policy that puts water, milk, and juice first, with other beverages available on request." All of these are welcome changes, yet, it remains true that most restaurant meals remain largely unhealthy.

### International Life Sciences Institute

The ILSI says, on its website, that it is "a global, nonprofit federation dedicated to generating and advancing emerging science and groundbreaking research to ensure foods are safe, nutritious and sustainable, and that they improve planetary and human health and well-being in the 21st century." It claims:

> ILSI convenes scientists at the forefront of research on nutrition, food safety and sustainability, and operates within a framework of the highest principles of scientific integrity. ILSI's trusted experts and volunteers around the world work synergistically and transparently across academia and the public and private sectors.

ILSI lists the following as its operating principles.

- Science for the Public Good. All ILSI scientific activities have a primary public purpose and benefit.
- Collaboration. Scientists from geographically diverse regions of the world can best address complex science and health issues by sharing their unique skills, insights, and perspectives.
- Shared Values. ILSI believes scientists from industry, government, and academia and other sectors of society can and should work together to identify and address topics of common interest.
- Transparency. All ILSI's activities are conducted in an open and transparent manner and all scientific outcomes are made available to the public to ensure confidence in the integrity of the scientific process. The purpose and funding sources for all ILSI-sponsored meetings; symposia; conferences; seminars; and workshops are fully disclosed. All publications list funding sources and sponsors. Speakers and authors sign disclosures of financial and other interests related to the contents of their presentations and/or articles.
- Lobbying and Advocacy. ILSI does not lobby, conduct lobbying activities, or make policy recommendations.

Yet, according to Nestle (2018, p. 41), ILSI "seems to never miss an opportunity to defend the interests of its four hundred or so corporate sponsors." She asserts that "ILSI has a vested interest in defending the scientific quality of industry-funded research." If true, this means ILSI is vested in the status quo, and the status quo is remarkably unhealthy and dangerous.

### Alliance for a Healthier Generation

The Alliance for Healthier Generation is heavily supported by industry, including ConAgra, Tyson Foods, Rich Products, and Kraft Foods. Its website says it does

> work with schools, youth-serving organizations, businesses, and communities across the nation to transform the places kids spend their time into healthier environments. Through our evidence-based programming and innovative resources, we empower those who work closely with children to help them develop lifelong healthy habits. We support a holistic approach to health promotion. In addition to physical activity and healthy eating, we address multiple, critical child and adolescent health issues, including social and emotional health, sleep, and asthma.

Imagine a group, allegedly interested in promoting health, that represents companies that sell meat and ultra-processed foods! Still, its founders include the AHA and Clinton Foundation and large donors include BlueCross BlueShield of South Carolina Foundation, the Centers for Disease Control and Prevention (CDC), and others.

Gray (2019a, p. 19) shows that one of the Alliance's sponsors, Tyson Foods, is "the world's largest processor of chicken, beef, and pork, with more than $40 billion in revenue in 2015. Some of its products are sold under different labels such as Jimmy Dean. Hillshire Farm, and Ball Park." Tyson acquired Cobb in 1994, and it "has a dominant presence in most regions of the world, with bedding facilities and partnerships in Latin America (notably Brazil), northern and southern Africa (Egypt and South Africa), Asia (China, Japan, Korea, and Thailand), the EU, and Russia" (Silbergeld, 2016, p. 69)

Gray (2019b, p. 192) goes on to say:

> It is the face of corporate food and has a record of felony environmental violations, price-fixing, lying about antibiotic use in chickens, and animal abuse. Tyson managers were also indicted for colluding with immigration officials to smuggle workers into the United States to work at their factories; the company itself escaped charges.

This certainly makes the company relevant for the study of food crime!

### Food Marketing Institute

The Food Marketing Institute (FMI), according to its website, "works with and on behalf of the entire industry to advance a safer, healthier and more efficient consumer food supply chain." FMI "brings together a wide range of members across the value chain—from retailers who sell to consumers, to producers who supply the food, as well as the wide variety of companies providing critical services—to amplify the collective work of the industry." So, it is clearly an organization working on behalf of the industry. It even claims:

> FMI is a champion for the food industry and the issues that make a difference to our members' fundamental mission of feeding and enriching society. The reach and impact of our work is extensive, ultimately touching the lives of over 100 million households in the United States and representing an $800 billion industry with nearly 6 million employees.

Its website has information on a huge variety of topics from asset protection to biotechnology to food prices to tax reform and workplace issues. Like other organizations, this front group also has its own PAC, FoodPAC, that contributes hundreds of thousands of dollars in each election.

According to Hauter (2012, p. 47), FMI "has been at the forefront of lobbying to: deregulate trucking; pass NAFTA; stop consumer-friendly labeling; weaken antitrust laws; weaken child labor laws; prevent health-care reform; weaken organic standards; irradiate foods; and weaken labor standards and worker protections." So, like other industry front groups, FMI works to beyond food-related issues and is vested in maintaining the status quo.

### Other Groups

Other front groups include the Alliance for Food and Farming, American Council on Science and Health (ACHS), Animal Agriculture Alliance, International Food Information Council, and recently formed groups such as Alliance to Feed the Future, America's Farmers, Center for Food Integrity, Global Harvest, Protect the Harvest, and US Alliance for Farmers and Ranchers.

Some industry groups, such as General Mills, create their own groups to create and publish research. The group by General Mills is called the Bell Institute of Health and Nutrition. See? There is no hint in its name that it is affiliated with or works in the interests of a private food manufacturer. Similarly, Coca-Cola created its own Beverage Institute for Health & Wellness, an organization that made clear on its website that it sought to increase understanding of the role that beverages can play in both developed and developing countries, a sure sign of its intent to continue the expansion of its products across the globe. Then there is the ConAgra Foods Science Institute, Gatorade Sports Science Institute, and Nestle Nutrition Institute.

The American Council on Fitness and Nutrition (ACFN) was "formed in January 2003 by a coalition of food and beverage companies, trade associations, and nutrition advocates to work toward comprehensive and achievable solutions to the nation's obesity epidemic." Of course, nothing the group promotes or supports has anything to do with reducing people's exposure to unhealthy foods produced by its members. According to Simon (2006, p. xxi): "ACFN publishes industry-friendly articles in both the academic press and the general media, usually without revealing its corporate backing." Similarly, the ACHS has opposed even efforts to remove sugary sodas from schools, writing "banning specific foods won't stop kids from eating other foods, and it's their total calorie intake . . . that matters, not a few 'evil' foods." Fair enough, but of course, this claim ignores the fact that sugary sodas are high in calories and completely empty on nutrition.

Even groups like the Academy of Nutrition and Dietetics (formerly the American Dietetic Association, or ADA) are sponsored by many giant food companies, including General Mills, Kellog's, Mars, PepsiCo, and SoyJoy, as well as "official partners" including Coca-Cola, Hershey's, and the National Dairy Council (Minger, 2013). ADA took money from and partnered up with a major chocolate manufacturer to get out its messages out about diabetes and obesity (Simon, 2006)! Is this not like an anti-drug agency partnering up with a major drug dealer to deliver anti-drug messages to consumers?

## Food Companies Are Culpable for Advertising Unhealthy Products

Nestle (2002, p. xiii) notes: "Food companies will make and market any product that sells, regardless of its nutritional value or its effects on health." As such, one could argue that food companies at times act like psychopaths in that they operate without regard to the negative impacts of their behaviors on others. Leon and Ken (2019) argue that "Big Food" operates with a paradoxical business model, one that strives to sell more and more of its foods while simultaneously serving unhealthy foods that have serious negative health outcomes for their own consumers.

Importantly, according to Hatanaka (2019, p. 11): "Research finds that advertising affects not only the brands of food consumers buy, but also their eating patterns." And she notes: "Food and beverage companies spend $10 billion to $15 billion a year marketing their products to children and adolescents; the vast majority of commercials are dedicated to food products that are high in sugar, fat, and/or sodium," as shown earlier. Herz (2018, p. 258) agrees, writing: "The media imagery and messaging about food that bombards us daily has a tremendous impact on our eating behavior. This is why advertisers spend so much money trying to turn our attention toward their brands." She notes that more than $1 billion was spent on snack advertising in 2014 in the United States alone, and that about $14 billion is spent on food advertising around the globe per year: "This is more money devoted to publicity and promotion than in any other retail sector except automotive."

When you consider the potential impacts of advertising, please remember that food companies are: (1) In the business of making money; (2) spending money on advertising, which is a cost or loss of money; and (3) not going to spend money on things that do not work. Take these three obvious realities together and you can be reasonably certain that food companies have evidence that their advertising works! And, as profit-making enterprises, "food companies are hardly about to encourage us to make fewer purchases" (Herz, 2018, p. 253). So, this is one way in which food companies are culpable for overeating in America as well as around the world.

Cohen (2014, p. 204) claims that the food industry uses aggressive marketing campaigns to get us to try (and like) their foods and that these "undermine our health behaviors." Cohen argues these efforts by food companies are not in the interest of even their customers:

> It is neither normal nor natural for us to be confronted multiple times each day with images of and access to unlimited foods, especially candy, cookies, soda, pastries, fries, and chips. The environment is engineered on behalf of a food industry that is so concerned with profits that it will do anything to get us to eat as much as

we can. A food industry that cared about us would not be pushing sugar, salt, and salt. It would not manipulate us to eat too much with extra-large portions and processed foods.

*(p. 207)*

She states that we are being manipulated by corporate America to make bad choices. Government agencies are culpable here, as well, for failing to stop this.

The success of advertising, as well as branding, explains why people prefer McDonald's hamburgers and Coke when they are labeled even over the same exact products when they are unlabeled. Kids even say carrots taste better when served on a plate with the McDonald's label; healthy foods with "fun labels" also taste better, according to kids in different studies.

Children are particularly vulnerable to advertising, given that their brains are not fully developed, leading to impulsivity and vulnerability to persuasion. And Nestle (2002, p. 178) writes that "it is easy to understand why children of any age present an irresistible marketing opportunity and why food companies spare no effort to each them." And given what kids tend to buy, it is no surprise they are targets. "Nestle shows that some of the main products kids buy are candy, chewing gum, soft drinks, ice cream, salty snacks, fast food, and cookies," hence mostly unhealthy "food" items.

Herz (2018) notes that about nine out of ten television commercials seen by kids and adolescents featured savory or sweet processed snacks, versus less than 6% for fruits and nuts. Companies are culpable for these realities, as well. With regard to commercials, Nestle (2002, p. 180) notes that "children lack the critical facility to distinguish commercials from program content." And she points out that kids who watch commercials tend to eat more calories, something that is bad for them, especially considering the nature of the calories they tend to consume. Studies show that the more kids watch TV, the more vulnerable they are to ads, meaning the more they prefer the foods they see advertised. Studies also show that kids who see food ads tend to eat more while viewing the ads: "food ads cue us to eat. And in the case of child-targeted ads it's all about fun with high-calories snacks, which propels desire in just one direction: 'I want some'" (Herz, 2018, p. 271). Some countries, including Sweden, Norway, Canada, and the United Kingdom, have banned or severely restricted food advertising to children, so this means it is a possibility for the United States, though the likelihood is small given the power of major food companies in the country and the unwillingness of government to do anything about it.

According to Moss (2021), kids are more vulnerable to marketing than adults also because external factors in their lives play a larger role in their decision-making than does internal judgment. Part of this owes to the fact that kids still watch a lot of TV, where they are exposed to a huge amount of advertisements. Hauter (2012, p. 305) claims that

Children ages two to seven see and average of twelve food ads a day—a total of 4,400 food ads annually, or nearly thirty hours of food advertising a year. Children ages eight to twelve see an average of twenty-one food ads a day—7,600 food ads, or over fifty hours of food advertising a year, And teenagers ages thirteen to seventeen see and average of seventeen food ads a day—more than 6,000 food ads, or over forty hours of food advertising a year.

Moss (2021, p. 74) notes that "kids ages two to eleven are still watching three hour and nineteen minutes *a day*, and during that time they'll see twenty-three ads for food that is high in sugar and fat." Moss uses the company Kraft to give an example. He writes:

The technologists at Kraft, like those at other companies, spent their careers pursuing kids, teens, adults—anyone who might eat their products. They didn't see this as nefarious. Their job was to maximize the appeal of their goods. In the labs, that meant working to hit the optimum *bliss point* for sweetness, the *mouthfeel* for fat, and the *flavor burst* of salt, as this chemistry was known in the industry. They engineered colors, textures, and smells to enhance the allure.

Then, they "were joined on the marketing side by people with a deep appreciation for the role of psychology in purchase decisions. Kraft, after all, was a company, not a charity" (p. 130). And recall, Kraft got this approach from Big Tobacco.

Moss (2021, p. 134) adds:

Kraft was hardly alone in targeting our vulnerabilities. In its records was a 1998 memo created by Nabisco to discuss a consultant's take on going after kids. The upshot: Teens were no longer seen as a young enough mark. Tweens were the industry's new bull's-eye, given how they at this earlier age would establish what they liked for the rest of their lives.

Evidence showed that Nabisco talked to employees with Pepsi, Frito-Lay, Burger King, and McDonald's, and all of them agreed that taste preferences occur early in life and thus must be addressed in ways that result in favorable outcomes for the companies. They would target tweens in movies, music, games, websites, and sports.

Sound like a conspiracy, where more than one party enters into an agreement to take an action? Whether true or not, food companies are clearly culpable for targeting kids, as well as for featuring unhealthy products in their ads rather than healthy ones. Booth et al. (2019, p. 369) put the crime in marketing to kids: "There are surely no greater crimes than those directed to the most vulnerable. On that basis, the exploitation of children's gullibility and credulity through promotion of unhealthy food products can be seen as dastardly and indefensible."

One train of thought when it comes to the culpability of food companies for bad health outcomes associated with diet is that, since we are targeted by companies with certain foods since at least early childhood (and, truth be told, into the adolescent years, as they work hard to build brand loyalty), their efforts to get sugar in our bodies through their products will ultimately teach us what food should taste like— that is, sweet. Perhaps this is one reason product manufacturers put sugar in so many different types of food products. Consider advertisements for sugary cereals, for example. Moss (2013) discusses a study that found "the sweetest brands were found to be the ones most heavily marketed to kids during Saturday morning cartoons." Another nine-month survey of weekend daytime TV found 3,832 ads for mostly sugary cereals, 1,627 for candy and gum, 841 for cakes, piers, and other desserts but only four for unsweetened foods such as meat, fish, vegetable juice, and so on. And none of this includes an online activity. Companies such as Kraft and so many others feature games as well as cartoon characters for small children that directly promote sugary foods including cereals and cookies. According to Moss (2013, p. 241): "Kraft had directed every last shred of its talent and energy to making its products as enticing as possible. Central to this mission were the formulations of salt, sugar, and fat that made the products attractive." Beyond that, marketing to kids and parents plays a major role in how they make people try their products in the first place.

Soda companies such as Coke work to "identify the ways in which both teens and adults can be made more vulnerable to persuasion" (Moss, 2013). For Coke, it is through there Coca-Cola Retailing Council. Knowing that 60% of grocery purchases are "complete unplanned" (according to a study by Coke), the company works hard to get their products as close to cash registers as possible (Moss, 2013, p. 113).

All advertising is intentional behavior. Food companies are culpable for intentionally advertising their unhealthy products to consumers and especially to vulnerable children.

## Companies Are Culpable for Deceptive Advertising

Deceptive advertising is any advertisement of a product that has the potential to mislead. Many examples of deceptive advertising in foods are provided in the following. Before getting into the examples, it is important to point out that food companies know what they are doing here. Consider this example. We will eat more high-calories snack foods when they are labeled with things such as "low fat," "no trans fat," or "organic" (Nestle, 2013). Sometimes, lower sugar simply means the companies have removed some of the added sugars from their products to create the impression that the still unhealthy products are healthy. Likewise, the term "free" as in the case of calorie-free or sugar-free or fat-free, does not actually mean no calories or no sugar or no fat, just really low amounts. And the term "real," as is on fruit juice products, tricks consumers into believing they are getting just fruit juice, as opposed to added sugars and artificial ingredients in some cases.

Food manufacturers have a long history of using images and language related to fruit to mislead consumers, even in sugary drinks with no actual fruit (Moss, 2013). Consider the drink Kool Aid, basically sugar water, which would promote the fact that each bottle contained a splash of real fruit juice ("barely half a tablespoon of juice, a mere 5 percent of the total formula"), and the product Tang (also sugar water) adding two tablespoons of real fruit juice, "decorated . . . with pictures of fresh oranges and cherries" and rebranded "Tang Fruit Box" (Moss, 2013, pp. 125–127). Kool Aid went further and

> engineered to evoke the image of fresh fruit in as many ways as possible: They were made in a variety of imitation fruit flavors, including cherry, grape, orange, and tropical fruit, and they were given the most enticing imitation aromas that lab technicians could devise so that when the bottles were opened, they emitted powerful fruity smells. Even the bottles promulgated the mythology of health: Their plastic sides were embossed with the shapes of fruit.
>
> *(p. 128)*

Recall that these products were, for some time, controlled by tobacco companies. The companies that own them now merely continue the same industry-wide technique of making misleading claims about products.

Note that drinking fruit juice is generally not a good idea anyway, for the real value of fruit comes from eating it, including the fiber contained within. Moss (2013, p. 134) explains:

> Juice concentrate is made through an industrial process that is highly variable, including any or all of the following steps: peeling the fruit, thereby removing much of the beneficial fiber and vitamins; extracting the juice from the pulp, which loses even more of the fiber, removing the bitter compounds; adjusting the sweetness through varietal blending; and evaporating the water out of the juice. At its extreme, the process results in what is known within the industry as "stripped juice," which is basically pure sugar, almost entirely devoid of the fibers, flavors, aromas, and or any of the other attributes we associate with real fruit. In other words, the concentrate is reduced to just another form of sugar, with no nutritional benefit over table sugar or high-fructose corn syrup. Rather, its value lies in the healthy image of fruit that it retains.

Incredibly, when this process is used, product manufacturers can put the words "made with real fruit" right on the packaging, creating the illusion of health when literally none is to be found.

The company Capri Sun, purchased by Phillip Morris tobacco in 1991, was sweetened with high-fructose corn syrup and such fruit juice concentrate and was packaged with the claim "Natural fruit drink. No artificial ingredients." But then, what would you expect from a tobacco company? According to Moss (2013, p. 139), Kraft foods "had its own secret weapon, borrowed straight from Phillip Morris . . . they used targeted lists prepared by the tobacco company in selling cigarettes" to make targeted phone calls to get their products in stores. And many of its efforts, such as that associated with its well-known Lunchables product, were to highlight "the fun, the cool, and most of all, the feeling of power it brought to their lives" (p. 203), just like with tobacco targeting of children. But, in the case of food, not only did they target kids with the idea of fun, they also targeted moms, by letting them empower their kids to assemble their lunches in any way they want. A totally predictable outcome of feeding kids these kinds of foods, however, is a rise in type 2 diabetes as well as obesity (p. 206). But the companies do not care, as the costs are externalized, meaning, falling on the public. Food corporations remain culpable, however.

Another example comes from ice cream! The company Unilever, based on a study of only eight people who were put into an MRI for brain scans as they tasted ice cream, would claim "Ice Cream Makes You Happy—It's Official!" This ridiculous claim was based on the fact that people's pleasure centers lit up when eating ice cream.

Moss (2013, pp. 90–93) notes the following examples of cereal companies making obviously false claims about their products:

1. Kellogg's claimed in an ad campaign that Frost Mini-Wheats were a brain food and claimed that eating their cereal would help kids get an A for attentiveness.
2. Kellogg's claimed that Rice Krispies with added vitamins and antioxidants would bolster children's immunity from disease.

The claims of yogurt companies about the health benefits of their products really stand out. For example, there is the claim that natural probiotics in yogurt offer tangible health benefits. Nestle (2018, p. 70) shows that even though this claim really has stuck or taken hold in society, the evidence does not hold up to scrutiny. Yet:

> Yogurt-funded studies show that yogurt is associated with another wide array of benefits: reduced risk of metabolic syndrome, type 2 diabetes, weight gain, and obesity; higher bone density in older adults; and better digestibility by people intolerant to the lactose sugar in milk.

Meanwhile, at one point, chocolate milk companies made the claim that "milk alleviates symptoms of concussion in high school football players" (Nestle, 2018, p. 71). The study, announced by the University of Maryland, claimed that new, high-protein chocolate milk "helped high school football players improve their cognitive and motor function over the course of a season, even after experiencing concussions!" Amazingly, there was no published, peer-reviewed scientific article on which these claims were based. Nestle lays out all of the weaknesses of the study and notes that the sponsoring company in question "had provided support for salaries and research materials, and a Pennsylvania dairy company had contributed $200,000 to" the professor who conducted the study! This strikes me as highly unethical, to say the least.

Another example of a product that some allege to be misleading relates to claims of Gatorade being a hydration drink, when, in essence, it is mostly just sugar water (Simon, 2006, p. 110). And Kraft's Capri-Sun once claimed, right on the box, that its product hydrated better than water, a claim lacking in any scientific evidence other than kids are more likely to drink more of the sugary drink than they are water because it tastes better.

Nestle (2018, p. 76) discusses the case of Royal Hawaiian Macadamia Nut, Inc. which, in 2015, petitioned the FDA to be able to advertise the idea that daily consumption of macadamias may reduce the risk of heart disease, as part of a healthy diet. The FDA ruled it would permit this language:

> Supportive but not conclusive research shows that eating 1.5 ounces per day of macadamia nuts, as part of a diet low in saturated fat and cholesterol and not resulting in increased intake of saturated fat or calories may reduce the risk of coronary heart disease.

Then there is the Wild Blueberries of North America, who wants us to know that "Wild Blueberries have twice the antioxidant capacity per serving of regular blueberries." They claim: "A growing body of research is establishing Wild Blueberries as a potential ally to protect against disease such as cancer, heart disease, diabetes, and Alzheimer's" (Nestle, 2018, p. 77). Nestle shows that their subsequent marketing campaign, linking blueberries to health, was highly effective. But she points out that,

> they are a fruit like any other. Their antioxidants may counteract the damaging actions of oxidizing agents (free radicals) in the body, but studies of how well antioxidants protect against disease yield results that are annoyingly inconsistent. When tested, antioxidant supplements have not been shown to reduce disease risk and sometimes have been shown to cause harm.

She adds: "The US Center for Complementary and Integrative Health at NIH judges antioxidants as having no special benefits" (p. 78).

With regard to pecans, the industry obtained a federal marketing order (FMO from the USDA—a means of helping produce the product and other specialty crops):

> The FMO for pecans applies to pecan growers in fifteen states; it authorizes the pecan industry to collect data, recommend quality standards, regulate packages and containers, and conduct research and promotional activities. It also establishes an American Pecan Council to work with the USDA to decide how the funds will be used and to oversee these activities.
>
> *(Nestle, 2018, pp. 79–80)*

FMOs are essentially regional checkoff programs. Nestle acknowledges that nuts, as generally unprocessed whole foods, are generally considered healthy, and people who regularly eat them tend to be healthier. Yet, a poster at a national conference—the ASN—claimed that eating pecans we associated with "reductions in fasting insulin, glucose, blood lipids, systolic blood pressure, and inflammation, and longer lag time for LDL oxidation . . . although none of these changes were statistically significant" (Nestle, 2018, p. 81). Two things: (1) Not statistically significant means that any of these results could be due to random chance; and (2) the study was funded by the National Pecan Shellers Association.

Then there are the incredible claims by Pom Wonderful, a pomegranate company:

> Since at least 2001, POM Wonderful had been investing in studies designed to show that daily intake of eight ounces of pomegranate juice of a supplement of pomegranate polyphenol extract would produce higher levels of antioxidants in the body and would therefore reduce risks for cardiovascular disease, type 2 diabetes, prostate cancer, and erectile dysfunction. . . . It claimed these benefits in advertisements.
>
> *(Nestle, 2018, p. 83)*

One ad, reprinted in Nestle (2018, p. 84), claimed, in bold text: "Cheat death. POM Wonderful. The Antioxidant Superpower." The ad appeared the day after a district court judge actually ruled that the health claims by POM were not scientifically substantiated. Yet, the ad actually said: " 'Natural Fruit Product with Health Promoting Characteristics.'—FTC Judge." Imagine intentionally misquoting a judge to mislead customers. Talk about clear culpability for potential harmful outcomes associated with the consumption of a product. Other ads claimed pomegranate's benefits in terms of reducing heart disease, prostate cancer, and erectile dysfunction.

The US FDA, in 2010, held that POM advertisements were "positioning the juice and supplements as drugs and were therefore illegal under current provisions of the laws that govern FDA' activities." Then the FTC instructed POM to stop making unproven claims. In response, POM sued on First Amendment grounds, but the court upheld the FTC position. Not to just take defeat lying down, the next day, POM ran a full-page ad in the *New York Times*, with the title, "FTC v. POM: You be the judge." The ad featured some quotes from the judge's decision, one being: "Competent and reliable scientific evidence shows that pomegranate juice provides a benefit to promoting erectile health and erectile function." Yet, the ad omitted the sentence that followed: "There is insufficient competent and reliable scientific evidence to show that pomegranate juice prevents or reduces the risk of erectile dysfunction or has been clinically proven to do so" (Nestle, 2018, p. 86). Again, the company is culpable both for trying to mislead consumers about what a judge said as well as about its product. This is intentional behavior, making them morally if not legally responsible for it.

Other amazing health claims come from chocolate companies! And who does not love chocolate? The Hershey Corporation said that chocolate can "inhibit naturally occurring deactivation of the brain during mundane and less interesting tasks (translation: it can keep you alert when you are bored)" (Nestle, 2018, p. 54). And Mars claimed that its candy bars help increase blood flow, lower blood pressure, and reduce the risk of heart disease and said you should eat two bars per day (Nestle, 2018, p. 56)! When confronted by the FDA over its claims, Mars began marketing its CocaVia product in pills and powder as dietary supplements, meant to reduce LDL (or bad) cholesterol in the blood, which is good for your heart. This could thus pass the requirements of the Dietary Supplement Health and Education Act of 1994, allowing companies to make claims that their products help with some structure or function in the body. Nestle notes that, after losing several cases to food companies on First Amendment grounds, the FDA had to back off of pursuing faulty claims by food manufacturers in favor of only cases that obviously violate the law.

Mars funded studies to show that cocoa flavanols do what they claimed they did. The self-regulatory National Advertising Division urged Mars to back down from its claims since the findings it cited and advertised in at least one national newspaper came from studies it funded. Ultimately, Mars backed down and admitted that chocolate is not a healthy food. Further, according to Nestle (2018, pp. 58–59), the company

disclosed genetically modified ingredients on M&M's labels, supports the FDA's proposals for voluntary sodium reduction and labeling of added sugars, promises to label its candies as "everyday" or "occasional" options and to stop using artificial dyes, and is so critical of the industry-funded ILSI for sponsoring efforts to attack the dietary guidelines and other such position that it withdrew from this group in 2018.

So, it is possible for companies occasionally to do the right thing! Keep this in mind should you believe that change is not possible.

With all these claims, it is important to be reminded of this from Nestle (2018, p. 76): "Foods are not drugs" (and she means this in the way that they do not heal disease; evidence presented in this book shows that many ultra-processed and fast foods are indeed addictive as drugs). The term "superfood," invented by food companies, has no medical meaning and is simply "an advertising concept" or a marketing tool.

In the following, I show several more examples of misleading products produced by food companies. They include low-fat foods, "natural," "real," and "free" foods, as well as "whole grain" breads. Before this, consider three types of health claims on products that are seen in today's food products. Health claims about products must be backed by "significant scientific agreement." Yet, *qualified health claims* require less quality evidence and often appear with an asterisk and explanation. *Structure/function claims* merely describe how food can assist in a function or process in the body. The latter case includes a wide variety of claims by food makers (as well as vitamin and supplement companies), and literally, none of them are evaluated by the FDA, at least as noted on the packaging of these products. Here, food companies are culpable for making faulty claims, and the FDA is culpable for failing to regulate those claims.

Herz (2018) notes that low-fat foods are generally not necessarily healthier than their full-fat alternatives. She writes, "most 'reduced fat' foods contain nearly the same calories and sugar as their full-fat versions" (p. 189). Even foods labeled healthy, often only because they contain something of value such as fiber or vitamins and minerals, may lead people to eat more than they otherwise would, and, in the case where the foods contain unhealthy ingredients, this can lead to bad health outcomes.

Moss (2021, p. xxvii) calls many diets foods "hardly distinguishable from their non-diet food cousins. He shows that one Lean Pockets product has 30 percent less calories than a similar Hot Pockets product, and Velveeta Light has only 13 percent fewer calories than regular Velveeta." And he notes:

Fat has more than twice the calories as sugar, and so, by changing up their products to reduce the fat, the processed food companies could see these new diet versions of their products as perfect for two kinds of dieters: those who wanted to better their cardiovascular health and those who were trying to lose weight.

The term "natural" in food products is meant to suggest that there are no artificial or synthetic additives. Yet, Cargill uses the term natural for Stevia, an artificially manufactured sweetener made from a plant. The company was sued and it admitted no wrongdoing, nor did it agree to stop using the term in the future. When the FDA sought information on what the meaning of "natural" was when it came to food products, the ASN took the side of the industry and argued for the inclusion of "synthetic" vitamins in products under the term natural. Nestle (2018, p. 137) writes: "Marketing vitamin-fortified foods as 'natural' opens the door to other 'healthy' synthetic additives. Once again, ASN was supporting industry marketing interests against what some members of the society might prefer 'natural' to mean."

Similarly, the Academy of Nutrition and Dietetics also has many corporate sponsors. Perhaps not surprisingly, then, it developed a labeling program to put an "Academy's Eat Right" seal on products that met certain nutrition standards. The first product to get this seal? Kraft singles! According to Nestle (2018, p. 147): "Kraft singles are 'cheese' slices, in quotation marks because they are not cheese. They are 'pasteurized prepared cheese product' with a list of ingredients so long that one reporter said 'it read like a novel.'" The Academy claimed the seal was not an endorsement but most people thought it was, including comedian Jon Stewart of The Daily Show. After pointing out that Kraft cannot legally call the product cheese, he said: "Turns out the Academy of Nutrition and Dietetics is an Academy the same way this is cheese."

Today's breads are, by in large, unhealthy. With new grain milling technology that removed the bran and germ from grains, breads with higher calories were created that also had less nutrients and fiber. With the addition of bleaching agents, shelf-life went up. Nutrients had to be added back in to avoid certain diseases. It would make sense to going back to serving all whole-grain breads, but "consumers and producers alike have grown accustomed to the taste and convenience of white flour. People prefer its softer texture, how easily it bakes, and how tastily it pairs with everything from fruit preserves and peanut butter to lunch meat." The result is that white-flour grain products now make up about 22% of all calories consumed. White breads are linked to outcomes such as obesity.

Walker (2019) claims that products made with white flour still contain claims like "whole wheat," multi-grain," "100% wheat," "cracked wheat," "bran wheat," and more.

This raises the issue of false advertising in product marketing and sales. "Refined" means the fiber is stripped of fiber and nutrients, and "enriched" means that some nutrients were added back in: "Dressing up inexpensive food ingredients to appear wholesome and nutritious is one of the food system's most tried-and-true strategies. The only label that ensures the bread contains the entire seed and all its fiber is . . . *100% whole wheat*"; yet, this "does not preclude the addition of sugars, fats, and salt" (p. 33).

Misleading claims about unhealthy foods or false health claims about food generally are immoral. They are intentionally made by companies and, as such, the companies are morally if not legally responsible for them.

### Food Companies Are Culpable for Food Fraud

Food fraud is defined as "deliberately placing food on the market, for financial gain, with the intention of deceiving the consumer." It is the "deliberate and intentional substitution, addition, tampering, or misrepresentation of food, food ingredients or food packaging; or false or misleading statements made about a product for economic gain" (Manning & Soon, 2019, p. 127). Manning and Soon claim the following foods are especially vulnerable to food fraud:

- Foods of high value, where substituting an alternative ingredient in the food can provide significant financial gain
- Foods associated with an ideological grouping, whereby adulteration of that food can be seen as an ideological or ethnic challenge to that group itself
- Foods produced, manufactured, and stored in readily accessible or poorly supervised areas, or where staff have little awareness of the potential for food fraud
- Foods susceptible to tampering or interference where this can go undetected by the inspection and testing routinely undertaken in the food supply chain
- Widely disseminated or distributed foods with complex interactions in the supply chain.
  *(p. 128).*

Food fraud can include adulteration, defined as "the specifically motivated, intentional replacement of an ingredient for economic or ideological gain" (Manning & Soon, 2019, p. 129). Economically motivated adulteration is defined by the US FDA (USDA) as "the fraudulent, intentional substitution or addition of a substance in a product for the purpose of increasing the apparent value of the product or reducing the cost of its production." Fish mislabeled to be something it's not is a good example of this issue. Another is the labeling of sugary substances that are not actually honey as honey (Hyde & Savage, 2019). A study of food fraud in the European Union from 2002 to 2015 found fraud in fish and seafood products at 17.4%, nuts at 16.4%, meat and meat products at 16%, and fruits and vegetables at 8%.

There are at least two kinds of food fraud cases in the world. One is selling foods and making false claims about them, or misrepresenting what is being sold in the first place. Take as an example, the case of olive oil fraud. The Joint Research Center, an internal scientific service of the European Commission, found 32 cases of olive oil fraud between September 2016 and December 2019. They included the following:

- Sixteen of the cases involved the substitution of olive oil with other oils.
- Eleven cases concerned the mislabeling of olive oils.

- Four cases involved the false use of a geographical indicator.
- Five cases concerned the distribution of counterfeit products.
- Six related to the dilution of olive oils with other oils or inferior grades.
- One case involved theft.

Twenty of the 32 cases occurred in Europe and: "The most common infringement practices were marketing virgin olive oil as extra virgin and selling blended olive and vegetable oils as pure olive oil." Then, outside of Europe, "olive oil fraud usually included dilution and substitution of oils." For example, in Brazil, the mixing of olive oil with lampante or soybean oil was the most common fraudulent practice. A 2017 example comes from Denmark where "only six of the 35 sampled extra virgin olive oils sold in Danish supermarkets were extra virgin." Meanwhile, in Greece "the police arrested seven people and charged them with adding green dye to sunflower oil and marketing it as olive oil." Finally, in Spain, "the world's largest olive oil cooperative was fined in 2018 for failing to pay the tariffs on imported olive oil from Tunisia, which was then blended with lower quality olive oil and exported to the United States as virgin olive oil" (Olive Oil Times, 2021). All food fraud is intentional behavior, making the involved companies culpable for it.

Then there is the stunning case of Ireland, where in 2012, horse DNA was found in over one-third of all beef burger samples (pig DNA was also found). Investigations suggested different kinds of adulteration were taking place, from using additives as fillers to outright fraud in labeling food products. Another round of tests found up to 100% of meat products such as frozen lasagna and spaghetti Bolognese meals contained horse meat!

According to Manning and Soon (2019, p. 133),

> 81 cases were due to adulteration of food with horsemeat. Fifty-five of the horsemeat incidents were found in the meat and meat products (including poultry) category, which 26 were reported in prepared dishes and snacks. The other 52 per cent of fraud/adulteration incidents were due to mislabeling, illegal trade and others, such as an absence of declaration of compliance, fraudulent health certificate and absence of a certified analytical report.

Corini and van der Meulen (2019, p. 165) point out that "two of the slaughterhouses that committed the fraud are located in the Netherlands." Since the products were considered unsafe, they were recalled there. In Italy, however, the issue was not viewed as one of food safety but rather one of commercial fraud, so no recall was issued.

In 2020 and 2021, horse slaughter for meat continues, where about 3,000 horses were slaughtered for meat: A "total of 1,549 thoroughbreds with passports issued by horseracing conglomerate Weatherbys were slaughtered in 2020, followed by another 1,105 in 2021." Some of this meat allegedly ends up in the human food chain: "Consumption of horse meat has been growing globally since the 1990s. It is considered a delicacy in parts of Italy, Holland, Switzerland and Belgium, and is also commonly served in China, Russia, Central Asia, Mexico, Argentina and Japan" (The Journal, 2022). There is no evidence this is still being sold as beef, although this remains a possibility due to the deregulation of the industry in the United Kingdom and the re-emergence of self-policing regulatory schemes there.

The UK Food Standards Agency handles cases of food fraud. Its "food crime" website defines food crime as "serious fraud and related criminality in food supply chains." Types of crimes the agency handles include the following:

1. Illegal processing—slaughtering or preparing meat and related products in unapproved premises or using unauthorized techniques
2. Misrepresentation—marketing or labeling a product to wrongly portray its quality, safety, origin, or freshness
3. Waste diversion—illegally diverting food, drink or feed meant for disposal, back into the supply chain
4. Substitution—replacing a food or ingredient with another substance that is similar but inferior
5. Document fraud—making, using, or possessing false documents with the intent to sell or market a fraudulent or substandard product
6. Theft—dishonestly obtaining food, drink, or feed products to profit from their use or sale
7. Adulteration—including a foreign substance that is not on the product's label to lower costs or fake a higher quality.

Other cases that made the news included a falsification of official health documents in seven European Union countries, fraudulent ketchup being sold as Heinz, poultry meat covered in feces and slime and cleaned with chemicals to be sold, meat from rats, minks, and foxes marked as beef and mutton, shrimp injected with gelatin, contaminated pork meat, cabbage treated with formaldehyde, and sewage oil treated with chemicals sold as cooking oil (Walker, 2019)! Walker (2019, p. 156) also notes the "egregious criminal fraud leading to a half-billion eggs needing to be recalled, and pathogen-tainted peanuts knowingly shipped to forty-six states."

Fraud is already illegal, a true food "crime." It is also committed with intentionality, making offenders legally culpable for such behaviors.

## Food Companies Are Culpable for Product Shrinkage

One way food product producers sell more products is keeping prices as low as possible. In situations where the cost of ingredients rises, manufacturers will intentionally design products to hold less volume and then sell them at the same price, so that you get less product for your money—a practice known as *product shrinkage*.

Product shrinkage occurs when less and less of a product (e.g., potato chips, baked beans, soup, toilet paper, and virtually any other product) is packaged in the same size container over time. That is, the product is shrinking whereas the size of the package is not. Here are several real examples. One major company that produces tampons stopped offering 40 tampons in a box and began offering only 32 tampons in a box (for the same price). Women noticed, complained, and ultimately the company began offering 40 tampons again. Only, the corporation added a label to the box, saying "Now 8 more free!" Of course, women were not getting 8 free tampons, they were instead receiving the same 40 that they used to receive before the company shrank the amount of the product offered to consumers. This example of product shrinkage was noticed because it was so blatant.

Another example of product shrinkage is a package of toilet paper that says "Now Even Better." In fact, the product is exactly the same, only the new package contains less toilet paper. Perhaps the claim that the product is better is accurate because it is better for the company (which will profit more by selling less of the product for the same price). The company went from offering a four-pack of toilet paper that contained 400 two-ply sheets per roll to only 396 two-ply sheets per roll (meaning in a four-pack, consumers would receive 16 less two-ply sheets of toilet paper, hardly noticeable). Instead of announcing that the company was providing less product for the same price, the company said nothing and instead added an additional label to its package, saying: "Now even better!"

Usually, corporations say nothing about product shrinkage—they make no changes to their packaging other than the fine print on the back of the product which indicates how much of the product is included (e.g., ounces). This practice is common in virtually every product in your grocery store. It is, simply stated, accepted as a legitimate business practice.

Another example is a potato chip company that advertised its popular product as "now even cheesier!" Expecting to experience a new, even better potato chip, consumers were probably disappointed to learn that the company really just included fewer chips in every bag (but for the same price). How did the chip get "even cheesier" then? The company put fewer chips in the bags but the same amount of cheese. Fewer chips and the same amount of cheese topping equals more cheese topping per chip (but still less product overall, and for the same price).

It is important to point out that product shrinkage is ubiquitous in food products. Next time you are in the grocery store, take special note of product sizes, from beans to soups to sauces, and so on. Ask yourself why some products come in sizes like 15.4 ounces instead of 16 ounces or 11.6 ounces instead of 12 ounces. These are sure signs of product shrinkage!

Such behavior is intentional on the part of companies—intended to deceive consumers. Companies involved in it are both morally and legally responsible for product shrinkage.

## Food Companies Are Culpable for Deleterious Health Outcomes and Other Harms of the System

Food companies are culpable for all of the detrimental health outcomes associated with the typical American diet (discussed in Chapter 6, such as obesity and diabetes), as well as the harms associated with the system itself (discussed in Chapter 7, such as environmental degradation). There is no need to revisit those topics here, other than to remind you of the enormous harms these outcomes impose on society—costs in at least hundreds of billions of dollars. It is food companies that make and sell the products that produce negative health outcomes discussed in Chapter 6. And it is food companies who create the conditions outlined in Chapter 7 that endanger animals, workers, and the environment. Food companies are culpable for these outcomes.

## Consumer Responsibility

The "simple principle" asserts that if it is wrong to produce something, it is also wrong to consume it (Barnhill et al., 2018). If this is true, it is us, the consumers, who are ultimately responsible for the conventional food system. Take factory farming, for example. It is us, the consumers, through our demand for factory-farmed meat, who facilitate and incentivize factory farming (Martin, 2016, p. 212). At the same time, the industry "very aggressively creates a favorable sales environment for its products. It does so by lobbying political representatives to eliminate or not enforce unfavorable regulations; by co-opting nutrition experts by supporting favorable research; and by marketing and advertising" (Kaplan, 2016, p. 266), as well as all the other mechanisms identified in this chapter. Kaplan concludes thus: "The success of the food industry is due less to consumer demand than to its own efforts, even force. The deck is stacked in their favor."

Simon (2006, p. 12) agrees about food company control of experts, writing:

> Food companies claim to welcome . . . scrutiny and insist that their expert advisory boards set "objective" standards for their policies. But . . . these so-called experts are in fact bought and paid for, and their conditions certainly are not subject to anything like public debate and discussion.

Further, the industry pays researchers "to produce industry-friendly findings for publication in prestigious academic journals—which sometimes neglect to make those biases known" (p. 169). Finally, food corporations form "partnerships with professional health association and disease foundations. By creating such alliances, manufacturers can ensure that these 'respectable' groups remain complicit with their take on the relevant science, while scoring valuable PR points" (p. 182).

The point of all this is that actors in the conventional food system wield enormous power over us, most of which is unseen and unknown. And, unless we all start growing our own food—enough to sustain us and our families—we must rely on the conventional food system for sustenance.

Still, "we as consumers are essential components of the food and agriculture system, and hence, we are also partly responsible for the problems associated with it" (Hatanaka, 2019, p. 3). Hatanaka (2019, p. 7) clarifies:

> Because most of us consume globalized, standardized, corporatized, processed, and disconnected foods, we play an important role in maintaining the status quo of the food and agriculture system, which increasingly imposes many challenges. . . . Among other things, these challenges include labor exploitation, resource depletion, unethical animal welfare practices, uneven access to healthy and nutritious foods by race and class, hunger, and increasing obesity.

Given that the system would not exist if not for us, it is up to us to change it. At least that much is clear—the conventional food system has no reason or incentive to change unless we demand it. And we have to put our money where our mouth is, for it is profit that drives the system. This is the fundamental consumer responsibility needed to force change in the system.

Walker (2019, p. 172) also seems to blame the consumer for the maladies of the conventional food system, writing that "American culture wanted convenient food, with little physical effort to get it. Food manufacturers were all too happy to oblige, turning out rich processed foods with captivating tastes, textures, aromas, appearance, and even sound." He notes that perfecting this formula was money in the bank for companies, while we end up having to figure out how to stop eating so much junk food. Moss (2021, p. 126) points out that food manufacturers "didn't invent the attributes that make their products so addictive, as much as they simply gave us what we innately want." I would argue this is a chicken or the egg situation, because, after all, which actually came first? This type of food or the desire for it? And then, of course, once one begins eating these foods, the issue of addiction kicks in, making quitting very difficult (see Chapter 5).

Schrempf-Stirling and Phillips (2019, p. 117) claim that corporate food "executives are faced with the brute fact that consumers very often choose the less healthy [food]—even when health information is relatively available and understandable." Schrempf-Stirling and Phillips (2019, p. 122) note:

> Competitive pressures limit corporations' capacity to make and sell healthy alternatives. . . . Companies that would like to offer healthier options face the prospect that their competitors will see this as an opportunity to steal shelf space and market share. If healthier ingredients are more expensive, making healthier foods will put manufacturers at a cost disadvantage due to collective action and challenges associated with being the first companies to make these changes.

In Chapter 3, I showed how we actually have less control over what we eat than we like to think we do. So we are not fully responsible for problems of obesity, diabetes, and so on. Still, it is us who choose the foods we buy and eat, and surely we have some control over that behavior. So, it should be pointed out that it is us—the consumers—who must change the system if we want to reduce the harms associated with it. As long as major food companies are profiting off of the status quo, we can be sure it will not change.

## Responsible Actions of Food Companies

While the focus of this book is on food crime—deviant and criminal acts of food companies committed with culpability—food companies also engage in responsible behaviors to try to make their foods healthier, provide alternative versions of their foods, and protect the environment. That is, food companies simultaneously engage in behaviors that may be deemed responsible by those who want to bring about needed change to the conventional food system.

Numerous multinational food and beverage companies (including Ferrero, General Mills, Grupo Bimbo, Kellogg's, Kraft Foods, Mars, Nestle, PepsiCo, the Coca-Cola Company, and Unilever) have worked within the context of the International Food and Beverage Alliance to reportedly increase their commitments to public health through means such as making products healthier and changing how products are marketed to children (Alexander et al., 2011). Yet, efforts by companies such as Kraft and Campbell soup to offer healthier versions of their products failed miserably, leading the CEO of the former company to be removed and the latter company to be downgraded on Wall Street (Walker, 2019). If we do not buy healthier versions of their products, we can be certain they will not keep making them.

Companies like McDonald's have a history of announcing moves that are deemed socially responsible in the media. For example, in 2002, McDonald's pledged to remove trans fats from its cooking oils. It also announced, in 2004, that it would stop serving supersize meals. The chain also offers salads and grilled chicken meals, both of which are supposedly less bad for you compared to other menu options (keep in mind that iceberg lettuce has no nutritional value and some salad dressings have as much fat and calories as the burgers the chain is famous for). According to Simon (2006, p. 74):

> McDonalds number-one motivation is to keep its customers addicted to its products, and lettuce covered with fried chicken [as in the case of one of its salads] furthers that goal. But touting its "premium salads" gives the false impression that the company sells healthy items.

Alexander et al. (2011, pp. 32–33) also note that

> most of the leading packaged food companies in China and India, and several companies in the US, South Africa, Brazil, Mexico and the UK are national companies without published evidence of significant health and wellness pledges. Additionally, numerous soft drink companies in each country as not engaged in significant health and wellness pledges.

Clearly, then, there is room for significant improvement in this area.

Companies are also fortifying foods—that is, adding micronutrients to foods. Even processed foods can be and are fortified with micronutrients, such as wheat flour and milk (Kimura, 2019). And now such products are often referred to as "functional foods" with statements such as "this product is a good source for" whatever vitamin or mineral is being promoted by product manufacturers. Yet, consider the danger of allowing companies to sell sodas that have claims consistent with functional foods, or sugar-coated cereals; one likely outcome of consumption of these products is weight gain, at the very least. The Business Alliance for Food Fortification includes companies such as Ajinimoto, Coca-Cola, Dannon, Heinz, Nestle, and Unilever and promotes food fortification in developing countries. Another group, Scaling Up Nutrition Business Network, aims to achieve the same goal (Kimura, 2019). Whereas this can be considered helpful and even philanthropic behavior, it also helps facilitate "the extension of corporate marketing to the untapped markets of rural and poor households" (Kimura, 2019, p. 73). Putting nutrients in processed foods creates the illusion that they are healthy for us, when in fact, they are not (Moss, 2021).

Other commitments come in the form of goals stated by major food corporations to, for example, dramatically reduce greenhouse gas emissions over time. Walmart, for example, reportedly has a sustainability initiative to reduce waste, source sustainable products, and monitor its product suppliers' labor practices (Bloom, 2019). PepsiCo, where more than half of its products start in the ground, aims to "reduce absolute greenhouse gas emissions by more than 40 percent by 2030 and to achieve net-zero emissions by 2040" (Klein, 2021). It also wants 100% of its packaging to be recoverable or recyclable by 2025. Similarly, Danone aims to reduce emissions by 50% by 2030. Further, at Kellogg's, where 76% of its packaging is recyclable and 86% of its products are plant-based, it is making its sub-brand Morningstar 100% vegan. It also "is aiming for a 15 percent per pound of food produced reduction in emissions by 2020 and 100 percent reusable, recyclable, or compostable packaging by the end of 2025" (Klein, 2021). The company wants to be at 100% renewable electricity by 2050.

At the same time: "Less than 10% of the world's largest food companies are aiming to reduce their emissions in line with climate science, while less than 13% are taking sufficient action to eliminate forced labour" (Edie, 2022). An assessment of 350 large businesses, "taking in all parts of the food value chain including agriculture, ingredient supply, manufacturing and processing, grocery retail and hospitality," awarded scores to major corporations for their "efforts on environment, nutrition, social inclusion and overarching governance" (Edie, 2022). The worst scores were for Subway, Sunkist Growers, Jollibee, Schreiber, JCB, Koch Foods, and American casual dining chains Bloomin' Brands and Darden Restaurants. In total, 100 companies scored less than 10 out of a possible 100. Further, not a single company scored more than 72 out of 100, but the highest scores went to Unilever, Nestle, Danone, OCP, and Anheuser-Busch InBev. Edie (2022) notes that 227 out of the 350 businesses have emission cut targets, "just 26 percent have set science-based targets to reduce emissions from their direct activities in line with the Paris Agreement's 1.5C pathway" (the goal to hold global temperature rise to less than 1.5°C).

When it comes to forced and child labor,

> 304 of the 350 companies—almost 87%—are not taking action to eliminate adult forced labour in their supply chains in a manner consistent with the UN's Guiding Principles on Business and Human Rights. Additionally, 2020 of the companies—57%—do not explicitly prohibit child labour in their supply chains.

## Conclusion

In this chapter, you learned about specific actions and behaviors for which food companies are culpable. At a minimum, this includes producing excess calories, putting food in non-food environments, funding research to create one-sided studies, using front groups to confuse consumers, advertising unhealthy products, engaging in deceptive advertising, selling fraudulent foods, and shrinking products over time (i.e., selling less products over time in same size containers).

I made the argument that food companies are morally if not legally responsible for these behaviors. From the perspective of a criminologist, some of these things could very well be made illegal and thus would actually be "food crimes," yet, I do not argue to criminalize any behavior in this chapter (other than food fraud and possibly product shrinkage). Instead, I would urge and expect government regulatory agencies to crack down on food companies to make sure they stop engaging in such acts.

On top of the behaviors identified in this chapter, keep in mind that elsewhere, I showed that food companies are also culpable for food addiction (Chapter 5), deleterious health conditions such as obesity and diabetes (Chapter 6), as well as a whole host of harms associated with the food system itself (Chapter 7). It is past time for government agencies that have supervisory capacity over food corporations to hold them accountable for such acts. To the degree they fail to do this, it makes government regulatory agencies culpable, as well. And finally, the consumer shares some blame, for it is us who make the choices to continue to support companies who behave unethically and/or illegally, or not.

# 9

# ECONOMIC BENEFITS OF THE FOOD SYSTEM

## Introduction

The conventional food system comprises a significant portion of the economic value in the country (approximately $776 billion of GDP in 2012, or 5% of the total) (Nesheim et al., 2015); this is another demonstrable benefit to society. By 2017, the number had topped $1 trillion (USDA, 2022). In this chapter, I focus specifically on the economic benefits associated with the conventional food system. The ultimate goal is to encourage a costs–benefits analysis of the conventional food system related to how it operates and the outcomes it produces for society. While it is expected that the benefits of the system will outweigh its costs, the hope is that this analysis will lead to possible reforms in order to reduce the costs associated with it.

The costs of this system were, in essence, discussed in Chapter 6 (harms associated with the foods we eat) and Chapter 7 (harms associated with the system itself). As you saw, costs associated with the conventional food system are enormous in terms of illness, death, as well as financial costs produced by deleterious health outcomes and environmental degradation. An extensive costs–benefits analysis of the system is beyond the scope of this book, but a preliminary one is offered in this chapter in hopes of encouraging future work in this area.

## Economic Benefits of the Conventional Food System

There are many measurable benefits associated with the conventional food system. In this section, I discuss some of the outcomes identified by research. The supply-side food system assures food for most people, contributes to GDP, increases overall employment, generates tax revenue through salaries, helps people achieve access to health care, contributes to the quality of life in rural areas, and provides benefits to other industries including agriculture and technology.

### Access to Food

The primary benefit of the food supply chain is providing "abundant," "safe," "high-quality," and "affordable" nutrition to more than 300 million Americans, as well as

DOI: 10.4324/9781003296454-9

even more people around the world (Nesheim et al., 2015, p. 195). This is claimed to be "the primary indicator of social and economic success in any food system." Given that the demand for food is constant, the contributions made to society by the conventional food system are among the most stable of all industries. The conventional food system provides ample, diverse food products for a large share of the population, and it does so in a relatively affordable way. Of course, as shown in this book, the system also provides an enormous amount of junk food; up to 60% of all food produced in any given year is junk.

Data show that the food supply chain Americans depend on for food provided $1.4 trillion worth of food in 2014. CED (2017, p. 19) notes: "Total expenditures on food were the third most important category in total consumer expenditures, after health care ($1.9 trillion) and housing ($2.1 trillion)." Yet, Americans only spend about 10% of their disposable income on food (CED, 2017, p. 19), compared with 40% in 1900 and 30% in 1950 (Lorr, 2020). According to the CED (CED, 2017, p. 7): "A global comparison shows that US consumers spend the smallest share of income on food, and much less than consumers in other countries with comparable income levels." Lorr notes that "we spend less money than almost every other country in the world on food and we spend less time gathering that food than at any time in history" (p. 6).

There is thus little question that the conventional food system significantly contributes to the general welfare of society, assuring that Americans are both happy and that most people meet their basic needs. However, the system does not do an equitable job in this area, meaning that access to food and its affordability to people vary widely across the population. According to the US Department of Agriculture (2018): "In 2017, households in the middle income quintile spent an average of $7,061 on food, representing 14.3 percent of income, while the lowest income households spent $4,070 on food, representing 34.1 percent of income." This means food places the heaviest burden on those with low incomes. Figure 9.1 depicts these data.

These data show that, although many are accessing food affordably, others are spending close to half of their earnings to purchase food. In reality, lower income earning Americans are struggling to balance food affordability and accessibility with low wages and the opportunity to access sources of healthy food at home (FAH). One potential consequence is increased reliance on food away from home (FAFH) sources primarily found in the fast-food industry; these sources tend to be largely unhealthy, thereby contributing to deleterious health outcomes discussed in Chapter 6.

### Gross Domestic Product

The Committee for Economic Development (2017, p. 5) claims:

> The food sector plays an essential role in the US economy, accounting for about 5 percent of gross domestic product [GDP], 10 percent of total U.S. employment, and 10 percent of US consumers' disposable personal income (DPI). The food sector has total sales of $1.4 trillion, including food consumed at home and away from home.

Nesheim and colleagues (2015, pp. 37–38) show government estimates that suggest the contribution to GDP was $776 billion in 2012, which is roughly 5% of the total (2% from

**Food spending and share of income spent on food across U.S. households, 2021**

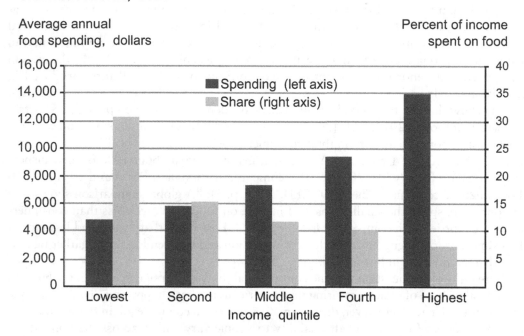

Source: USDA, Economic Research Service using U.S. Department of Labor, Bureau of Labor Statistics, 2021 Consumer Expenditure Survey data.

**FIGURE 9.1** Food Spending as a Share of Income Declines as Income Rises

*Source:* www.ers.usda.gov/data-products/chart-gallery/gallery/chart-detail/?chartId=58372

food processing, 2% from manufacturing, and 1% from agriculture). Again, this value now tops $1 trillion (USDA, 2022).

Figure 9.2 shows the contribution of different sectors to the nation's GDP. Although GDP is generally used as a measure of a positive contribution to the economy, note that some of it comprises negative contributions. For example, the food service industry largely serves foods that are high in fat, sodium, and sugar and absent much nutritional value, leading to illness and death; the textile and leather industry leads to widespread suffering of animals; the tobacco and beverage industries produce products that kill far more Americans than all illicit drugs combined; the fishing industry kills hundreds of billions of fish every year, as well as other species not even targeted for capture; and farms pollute the air and water and utilize harmful chemicals, and raise the animals (i.e., cows, chickens, turkeys, hogs, etc.) that are slaughtered by the billions each year. Many of these costs were discussed in Chapter 7.

A review of the top 500 companies in the world shows that they included 39 food companies in 2013:

Annual revenue of these 39 food companies ranged from $6.5 to $469.3 billion. . . . The most numerous firms represent the food manufacturing and retail food sectors.

**TABLE 9.1**  Top Ten Revenue-Generating Food Companies in 2017

1. Archer-Daniels-Midland Co. ($62.3 billion in 2016 revenue, $1.3 billion in 2016 net income)
2. Bunge Ltd. ($42.7 billion in 2016 revenue, $745 million in 2016 net income)
3. Tyson Foods ($36.9 billion in 2016 revenue, $1.8 billion in 2016 net income)
4. Kraft Heinz Co. ($26.5 billion in 2016 revenue, $3.6 billion in 2016 net income)
5. Mondelez International Inc. ($25.9 billion in 2016 revenue, $1.7 billion in 2016 net income)
6. General Mills Inc. ($16.6 billion in 2016 revenue, $1.7 billion in 2016 net income)
7. Kellogg Co. ($13 billion in 2016 revenue, $694 million in 2016 net income)
8. Conagra Brands Inc. ($11.6 billion in 2016 revenue, $677 million in 2016 net income)
9. Hormel Foods Corp. ($9.5 billion in 2016 revenue, $890 million in 2016 net income)
10. Campbell Soup Co. ($8 billion in 2016 revenue, $563 million in 2016 net income)

In general, the largest profits are found in the food manufacturing sector, primarily among large multinational companies and in the food service sector.

*(Nesheim et al., 2015, p. 194)*

The top food companies from 2021 are shown in Table 9.1. As you can see in the table, the top companies include PepsiCo, Nestle, JBS, Anheuser-Busch InBev, Tyson Foods, Mars, Archer Daniels Midland Company, Coca-Cola, Cargill, and Danone.

## Employment

About 1.3 billion people around the world work in agriculture, roughly one in three of all jobs. Further: "In developing countries, agriculture remains the biggest employer of the rural poor, and is one of the largest contributors to national economic revenues" (Nesheim et al., 2015, p. 93).

There are several sectors of the conventional food system in which people are employed. They include production, processing, distribution, retail, and service. Production workers grow food and tend to animals on farms. Only 1.4% of people today work on farms (Walker, 2019). And, according to Gray (2019b, p. 185):

Today's US farmers are highly reliant on, if not addicted to, cheap labor. Because most US citizens are reluctant to do hard manual labor for low wages, farmers have sought out vulnerable workers who are willing to take these jobs. Present-day farmworkers are mostly immigrants; 72 percent of US crop workers are foreign born and 46 percent are undocumented.

On East Coast farms, the primary worker is Latin American.

The term, "exploitation of labor," may not have a specific meaning, but here is a pretty good example of it when it comes to the conventional food system: "At least half of the 10 lowest-paid jobs in the United States are in the food industry, and they rely disproportionately on federal benefits. Walmart and McDonald's are among the top employers of beneficiaries of food stamps and Medicaid" (The Guardian, 2022). One could argue that it is unethical for vastly profitable companies to be built on the backs of underpaid employees. It appears that the success of these companies literally requires low-paid workers, suggesting a systemic problem with the industry. These companies are unethical if not unsustainable, assuming we want people who work to actually be able to afford to live. Companies are culpable for the low pay of their workers.

Then there is the nature of the work itself, with often long hours and serious dangers facing workers. In farms, "low paid workers have little protection from long hours, repetitive strain injuries, exposure to pesticides, dangerous machinery, extreme heat and animal waste." And the industry relies heavily on immigrants: "Between 50% and 75% of the country's 2.5 million farmworkers are undocumented migrants who have few labor rights and limited access to occupational healthcare" (Lakhani et al., 2021).

Processing workers transform farm products into food products. Distribution works transport food throughout the system as well as across the country. The retail sector includes grocery store workers and others involved in stocking shelves and prepping products for sale. Finally, the service sector, the largest employer of food workers, is where workers serve food, for example, at restaurants.

According to Gray (2019b, p. 184): "The food industry is the largest employer in the United States, accounting to 14 percent of all US jobs." However,

> the jobs offered are not well paid. . . . Compared to other industries, food workers earn the lowest median hourly wages ($10 an hour versus $17.53 in all industries), rely more on food stamps (2.2 times the use of all other industries), and are more food insecure (20 percent compared to 13% of the US population). These problems of the conventional food system can and must be solved by the major food companies that comprise the system.

Taken as a whole, Gray (2019b, p. 188) characterizes food work as "unstable, temporary, low paid, with unpredictable schedules and difficult hours." Then, much of the work is very hard and unsafe, for example, in meat and poultry processing. Gray points out that, perhaps for this reason, more than half of the workforce in poultry processing is immigrant "and employers have actively recruited them to take the place of local workers in an attempt to find cheaper, more docile workers" (p. 190). According to Konefal and Hatanaka (2019, pp. xxxi–xxxii),

> the agriculture sector . . . exploits and reproduces inequalities in society. Most notably, Latinos continue to do the bulk of farm work in the United States, which is low paying and dangerous. Similarly, racial and ethnic minorities are also disproportionately in meat processing plants in the United States, again some of the lowest paying and most dangerous work. In other words, agriculture and food processing lock many poor racial and ethnic minorities into low-paying and dangerous work that offers little possibility of social mobility.

Notice that, even in this discussion of the benefits of the system, costs emerge. It is large food companies that are responsible for the realities identified earlier.

Myers (2019) shows that about 30% of the 20 million food chain workers in the United States are food insecure, and he claims this is twice the rate of the whole US workforce. Further,

> farmworkers have rates of food insecurity over three times greater than that of the general US population because they are excluded from the basic labor protections and

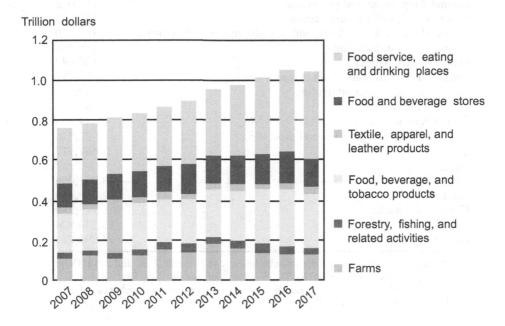

Value added to GDP by agriculture and related industries, 2007–17

Trillion dollars

Legend:
- Food service, eating and drinking places
- Food and beverage stores
- Textile, apparel, and leather products
- Food, beverage, and tobacco products
- Forestry, fishing, and related activities
- Farms

Note: GDP refers to gross domestic product.

Source: USDA, Economic Research Service using data from U.S. Department of Commerce, Bureau of Economic Analysis, Value Added by Industry series.

**FIGURE 9.2** Contribution to GDP

*Source:* www.ers.usda.gov/data-products/chart-gallery/gallery/chart-detail/?chartId=92983

worker rights that many Americans take for granted, such as collective bargaining rights, the right to join a union, and overtime pay.

Then there is the restaurant industry, employing 10 million workers, where "almost half of workers [are] earning at or below the federal minimum wage." Thus, about 17% of them are below the federal poverty line, compared to only about 6% of other workers. In addition to poor pay, there is a general lack of quality health care alleged in the industry, and unionization efforts are often blocked by food companies. It is up to food companies to change the aforementioned conditions.

Data from the US Census Bureau (2018) are shown in Table 9.2. These data illustrate the total number of people employed in different facets of the conventional food system.

Approximately two million farms are involved in producing the food we eat. According to Nesheim et al. (2015, p. 168): "About 40 percent of the U.S. land area is used for farming, with 2.1 million farm operations generating nearly $400 billion in sales (55 percent from crops and 45 percent from livestock) and more than $100 billion in net farm income in 2013." These farms employ millions of people, somewhere between two million and six

**TABLE 9.2** Employees in the Conventional Food System (2012)

| *Manufacturing and processing* | |
|---|---|
| Animal slaughtering and processing | 487,072 |
| Bakeries and tortilla manufacturing | 258,345 |
| Fruit and vegetable preserving and specialty food manufacturing | 166,050 |
| Other food manufacturing | 159,296 |
| Dairy product manufacturing | 130,980 |
| Sugar and confectionary product manufacturing | 69,717 |
| Soft drinks and ice manufacturing | 59,942 |
| Grain and oilseed milling | 52,766 |
| Animal food manufacturing | 44,873 |
| Seafood product preparation and processing | 32,876 |
| Total | 1,461,917 |
| SALES | |
| Food services | 9,701,096 |
| Food stores | 2,750,459 |
| Grocery wholesalers | 779,251 |
| Total | 13,230,806 |
| GRAND TOTAL | 14,692,723 |

*Source:* Table 1. Post-farm employees, establishments, and payroll, 2007 and 2012. Committee for Economic Development (2017). *Economic contribution of the food and beverage industry.* Downloaded from: https://www.ced.org/pdf/Economic_Contribution_of_the_Food_and_Beverage_Industry.pdf

million workers depending on the source (Nesheim et al., 2015, p. 174). Yet, only about half of farmworker positions are full-time and about 20% are temporary contract workers. Further, farm work is highly seasonal, and pay levels are very low, especially given the hard and potentially dangerous levels of work.

According to CED (2017, p. 22):

> Most farms are family owned and operated, and most also rely on nonfarm earnings to support household income. About 90 percent of farms have gross cash farm income of less than $350,000 annually. Most production—68 percent—occurs on the 9 percent of farms classified as midsize or large-scale family farms, which also account for 51 percent of all farmland. Farm and ranch families make up just 2 percent of the US population.

The smallest farms are not profitable and tend to rely on non-farm income to make ends meet. Data show that "57 percent of U.S. farm operations in 2012 had gross farm sales below $10,000, and these operations typically reported net losses from their farming business" (Nesheim et al., 2015, p. 171). For those farms that had less than $250,000 in gross sales, "nearly all of the roughly $70,000 gross average household income comes from off-farm employment and unearned income" (Nesheim et al., 2015, pp. 171, 173). This can be contrasted with larger, commercial farms with sales of more than $350,000, where the average farm household "made more than $200,000 in total income, with nearly 75 percent of this total accounted for by net farm income" (Nesheim et al., 2015, p. 173).

The raw ingredients produced by farms are transformed into foods and beverages that are sold at roughly 620,000 stores (e.g., grocery stores) and outlets (e.g., restaurants).

Millions more people work in these facilities. The food and beverage industry employs at least 1.5 million people at 27,000 establishments (CED, 2017). In addition, more than 13 million employees work in food sales. According to CED (2017, p. 22):

> Grocery wholesalers employ 0.8 million workers in 33,794 establishments and pay out $34 billion in payroll. By far the largest industry footprint is found in food retailing. Food stores serving the [Food at Home] FAH market employ 2.8 million workers earning $58 billion, and number 124,014. Foodservice for the [Food Away from Home] FAFH market employs 9.7 million workers earning $141 billion in 556,882 establishments.

### Average Incomes of Farmers and Food Workers

Workers in the food industry earn approximately $83 billion in "total salary and benefits," pay about $10 billion in taxes to the government, and account for $62 billion in "property income" as well as $9 billion for "imported inputs" (CED, 2017, p. 5). This sounds impressive until you remember the number of people involved in food production. According to Nesheim et al. (2015, p. 193):

> Poverty and injustice in the food system has been described in the literature for centuries. . . . Evidence shows that 40 percent of food industry jobs provide a wage at the federal poverty level; only 13.5 percent of the jobs provide wages that yield an annual income at 150 percent of the poverty level.

In essence, food production in the United States is possible only because of very low wages paid to workers. As one example, the average worker at fast-food restaurants is about 29 years (Chandler, 2019), and about 40% of workers live in near poverty! And the truckers who bring us our food from across the country are paid terribly and have high rates of turnover in their industry (Lorr, 2020).

Research suggests that about one-third of the nation's small farm workers are not held to federal laws enforcing the minimum wage. For example, if workers are hired through crew leaders or farm labor contracts, employers are not required to follow state and federal employment laws. Labor contractors typically promise candidates an increased quality of life and stable employment with decent wages but ask for a recruitment fee first. Usually, the promises made fail to meet the new employees' expectations, leaving them indebted with lower wages than planned, poor working conditions, long tedious hours, and a binding contract. Employers are able to distribute lower pay by paying under the table in cash, an illegal but simple act to execute. Wage theft also results from poor oversight in this industry where supervisors or employers steal a portion of the workers' wage or salary before distributing pay. The agriculture and food industry rank as two of the main places of employment for workers to suffer from wage theft.

Nesheim and colleagues (2015, p. 174) point out that farm workers earn low wages, work fewer days a year than other workers, and experience "chronic levels of underemployment, unemployment, and poverty." And in 2010, "the average hired crop worker earned less than $10 per hour, and median weekly earnings were about two-thirds of the average U.S. wage or salary worker's," leading to very high poverty rates. They also tend to live in substandard housing with high levels of family disruption and poor education for their children (Nesheim et al., 2015, p. 176). Again, the conventional food system is only profitable due to these conditions, and it is large food corporations who are responsible for these realities.

Of the total 23 million people working in food system jobs of some kind, the average salary is just over $19,000 per year, which was "less than half the average annual income of all workers in the United States in 2007." Only two subsectors—distribution/wholesale and waste recovery—boast average payrolls above the national average income of $41,525 (Nesheim et al., 2015, p. 183). The former average $57,000 per year. Farm workers average about $30,000 per year. Pay in food processing varies by sector, with an average salary of $41,000 among meat workers and $29,000 for poultry workers. Full-time workers in food manufacturing average about $27,000 per year.

The average salary in the food retail industry is only about $25,600 per year. Here, employee turnover is quite high, especially given the high rates of "wage theft (e.g., not receiving overtime payments, tip misappropriations)" (Nesheim et al., 2015, p. 192).

In addition to these obstacles faced by food workers, they are also subjected to double the average unemployment rate. This is often due to crop turnover, weather, winter freezes, a lack of job security, and no benefits or protection from labor laws. Many farm workers adopt the "migrant lifestyle" and follow locations of ripe harvests for work in coordination with the season. This lifestyle still presents obstacles of instability and a lack of job security.

### Access to Health Care

Although people working in the food industry often have the same problems with health care as other Americans, the portion of people living in farm-operator households has an uninsured rate of only about 9%, much lower than the overall population (Nesheim et al., 2015, p. 178). This is definitely one significant benefit of the conventional food system.

However, rates of non-coverage are much higher in the largest part of the food system, the food service industry. In this sector, a large portion of employees works part-time and rely on tips for income. In addition, since about half of all farmworkers are migrants and seasonal workers, rates of non-coverage are extremely high among these populations. They also tend not to have workers' compensation protection, paid sick days, vacation days, and so on. How we tolerate these realities is honestly beyond my understanding.

### Multiplier Effects

According to the Committee for Economic Development (2017, p. 6) the "relatively labor-intensive nature of food processing leads to strong economic multiplier impacts on local economies" (CED, 2017, p. 6). *Multiplier effects* include *direct effects* (such as when a company opens a business, employing and paying people, generating local economic benefits). It also includes *indirect effects* (such as when the new company buys products from other businesses, including raw commodities used in their product, electricity, transportation services, etc.). Then, there are *induced effects*, whereby employees spend their salaries at other local businesses (including, ironically, restaurants and grocery stores). In these ways, the conventional food system contributes to the quality of life in rural areas where most farming occurs. The effect, however, is not always good:

> The economic performance and quality of life for farm operators and hired farm workers can be an important contributor for community life and well-being, particularly in rural

areas where farming is a major driver of local social and economic activity. Researchers know that rural communities that rely most heavily on farming for their local economic base are more likely to experience economic stagnation and population declines.

*(Nesheim et al., 2015, pp. 177–178)*

Moreover, research also shows that rural communities with large populations of farm workers tend to have some of the lowest per capita incomes in the nation, face shortages in vital social services, and live in some of the "most stressed local fiscal conditions of any rural communities in America" (Nesheim et al., 2015, p. 178).

Still, according to CED, every dollar of food and beverage industry output

generates between $0.40 and $1.35 of additional economic activity. Additionally, every job in the food and beverage industry generates between one and three additional jobs in local and regional economies as employees spend their wages on local goods and services.

Then, at the national level, "economic multiplier impacts are larger, as these indirect impacts ripple through the economy," producing as much as "$4.00 of additional value added in the economy for each dollar of value added in the industry and five additional jobs for every industry job" (CED, 2017, p. 6).

The term *value added* refers to all "the proceeds from the sale of outputs minus the outlays for commodities purchased from other establishments" (CED, 2017, p. 28). According to CED (2017, p. 28): "The food and beverage industry accounts for nearly $164 billion in total value added, out of total food expenditures of $1.1 trillion for all domestically produced food and $1.2 trillion for food expenditures including imports used in food production." The largest contributor to value added in the food industry is the retail FAFH or restaurant industry, with $351 billion in value added. This is followed by farm and retail (which includes food processing and packaging, transporting, and whole selling) at $322 billion. CED (2017, p. 30): "Retail trade and farm production are third and fourth, with $138 billion and $112 billion, respectively." These benefits must be weighed against their costs—for example, the health costs associated with unhealthy FAFH.

In terms of how each part of the conventional food system contributes to added value,

farming and agribusiness input firms [were] responsible for just 12 percent of total economic value created in the food system (9.7 percent and 2.4 percent, respectively). Food processing and packaging together represent[ed] roughly 19 percent of value, while the food retail and food services sectors contribute[d] the greatest economic value added, with more than 44 percent of the total in 2012.

*(Nesheim et al., 2015, p. 39)*

## Costs–Benefits Analysis of the Conventional Food System

A costs–benefits analysis of the conventional food system is difficult for the simple reason that it is difficult, if not impossible, to put a dollar figure on some of the benefits (and especially the costs) of the system. The following effort, shown in Table 9.3, should be considered a rough first effort.

**TABLE 9.3** Costs and Benefits of the Conventional Food System

| Benefits | Costs | Costs | Costs |
|---|---|---|---|
| Food produced | $1.4 trillion | Poor access to health insurance | |
| GDP | $776 billion | Low wages | |
| Farm sales | $400 billion | Employment of undocumented immigrants* | |
| Company revenues | $100s billions | Food insecurity | $90 billion |
| Employee pay | | Death | 395,000 ($22 billion) |
|    Restaurants | $141 billion | Foodborne illness | 48 million |
|    Farms | $100 billion | Foodborne deaths | 3,000 |
|    Food stores | $58 billion | Diabetes | $245 billion |
|    Grocery | $34 billion | Obesity | $147 billion |
| Taxes paid | $10 billion | Lost productivity | $10s billions |
| Exports | $70 billion | Salary loss | |
| Value added | | Obesity | $17 billion |
|    Restaurants | $351 billion | Diabetes | $4 billion |
|    Processing | $322 billion | | |
|    Food/beverage | $164 billion | | |
|    Retail | $138 billion | | |
|    Farms | $112 billion | | |

*Source:* * Barnhill et al. (2018, p. 17) assert that "between 25% and 70% of farm labor comes from illegal immigrants."

As you can see in the table, the benefits of the food system are enormous and appear to outweigh its costs. However, many of the benefits of the system end up producing significant costs (e.g., feeding billions of people leads to obesity, diabetes, illness, and death, as well as environmental degradation which also imposes its own costs). These costs can surely be reduced through more sustainable and ethical production of food.

Another analysis of the food system attempted to establish its costs and benefits. According to Van Nieuwkoop (2019): "[Food and Agriculture Organization of the United Nations] estimates the gross value of global (primary) agricultural production at just over $5 trillion. The World Bank estimates (primary) agricultural value-added at about $3.2 trillion." After estimating that the food system generates between two and five times as much value as farm production, Van Nieuwkoop (2019) estimates that "the value of the global food system at roughly $8 trillion, or 10 percent of the $80 trillion global economy." He then goes on to claim that "negative impacts associated with the way the current food system operates are at least $6 trillion," although this is a conservative estimate that does not take into account many costs that studies have yet to attach amounts to. These include biodiversity losses, health-care costs associated with pesticide and chemical exposures, and costs associated with rising anti-microbial resistance. And Van Nieuwkoop (2019) writes:

> Note that the huge costs associated with biodiversity and ecosystem losses are not fully accounted for in the table above. The loss of pollination services which are essential for nearly 70 percent of global food production would be catastrophic. By some estimates, these costs would completely swamp the current global GDP.

So, the costs of the system likely outweigh its benefits, once everything is considered!

## Conclusion

In this chapter, I made an effort to establish the economic costs and benefits of the conventional food system in the United States. It is enormous and global companies that produce the foods we consume. The analysis shows that the benefits of the system are enormous and may outweigh the costs of the system. Yet, the costs of the conventional food system are enormous and not sustainable. Not only are people getting sick and dying at alarming rates based on what they eat (and do not eat), the environmental costs of our way of life may one day literally end life on Earth.

The costs of the conventional food system can be reduced through efforts of food companies to produce and deliver healthier food and to more sustainably do so. This is the key challenge of the multi-national companies who produce our food, and it is up to the state (i.e., national and international governmental bodies) to hold the key actors in the food system accountable for this outcome. Just as drug dealers are culpable for the drugs they provide to willing users, the companies who produce our foods are culpable for the foods (and beverages) we willingly consume. As one example, a 2006 Institute of Medicine report concluded that "food and beverage marketing practices geared to children and youth are out of balance with healthful diets, and contribute to an environment that puts their health at risk" (Bread for the World, 2016). Chandon and Wansink (2012, p. 573) also proposed that food marketing has contributed to obesity by increasing the accessibility to large portions of inexpensive, tasty, and calorie-dense food.

The governments and government agencies also share some culpability for the harms of the conventional food system. In this way, the harms associated with the conventional food system can be considered state-corporate crimes and are thus relevant for criminology.

# 10

# SUMMARY AND THE FUTURE

## Introduction

In this final chapter of the book, I offer a summary of the book and then address some potential reforms we can pursue to (1) make food healthier and (2) hold food companies responsible for their deviant and criminal behaviors. Though I am a fan of local, organic, sustainable food, it is no panacea. Moreover, though my reforms may be charged with merely tinkering around the edges of real systemic reforms, I do not advocate for toppling the conventional food system and replacing it with something different; I just do not see this as realistic, nor it is likely to be effective. As some food scholars have noted, it is simply not possible to feed the world through a food system that is exclusively local, for example. Instead, I offer suggested reforms that aim to change the system to make it more effective at providing better nutrition as well as more just for consumers, workers, and even animals.

## Summary of the Book

In Chapter 1, you learned that food crime refers to both deviant and criminal acts committed by key actors of the conventional food system. It includes everything from food fraud (criminal) to harmful food (deviant) and even environmental harms associated with the system itself. Food crime exists in the context of cheap capitalism, where profit-seeking underlies virtually every behavior discussed in this book. Contemporary dietary patterns now kill more people each year than murder, as well as all illicit drugs combined. Given the involvement of key government agencies in the production and sale of food, plus its (lack of) regulation for safety and health, the concept of state-corporate crime can help us frame the issue of street crime as an act of elite deviance (behaviors committed by the powerful, some illegal, some not), conducted by corporations for profit, with the assistance, approval, reward, and/or complicity of government. As shown in this book, there is culpability among both food companies and government agencies when it comes to matters of food crime. Even in cases where there is no intent to harm people or animals through food crime, there is often negligence, recklessness, and knowing behaviors involved in food crime.

DOI: 10.4324/9781003296454-10

In Chapter 2, you learned about the conventional food system, the term used to describe all of the actors and organizations involved in the growth, production, and sales of food products. The system is highly industrialized (Chickenized and McDonaldized), corporatized, monopolized, and globalized. The major implication of these realities is that the great bulk of foods being produced are largely unhealthy and the same nearly everywhere around the world. They are grown, produced, marketed, and sold by companies that are far less concerned with quality or healthiness than they are with turning a profit for Wall Street and shareholders. This has enormous health consequences for Americans as well as people around the world.

Chapter 3 showed what a healthy diet is comprised of, based on the government guidelines for food intake. A comparison of that with what people eat shows that Americans eat too much fat, saturated fat, sugar, and salt, plus lots of additives, and far too few vegetables, fruits, vitamins, and minerals. This reality exists for at least two major reasons. First, food companies historically played a large role in hindering the release of adequate food guidance for consumers. Hence, the major problems with mechanisms such as the food pyramid, MyPyramid, and MyPlate. Further, major food corporations latched onto the concept of energy balance to promote exercise as the solution to the many problems actually largely caused by poor diet! Second, the foods being produced by major food corporations are largely unhealthy. In particular, the cheapest and most widely available foods are the ones most consumed by Americans, as well as people around the world.

Chapter 4 illustrates that it is the food itself that is the first crime. That is, the food itself is largely unhealthy, with high levels of fat, saturated fat, sugar, and salt, plus potentially unhealthy additives. The use of antimicrobials in animal feed is also a significant problem within the conventional food system as it is leading to increasing cases of antibiotic resistance in humans. I do not argue that any food should be made criminal and thus am not actually calling on lawmakers to make unhealthy foods food "crimes," but I am instead calling the attention of regulators to better regulate the production of these foods and to greatly restrict their marketing, especially to vulnerable children.

Chapter 5 demonstrates that many of these foods can be considered drugs in that they produce changes in the brain, lead to cravings, produce tolerance and withdrawal, and are habit-forming and lead to compulsive use even in the face of serious deleterious outcomes associated with their consumption. In particular, ultra-processed and hyperpalatable foods seem to be the most addictive, and these are the foods being most produced in the United States and around the world. Surely food companies are culpable for the harmful outcomes related to food addiction produced by the foods they create and sell.

In Chapter 6, you learned about the harms associated with the standard American diet (or SAD). This includes major illnesses including obesity, diabetes, plus mental health, and other conditions. Nearly 400,000 Americans die every year from the foods they eat (and do not eat enough of). Health-care costs associated with poor diet now dwarf those direct harms associated with street crime every year. You also saw what a healthier diet looks like and were confronted with the issue of whether people actually have a full choice when it comes to the foods they eat. I argued that the notion of free will is a myth, particularly when it comes to what we eat; we eat what we like but we do not choose to like it, at least not freely.

In Chapter 7, you learned about the harms associated with the conventional food system itself. These include harms associated with killing and eating animals, including animal welfare concerns stemming from factory farming and CAFOs, pathogens in our food and environment from food production, hazardous working conditions, significant environmental damage, plus the confusing dual reality of food insecurity and food waste. The issue of environmental damage may stand out the most, given the enormous and serious consequences of issues such as climate change. Food companies must do more to reduce their environmental harm footprint, and consumers must insist that they do.

Chapter 8 dealt with the issue of the culpability of food companies for specific behaviors. Those include producing excess calories than needed by consumers to be and stay healthy, putting food in non-food environments so that we are surrounded by junk food nearly everywhere we go, funding research to create one-sided studies, using front groups to confuse consumers as part of the industry playbook to keep regulators and lawmakers at bay, advertising unhealthy products, deceptive advertising, food fraud, and product shrinkage, not to mention all the deleterious health outcomes documented in Chapter 4 and other harms of the system documented in Chapter 5. There is great recklessness in food companies when it comes to creating more food than is needed by the population, especially when so much of it ends up going to waste. Yet, of greater importance is the recklessness associated with surrounding us with especially unhealthy foods essentially wherever we go. There is likely specific intent to mislead when it comes to funding research and using front groups to make claims and defeat honest efforts to make food healthier and less harmful. There is also likely intent involved in the misleading claims made by food companies like those discussed in the book, advertising unhealthy products just to boost sales, and shrinking the size of packages over time to create the false impression that people are getting more food than they really are when they make purchases.

Finally, Chapter 9 shows that there are significant economic benefits of the conventional food system, including incredible access to food across the nation and globe, expansive contribution to GDP, significant employment, access to health care, as well as enormous multiplier effects. The contribution to GDP is more than $1 trillion per year, and tens of millions of people are employed in some part of the system. You saw, however, that wages associated with food system work tend to be quite low relative to other jobs and that benefits are often lacking completely in the case of part-time and seasonal workers. Though a full costs–benefits analysis of the system is beyond the scope of the book, you saw that the benefits of the system may very well outweigh its costs, although the costs of the system are still enormous and need to be reduced.

Clearly, there are significant problems associated with food. Many of these make up the focus of the food crime literature, the subject of this book. Where this book differs from other scholarship in the field is that this book explicitly focuses on harms associated with both food and the conventional food system, as well as identifies the culpable harms that comprise food crimes. Stated simply, we must hold food corporations responsible for the culpable acts they commit. Similarly, governments must be held responsible for these harms as well, since it is their obligation to regulate corporations in the public interest.

## Suggested Reforms

According to Winson and Choi (2019, p. 51), a broad spectrum of actions has started in resistance to industrial food production and sales,

> from local school boards banning the sale of soft drinks in schools (in the United States) to sweeping new guidelines on what can be served in schools (Canadian provinces),

nationwide taxes on soft drinks (Mexico), and bans on advertising nutrient-poor edible products on television programs aimed at children and teens (Britain).

Given the wide range of terrible outcomes associated with the contemporary western diet, much more can and should be done in the future (and present) to restrict the activities of food companies aiming to maintain the status quo. Some potential reforms are suggested shortly.

For now, consider some recommendations from the WHO. WHO produced a report in 2016, issuing three non-binding conclusions and recommendations: "Countries have a duty to ensure that food, beverage, and supplement companies do not exert undue influence over public health missions" (Nestle, 2018, pp. 208–209). This is a no brainer, for, why would we even let private, for profit businesses impact our public health goals? Stopping the impact of food companies on government dietary advice is a good place to start. Another helpful suggestion is to limit corporate donations to politicians and political parties, though this would take an act of Congress.

WHO has also issued guidance for foods being marketed to infants and young children:

WHO recommends that member states set policies that do not allow infant-formula companies to donate free samples to health facilities; host events, contests, or campaigns in health care facilities; provide gifts or coupons to health care staff, parents, or families; host events, contests, or campaigns in health care facilities; distribute educational materials to parents; or sponsor meetings of health or science professionals.

Shouldn't this be the standard for any food or beverage product, especially given that a large majority of available foods are so unhealthy?

Then there are suggested reforms stemming from the reality that our conventional food system is so industrialized. I agree with Silbergeld (2016, p. 245), who writes:

If industry has a right to be industrial, we have a right to treat it accordingly, and to expect it to comply with the rules and regulations that constrain the operations of other industries, including restraint on economic concentration and legal responsibility for currently externalized costs.

Restraining economic concentration likely refers to breaking up food monopolies, something that would likely be vastly unpopular, politically speaking. But, I support applying anti-trust laws to all corporations, as well as making food companies legally responsible for the costs they impose on society.

Silbergeld also insists that industrial food production abide by anti-monopolization laws (as just noted), "disclose the presence and use of hazardous materials in their operations," "provide evidence, on the public record, of the safety of all materials used in production," and "bear full liability for unsafe products and the burden of guaranteeing safety of their products." The government's role here is to make sure this occurs, through the use of effective regulation. All of this simply refers to the expectation that food companies follow the law as well as be good, moral citizens of society. Silbergeld adds that governments "institute and maintain adequate programs of oversight and inspection to assure worker and food safety within the industry" and "comply with statutory requirements to devise and implement adequate and transparent systems of inspection and regulation," all suggestive of adequate regulatory capacity. She also suggests that government must "approve any proposed change by any industrial enterprise . . . that could affect worker safety, food safety, and the environment," "collect and publish data on industrial operations in terms of hazardous materials, waste disposal, worker health and safety, and violations of food safety standards," and "promulgate biding regulations on industrial processes that ensure protection of human health, the food supply, animal welfare, and the environment" (pp. 247–248). I support all of these ideas.

Silbergeld continues, saying food companies should "be responsible for the entire life-cycle of production, including their external impacts, especially regulation and reduction of waste streams," "comply with local and state regulations on zoning in terms of location and concentration," "obey the labor laws of the country, including providing a safe workplace for all, not employing children, permitting the lawful organization of workers into unions, and ensuring a minimum wage, "be subject to workers' compensation laws," and "permit on-site inspections by authorized personnel of local, state, and national agencies" (pp. 246–247). None of these are outlandish, and any decent company that cares about its employees, customers, and community should welcome these steps.

I definitely want to see more done in the area of reform of the conventional food system than just making modest changes. We can take the advice of Chandler (2019, p. 30) in order to revise our current food situation by engaging in what he calls "creative destruction." It "describes the chaos that ensues when a new industry built on a new technology emerges and leaves networks of outdated industries in its wake." Undoubtedly, new technologies will change the food system. Yet, we can apply this concept more broadly and ask the question, what changes need to be made to destroy the parts of the conventional food system that need to be left behind in order to create better outcomes in our nation and around the world? In the following, I offer some suggestions; some of them, at least, aim for systemic change.

Based on the research presented in this book, I offer the following 25 suggested reforms. Some are aimed at assuring more healthy eating, others at informing consumers, still others at protecting workers and the environment, and finally others are intended to hold food corporations responsible for their deviant and criminal acts.

1. Grow, produce, and sell more healthy foods
2. Make food healthier by reducing the number of processed and ultra-processed foods and reducing the amount of salt, sugar, and fat in food
3. Regulate food additives through a public health approach led by the government
4. Ban antimicrobials in animal feed
5. Encourage vegetarianism
6. Grow a much wider variety of crops
7. Subsidize fruits and vegetables
8. Reduce the number and size of factory farms, CAFOs
9. Inform consumers about the reality of food addiction and hold food corporations responsible for it
10. Turn the regulation of food pathogens over to the government
11. Sell imperfect fruits and vegetables to reduce food waste
12. Educate consumers about ways to reduce food waste

13. Feed the world to reduce food insecurity
14. Hold food corporations responsible for the environmental damage they create
15. Insist on safe workplaces for all food-based employees
16. Hold food corporations responsible for their deviance and criminality
17. Ban the sale of food in non-food environments
18. Insist that all food-related researchers identify which companies they work for or are paid by
19. Create a national, searchable database on food-related research
20. Create a national, searchable database on food-related front groups
21. Ban deceptive advertising and insist the government call out companies engaged in this behavior
22. Vigorously pursue cases of food fraud
23. Outlaw product shrinkage
24. Ban the advertising of unhealthy products to children
25. Counter the advertising of unhealthy products with truth-related ads showing their harms

Each suggested reform is briefly discussed in the following.

### Grow, Produce, and Sell More Healthy Foods

Food manufacturers must simply produce more healthy foods, and a greater portion of the foods produced must be of the healthy variety rather than of the processed and especially the ultra-processed variety. Much more focus should be placed on fruits and vegetables and greater efforts should be made to make this a higher proportion of our diets.

### Make Food Healthier by Reducing the Number of Processed and Ultra-Processed Foods and Reducing the Amount of Salt, Sugar, and Fat in Food

Processing is often necessary for the production of food. Ultra-processing is unnecessary for healthy foods. While many of the hyperpalatable foods we like are of the ultra-processed variety, we are advised to eat few of these foods anyway. So, they should not be so readily available and especially in such large amounts.

### Regulate Food Additives Through a Public Health Approach Led by the Government

No one knows to what degree additives in food pose threats to consumer safety, and that is largely because no one really knows what is being added to our foods. This is an important issue of public safety, and, as such, a government agency should be involved in the regulation of food additives. Moreover, it, as a representative of the people, should make decisions about what is allowed in our food in the first place.

### Ban Antimicrobials in Animal Feed

Antimicrobials have no place in our food. Nor do they need to be added to animal feeds. As such, they should be banned by the government.

### Encourage Vegetarianism

Vegetarians tend to be healthier than meat-eaters. While it may not be wise to advise people not to eat any meat, it makes plenty of sense to encourage people to eat vegetarian to whatever degree they can.

### Grow a Much Wider Variety of Crops

There exist a wide variety of plants that can be grown for food that currently are not. Instead of growing just wheat, soy, and corn, farmers should be encouraged to expand the variety of crops they grow and harvest. Consumers like more choices and an educational campaign to inform consumers of the benefit of additional food products in their diets would go a long way toward assuring that new foods would be purchased and eaten.

### Subsidize Fruits and Vegetables

Fruits and vegetables should not be treated as "specialty crops," especially when leading dietary advice is to make these foods the center of one's food intake. As such, the government has a vested interest in subsidizing the production of fruits and vegetables so that more of them can be grown and brought to market.

### Reduce the Number and Size of Factory Farms, CAFOs

When more people switch from meats to fruits and vegetables, there will be less need for factory farms and concentrated animal feed operations. Both types of facilities, though needed to keep up with consumer demand for animal food products, will be less necessary when people eat less meat.

### Inform Consumers About the Reality of Food Addiction and Hold Food Corporations Responsible for It

Some foods act as drugs on the brain and lead to food addiction, very similar to drug addiction in many important ways. Food producers act as drug dealers when they target people with especially hyperpalatable foods with no regard whatsoever to their health and well-being. Lawsuits against big tobacco went a long way to convincing people that tobacco is dangerous to one's health, and they produced clear, convincing, and even damning evidence of the culpability of tobacco companies. Similar lawsuits against big food corporations may produce similar evidence showing they know about the harms of their products yet they continue to produce and sell them anyway.

### Turn the Regulation of Food Pathogens Over to the Government

The current system of regulating food pathogens is controlled by industry, and it does not necessarily operate in the interest of workers or consumers. The government must get back involved in regulating producing foods, at all stages of production, to reduce the likelihood that dangerous pathogens are present in our food.

### Sell Imperfect Fruits and Vegetables to Reduce Food Waste

Food producers operate on the assumption that people do not want to buy imperfect foods. Yet, with proper information, consumers can be encouraged to appreciate bruised or odd-shaped fruits and vegetables in the interest of reducing the waste of food which contributes greatly to climate change.

### Educate Consumers About Ways to Reduce Food Waste

A national campaign should be launched to educate people about other ways they can reduce food waste. This might include learning more effective ways to use ingredients as well as recipes to create new meals from leftovers.

### Feed the World to Reduce Food Insecurity

International, national, state-level, and even local organizations already exist to provide food to people in need. Further, government programs assist those who most need help. Yet, all of these efforts must be stepped up if we want to eradicate hunger once and for all.

### Hold Food Corporations Responsible for Environmental Damage They Create

Environmental harm is the cost of doing all kinds of business, including the food business. Currently, corporations externalize these costs to the public. Governments must insist that corporations pay for the damages they cause. This is a sure way to get profit-seeking companies to reduce the costs that they must end up paying.

### Insist on Safe Workplaces for All Food-Based Employees

It is immoral to employ people in hazardous conditions when those hazards can be reduced if not eliminated. Just because some people are willing to do such work (e.g., poultry processing) does not mean they deserve to suffer painful and often debilitating injuries. Injuries to employees are often seen as the cost of doing business. This must change and companies must be held accountable for the harms they inflict on their employees.

### Hold Food Corporations Responsible for Their Deviance and Criminality

This is an easy one. We hold street criminals responsible for their crimes; there is an entire criminal justice system designed to do this very thing. We must use this same system to hold those in power accountable for any crimes they commit. Beyond this, any acts of deviance that rise to the level of great seriousness in terms of harms caused should be legislated as crimes.

### Ban the Sale of Food in Non-food Environments

This might be controversial since Americans tend to be very sensitive to government efforts to restrict their freedoms. Yet, when you start from the premise that food does not *need* to be present in so many areas where it is present (especially junk foods in non-food environments), it is not so unreasonable to suggest that this practice be stopped. When you consider the clear harms associated with the consumption of hyperpalatable, ultra-processed foods, the government has a vested interest in reducing people's exposure to such foods.

### Insist That All Food-Related Researchers Identify Which Companies They Work for or Are Paid by

Nutrition research is essential to consumers being able to make informed food choices. Yet, we tend to *not* know who is conducting the research or how that can impact the research focus, results, and conclusions. Anyone who conducts food-related research must be forced by law to identify who funds their work and discuss the implications of especially corporate sponsorships.

### Create a National, Searchable Database on Food-Related Research

Consumers should have one reliable source of information when it comes to health and food. A national, searchable database on each and every food, including what is known about it and its potential harms and/or benefits would assist consumers in making more informed food choices.

### Create a National, Searchable Database on Food-Related Front Groups

Another useful database of food-related front groups would enable consumers to search any front group who is making claims about any issue in public media. This would assist us with knowing the interests being served by the claims-making activities of such groups.

### Ban Deceptive Advertising and Insist the Government Call Out Companies Engaged in This Behavior

Deceptive advertising is illegal. And it includes any ad that has the potential to mislead. The government must start enforcing its laws in this area so that when people are tricked to buy products that make questionable or false claims, some meaningful consequences will follow. Ideally, enforcing such laws will reduce the likelihood that companies will make such questionable and false claims in the first place.

### Vigorously Pursue Cases of Food Fraud

Food fraud is illegal, and so it should be the focus of some law enforcement agencies (e.g., the Federal Bureau of Investigation). No one should ever face the possibility of being tricked out of their money through fraudulent activities of food corporations or any other entity or individual selling food products.

### Outlaw Product Shrinkage

Product shrinkage, where less and less of a product is offered to customers in the same size packages over time, is deviant behavior. It is unethical and immoral and should be banned.

Consumers should not be tricked out of their money, especially by companies who aim to get people's business by selling them products consumers want and love.

### Ban the Advertising of Unhealthy Products to Children

Other countries have banned or greatly limited advertising food products to children. Free speech rights should not apply to corporate speech meant to sell products. Children are especially vulnerable to advertising, and parents are often too busy or distracted by real life to adequately shield their kids from it. Here, the government can help by banning the advertisement of unhealthy food products to children.

### Counter the Advertising of Unhealthy Products With Truth-Related Ads Showing Their Harms

Anti-tobacco ads have been among the most effective prevention ads ever created. A similar approach with food could inform consumers with the truth about the products being advertised to them by food companies. Consumers should be informed about which products are healthy and which are unhealthy, and the government can counter the impact of certain food ads that use cartoon characters and the like by showing what is really in those products.

## The Future: Toward Food Justice

Should such reforms be enacted, we would be on our way toward greater food justice. Gottlieb and Joshi (2010, p. 223) note that food justice aims for

> justice for all in the food system, whether producers, farmworkers, processors, workers, eaters, or communities. Integral to food justice is also a respect for the systems that support how and where the food is grown—an ethic of place regarding the land, the air, the water, the plants, the animals, and the environment.

The term *food justice* could refer to so many things given the broad nature of justice itself. At a minimum, it includes issues related to earned wages and working conditions within the conventional food system, food insecurity and inadequate access to high-quality food, racism, sexism, and other isms within the food system itself, unequal participation and ownership in decision-making about food, environmental unsustainability, and indigenous people's rights to control and have access to their own food (Barnhill et al., 2018; Colas et al., 2018). In food justice, farmers and all food workers would be earning a living wage, all workplaces would be safe and managers would be truly concerned with assuring the safety of their workers, food would be healthy and sustainably grown, and much greater efforts would be made to diminish the environmental harm footprint of food production.

The USDA already has programs in place that are consistent with food justice, such as its "farm-to-institution programs that support the purchase of fresh, local food in cafeterias at hospitals, senior centers, and city offices" and its "Healthy Food Financing Initiative" (with the Department of HHS) that seeks "to address the absence of fresh food in low-income neighborhoods" by offering "financing and marketing support for grocery stores, farmers markets, and other food sources to enter these areas" (Alkon, 2019, pp. 359–360). A vast expansion of such programs at both the state and national level is necessary to achieve goals related to food justice.

A term often even used synonymously with food justice is "food sovereignty," defined as "the view that not only should food be both produced and consumed in local food-sheds . . . but that the local populations should also own and control these local food systems" (Budolfson, 2016, p. 177). Local food is a general principle associated with food justice, as well, but its feasibility for feeding the world is presumed to be not achievable. Food sovereignty also "rejects the status quo of a globalized food system with enormous power exercised by multinational corporations, along with the underlying so-called 'neoliberal global order' and its imposition of institutions of so-called 'free trade' " (Budolfson, 2016, p. 177). Here, it is important to recognize that, although some food system reformers want to replace the system with something different—stripping global food companies of their power and growing food through other, more sustainable means—the likelihood of achieving this is small to non-existent, and so the reforms offered earlier in the chapter are aimed at reforming the existing system to make it more effective and more just.

Whyte (2018, p. 363) defines food sovereignty as "the right of peoples and governments to choose the way food is produced and consumed in order to respect livelihoods." And Powers (2018, p. 392) notes it would "empower the poor by providing them with legal ownership rights that cannot be overridden by authoritarian governments and create legal protections of land and water from devastation by those who have little long-term stake in environmentally sustainable and socially beneficial enterprises." Silbergeld (2016, p. 26) notes that food sovereignty generally refers to "protection for special products, usually identified as foods that are culturally important rather than essential to national food independence."

Wittman (2019) shows how food sovereignty has been nationalized in some nations including Ecuador, where its Constitution calls it both an objective and obligation, and Brazil, through its "Zero Hunger" program, its amended Constitution that includes a right to food, and its food security and education programs. It is unlikely (dare I say impossible) that any national right to food or food sovereignty will ever exist in the United States, but efforts toward these goals are welcome. These are more likely to occur at the local level in places where more progressive politicians and policymakers are in place.

Even in the face of stiff resistance to needed change, it is clear the conventional food system is plagued by serious problems that it is not yet correcting. As noted by Holt-Gimenez (2019, p. 88): "The corporate food regime is not ending hunger or ensuring dietary health for the world's people. Nor is it effectively addressing the problems of climate change, environmental destruction, or resource depletion." He advocates replacing the conventional food system with *agroecology*, characterized by farmers:

- Using natural sources (e.g., animal manures, legumes, and cover crops) to provide nutrients for the soil
- Controlling weeds with cover-cropping, inter-cropping, and mulching
- Managing pests with predators, crop rotations, ally-cropping, trap crops, and repellant crops.

*(pp. 98–99)*

Agroecology essentially means ecologically friendly agriculture that is environmentally sustainable, as well as empowering of farmers and local communities.

Holt-Gimenez writes that "agroecology is anathema to capitalistic agriculture . . . [it] doesn't provide opportunities for agribusiness to sell seeds, fertilizers, or pesticides" (p. 99). Thus, such a program will not be popular among anyone with ties to the current conventional food system, including, unfortunately, lawmakers at both the state and federal levels in the United States. And one fundamental question that needs addressing is how already existing food organizations can find ways to implement agroecology as part of their routine operations.

A national if not international movement toward such an agricultural approach is apparently underway, through means such as the International Panel of Experts on Sustainable Food Systems, which holds that sustainable means of agriculture can in fact "grow all the food we need to feed a growing population" (Wise, 2019, p. 277). Even the UN Food and Agriculture Organization (FAO) held a conference titled, "Scaling up Agro-ecology" that made the same type of claims. The FAO has a website devoted to agroecology at: www.fao.org/agroecology/home/en/. That such practices potentially run counter to the economic interests of big agri-business, one might expect significant opposition and barriers to the successful implementation of agroecology at either the state or national level. Yet, I believe global food corporations have a vested interest in embracing and utilizing it for their own survival.

You have likely noticed the term sustainable as part of both food justice and food sovereignty movements. Hatanaka and Konefal (2019, p. 368) define sustainability as "the resilience of people, communities, and ecological systems over time." They also provide the United Nations FAO definition: "ensuring human rights and well-being without depleting or diminishing the capacity of earth's ecosystems to support life, or at the expense of others well-being." Hatanaka and Konefal (2019, pp. 368–370) identify different elements of sustainability. They include:

1. Environmental—maintenance of ecological systems over time
2. Economic—ensuring sufficient economic production and development
3. Social—maintaining social systems over time

Hatanaka and Konefal also outline the three types of efforts engaged in achieving sustainability. They include:

1. Community-based initiatives—investments in alternative agriculture, such as local food, food sovereignty, and food justice
2. Standard, metric, and certification initiatives—academic, professional, civil society, and environmental organizations develop goals and measures to assess performance, verification of compliance with standards, and communication to consumers
3. Intensification initiatives—increases in productivity and efficiency through technological innovations

How are we doing in this area? Oxfam releases its "Behind the Brand" scorecard to rank major food companies on their commitments to sustainability. The top-ranked company was Unilever, with high scores in climate control, farmers' and workers' rights, plus improvements in the treatment of women as well as water management. The second was Nestle, and the third was Coca-Cola (Oxfam, 2016). Next were Kellogg's, Mars and PepsiCo, Mondelez, General Mills, and finally, Associated British Foods Plc and Danone.

The highest scores overall were in the category of climate, with Unilever, Nestle, and Kellogg's leading the way in this area. Meanwhile, farmers' and workers' rights had the lowest scores, with several companies receiving poor ratings including Coca-Cola. Kellogg's and Unilever have made the most progress, at 30% and 26%, respectively, since the campaign began. Meanwhile, "the world's leading supermarket and fast-food companies are doing little to address the environmental and human rights abuses associated with beef production," according to E360 (2022). Scores are depicted in Figure 10.1.

## Company Scores

| Scoring Key | 0 - 39 | 40 - 69 | 70 - 100 |
|---|---|---|---|

| Company | Score |
|---|---|
| TESCO | 65/100 |
| MARKS & SPENCER | 62/100 |
| Carrefour | 61/100 |
| McDonald's | 54/100 |
| Walmart | 49/100 |
| Jerónimo Martins | 45/100 |
| Yum! KFC Pizza Hut Taco Bell | 44/100 |
| Sainsbury's | 43/100 |
| rbi restaurant brands international Burger King Popeyes | 41/100 |
| METRO | 36/100 |
| Costco Wholesale | 28/100 |
| Auchan | 24/100 |
| Ahold Delhaize | 19/100 |
| ALDI | 14/100 |
| REWE Group | 9/100 |

**FIGURE 10.1**   Company Scores From Mighty Earth

*Source:* https://e360.yale.edu/digest/most-global-food-brands-continue-to-have-a-dismal-record-on-beef-and-deforestation

The website E360.com notes:

> only four companies—Tesco, Marks & Spencer, Carrefour, and McDonalds—are taking steps to stop purchasing beef from destructive suppliers. The remaining 11 food companies surveyed, including U.S.-based Costco, are not taking the necessary steps to stop procuring beef from suppliers with a poor record of sourcing cattle from deforested areas in the Amazon and elsewhere.

This is troubling because it is organizations such as supermarkets and fast-food companies that can enforce sustainability standards by ceasing to get beef supplied from growers who devastate the environment through means such as deforestation in places like the Amazon and in countries such as Argentina and elsewhere. Food justice requires such actions.

Interestingly, some projections for the future of food are consistent with the food justice movement and the issue of sustainability. For example, an analysis by ADM outlines trends that will likely impact food and beverage companies as well as consumers. They include:

- Consumers wanting to and buying products that help sustain health and/or contribute to healthy functioning of the body and mind, things like "supporting immune systems, enhancing mood and sustaining energy"
- Consumers desiring and purchasing products that are sustainably produced
- Consumers being interested in products that target the microbiome that can help with "weight management, immune system support and better emotional well-being.
- Consumers investing their dollars in plant-based products
- Consumers demanding transparency around issues such as where their food comes from.

*(ADM, 2021)*

To the degree these are true, consumers apparently desire healthier foods, as well as foods that are sustainably grown and produced. All of this is consistent with the food justice movement.

Four additional opportunities to redesign the food supply chain include "diversifying ingredients; sourcing lower-impact ingredients; upcycling ingredients and shifting to regenerative methods of food production" (Edie, 2022). There are reportedly thousands of edible plant species that are not being harvested and produced for food purposes; these should be invested in to achieve some of these goals. Another important step would be "to shift away from animal products to lower-impact alternatives" (Edie, 2022). A move to vegetarianism would reduce many of the problems associated with the conventional food system, as noted earlier. Each of these is also part of the food justice platform.

Food justice is ideally not some "pie in the sky" concept that is not obtainable. It is something that, simply stated, can be viewed as the opposite of food crime. Those behaviors that are immoral and/or illegal and that are harmful and committed with culpability are unjust actions. Food justice requires not only that those who commit them be held accountable but also that we stop allowing them and even enabling them. Food justice demands this. It is past time to do something about this.

# REFERENCES

Access to Nutrition Initiative. (2022). *Global Index 2021*. Retrieved from https://accesstonutrition. org/index/global-index-2021/

Action on Sugar. (2017). *Sugar and Health*. Retrieved from https://www.actiononsugar.org/sugar-and-health/

ADM. (2021). *ADM Announces Global Trends Set to Drive Nutrition Innovation for 2020*. Retrieved from https://www.adm.com/en-us/news/news-releases/2022/11/adm-announces-global-trends-set-to-drive-nutrition-innovation-for-2023/

Alexander, E., Yach, D., & Mensah, G. (2011). Major multinational food and beverage companies and informal sector contributions to global food consumption: Implications for nutrition policy. *Globalization and Health*, 7, 26–34.

Alexander, L., & Ferzan, K. (2018). *Reflections on crime and culpability: Problems and puzzles*. Cambridge, MA: Cambridge University Press.

Alkon, A. (2019). Food and justice. In J. Honefsal & M. Hatanaka (Eds), *Twenty lessons in the sociology of food and agriculture*. New York: Oxford University Press.

American Diabetes Association. (2017). *Diabetes myths*. Retrieved from www.diabetes.org/diabetes-basics/myths/

American Psychiatry Association. (2020). *What is addiction?* Retrieved from www.psychiatry.org/patients-families/addiction/what-is-addiction

American Society of Addictive Medicine. (2020). *Definition of addiction*. Retrieved from www.asam.org/Quality-Science/definition-of-addiction

Anderson, E., & Durstine, J. (2019). Physical activity, exercise, and chronic diseases: A brief review. *Sports Medicine and Health Science*, 1(1), 3–10.

Anderson, L. (2017). *Deviance: Social constructions and blurred boundaries*. Berkeley, CA: University of California Press.

Aramark. (2018). *Leading food companies form the Global Coalition for Animal Welfare*. Retrieved from www.aramark.com/about-us/news/aramark-general-global-coalition-for-animal-welfare-2018

Archer-Daniels-Midland. (2018). *Our Company*. Retrieved from https://www.adm.com/our-company

Arencibia-Abite, F. (2020). Serious analytical consistencies challenge the validity of the energy balance theory. *Heliyon*, 10(6), 7.

Arencibia-Albite, F., & Manninen, A. (2021). The energy balance theory: An unsatisfactory model of body composition fluctuations. *MedRxiv*. Retrieved from www.medrxiv.org/content/10.1101/2020.10.27.20220202v10

Arumugam, B., Suganyu, A., Saranya, N., & Suveka, V. (2015). Fast food addiction—The junk enslavement. *International Archives of Integrated Medicine, 2*, 62–70.

Aschwanden, C. (2021, June 30). Prohibited, unlisted, even dangerous ingredients turn up in dietary supplements. *Washington Post.*

Asomah, J., & Cheng, H. (2019). Food crime in the context of cheap capitalism. In A. Gray & R. Hinch (Eds.), *A handbook of food crime: Immoral and illegal practices in the food industry and what to do about them.* Bristol: Policy Press.

Ayaz, A., Nergiz-Unal, R., Dedebayraktar, D., Akyol, A., Pekcan, A. G., Besler, H. T., & Buyuktuncer, Z. (2018). How does food addiction influence dietary intake profile? *PLoS One, 13*(4), e0195541.

Bakan, J. (2004). *The corporation: The pathological pursuit of profit and power.* New York: Free Press.

Barbarosa, C. (2019). Consumer reactions to food safety scandals: A research model and moderating effects. In A. Gray & R. Hinch (Eds.), *A handbook of food crime: Immoral and illegal practices in the food industry and what to do about them* (pp. 385–402). Bristol: Policy Press.

Barnes, G. (2020). Smithfield's largest slaughterhouse struggling to contain virus. *NC Health News.* Retrieved from https://www.northcarolinahealthnews.org/2020/05/01/smithfield-struggles-to-contain-covid/

Barnhill, A., Budolfson, M., & Doggett, T. (2018). *Oxford handbook of food ethics.* New York: Oxford University Press.

Basu, S., Yoffe, P., Hills, N., & Lustig R. (2013). The relationship of sugar to population-level diabetes prevalence: An econometric analysis of repeated cross-sectional data. *PLoS One, 8*(2), e57873.

Beirne, P. (1999). For a nonspeciest criminology: Animal abuse as an object of study. *Criminology, 37*(1), 117–147.

Benzerouk, F., Gierski, F., Ducluzeau, P., Bourbao-Tourois, C., Gaubil-Kaladjian, I., Bertin, E., Kaladjian, A., Ballon, N., & Brunault, P. (2018). Food addiction, in obese patients seeking bariatric surgery, is associated with higher prevalence of current mood and anxiety disorders and past mood disorders. *Psychiatry Research, 267*, 473–479.

Best, J. (2020). *Social problems.* New York: W.W. Norton.

Bhushan, N. (2011). Injuries, illnesses, and fatalities in food manufacturing, 2008. *US Bureau of Labor Statistics.* Retrieved from www.bls.gov/opub/mlr/cwc/injuries-illnesses-and-fatalities-in-food-manufacturing-2008.pdf

Blanchard, T., & Matthews, T. (2009). Retail concentration, food deserts, and food-disadvantaged communities in rural America. In C. Hinrichs & T. Lyson (Eds)., *Remaking the North American food system: Strategies for sustainability.* Lincoln, NE: University of Nebraska Press.

Bloom, J. (2019). Globalization of food: The world as a supermarket. In J. Konefal & M. Hatanaka (Eds.), *Twenty lessons in the sociology of food and agriculture.* New York: Oxford University Press.

Boone, R. (2013, May). GMO wheat lawsuit: Idaho farmers sue Monsanto. *Huffington Post.* Retrieved from www.huffingtonpost.com

Booth, F., Roberts, C., & Laye, M. (2012). Lack of exercise is a major cause of chronic diseases. *Comprehensive Physiology, 2*(2), 1143–1211.

Booth, M., Wilkenfeld, R., Pagnini, D., Booth, S., & King, L. (2008). Perceptions of adolescents on overweight and obesity: The weight of opinion study. *Journal of Paediatrics and Child Health, 44*(5), 248–252.

Booth, S., Coveney, J., & Paturel, D. (2019). Counter crimes and food democracy: Suspects and citizens remaking the food system. In A. Gray & R. Hinch (Eds.), *A handbook of food crime: Immoral and illegal practices in the food industry and what to do about them* (pp. 367–384). Bristol: Policy Press.

Bread for the World. (2016). *How the Twisted U.S. Food System Feeds and Epidemic.* Retrieved from https://www.bread.org/article/how-the-twisted-u-s-food-system-feeds-a-national-epidemic/

Breslin P. (2013). An evolutionary perspective on food and human taste. *Current Biology: CB, 23*(9), R409–R418.

Brown, J., Shepard, D., Martin, T., & Orwat, J. (2009). *The economic cost of domestic hunger: Estimated annual burden to the United States, 2007.* Retrieved from www.sodexofoundation.org/hunger_us/newsroom/studies/hungerstudies/costofhunger.asp

Brown, S., & Sefiha, O. (2017). *Routledge handbook on deviance*. New York: Routledge.

Brownell, K., & Warner, K. (2009). The perils of ignoring history: Big tobacco played dirty and millions died. How similar is big food? *The Millbank Quarterly, 87*(1), 259–294.

Budolfson, M. (2016). Consumer ethics, harm footprints, and the empirical dimensions of food choices. In A. Chignell, T. Cuneo, & M. Halteman (Eds.), *Philosophy comes to dinner: Arguments on the ethics of eating* (pp. 163–181). New York: Routledge.

Bunge. (2018). *Our History*. Retrieved from https://www.bunge.com/who-we-are/our-history, and *Global Scale, Local Insight*. Retrieved from https://www.bunge.com/where-we-are, and *Our Value Chain*. Retrieved from https://www.bunge.com/our-businesses

Bureau of Labor Statistics. (2018). *National Census of Fatal Occupational Injuries in 2017*. Retrieved from www.bls.gov/news.release/pdf/cfoi.pdf

Burmeister, J., Hinman, N., Koball, A., Hoffmann, D. A., & Carles, R. (2013). Food addiction in adults seeking weight loss treatment: Implications for psychosocial health and weight loss. *Appetite, 60*, 103–110.

Burrows, T., Hides, L., Brown, R., Dayas, C., & Kay-Lambkin, F. (2017). Differences in dietary preferences, personality and mental health in Australian adults with and without food addiction. *Nutrition, 9*, 285.

Cai, W., Ramdas, M., Zhu, L., Chen, X., Striker, G., & Vlassara, H. (2012). Oral advanced glycation endproducts (AGEs) promote insulin resistance and diabetes by depleting the antioxidant defenses AGE receptor-1 and sirtuin 1. *Proceedings of the National Academy of Sciences of the United States of America, 109*(39), 15888–15893.

Campbell, J. (2022, January 22). Who owns your food? Big tobacco execs, that's who. Now they push sugar. *Medium*. Retrieved from https://medium.com/in-fitness-and-in-health/who-owns-your-food-big-tobacco-execs-thats-who-67f5770d3522

Campbell Soup Co. (2018). *About Us*. Retrieved from https://www.campbellsoupcompany.com/about-campbell/, and *Our Brands*. Retrieved from https://www.campbellsoupcompany.com/campbell-brands/?regions=North%20America

Canning, V., & Tombs, S. (2021). *From social harm to Zemiology: A critical introduction*. New York: Routledge.

Caplinger, D. (2016). A short history of big tobacco's fling with food. *The Motley Fool*. Retrieved from www.fool.com/investing/2016/09/23/a-short-history-of-big-tobaccos-fling-with-food.aspx

Center for Responsive Politics. (2019). *Alphabetical listing of industries*. Retrieved from www.opensecrets.org/lobby/alphalist_indus.php

Center for Responsive Politics. (2022). *Political action committees*. Retrieved from www.opensecrets.org/political-action-committees-pacs/2022

Center for Science in the Public Interest. (2022). *Big food: Sounds a lot like big tobacco*. Retrieved from www.cspinet.org/big-food-sounds-lot-big-tobacco

Chandler, A. (2019). *Drive-thru dreams: A journey through the heart of America's fast-food kingdom*. New York: Flatiron Books.

Chandon, P., & Wansink, B. (2012). Does food marketing need to make us fat? A review and solutions. *Nutrition Review, 70*(10), 571–593.

Chandrasekhar, A. (2021). *Nutrition ranking exposes hollow claims of global food companies*. Retrieved from www.swissinfo.ch/eng/business/nutrition-ranking-exposes-hollow-claims-of-global-food-companies/46751468

Chenoweth, D., & Leutzinger, J. (2006). The economic cost of physical inactivity and excess weight in American adults. *Journal of Physical Activity and Health, 3*, 148–163.

Christen, C. (2021, March 17). Meat consumption in the U.S. is growing at an alarming rate. *Sentient Media*. Retrieved from https://sentientmedia.org/meat-consumption-in-theus/#:~:text=How%20Much%20Meat%20Is%20Consumed,by%2040%20percent%20since%201961

Clausen, R., Longo, S., & Clark, B. (2019). From ocean to plate: Catching, farming, and eating seafood. In J. Konefal & M. Hatanaka (Eds.), *Twenty lessons in the sociology of food and agriculture*. New York: Oxford University Press.

Coca, E., & Barbosa, R. (2019). Responding to neoliberal diets: School meal programmes in Brazil and Canada. In A. Gray & R. Hinch (Eds.), *A handbook of food crime: Immoral and illegal practices in the food industry and what to do about them* (pp. 347–364). Bristol: Policy Press.

Coe, S., & Coe, M. (2013). *The true history of chocolate*. London: Thames & Hudson.

Cohen, D. (2014). *A big fat crisis: The hidden forces behind the obesity epidemic and how we can end it*. New York: Nation Books.

Colas, A., Edwards, J., Levi, J., & Zubaida, S. (2018). *Food, politics, and society: Social theory and the modern food system*. Berkeley, CA: University of California Press.

Committee for Economic Development. (2017). *Economic contribution of the food and beverage industry*. Retrieved from www.ced.org/pdf/Economic_Contribution_of_the_Food_and_Beverage_Industry.pdf

Conagra Brands Inc. (2018). *Company Milestones*. Retrieved from http://www.conagrabrands.com/our-company/company-milestones, and *Our Brands*. Retrieved from http://www.conagrabrands.com/brands

Constance, D. (2019). The industrialization of agriculture. In J. Konefal & M. Hatanaka (Eds.), *Twenty lessons in the sociology of food and agriculture*. New York: Oxford University Press.

Constance, D., Konefal, J., & Hatanaka, M. (2019). *Contested sustainability discourses in the agri-food system*. New York: Routledge.

Corini, A., & van der Meulen, B. (2019). Regulating food fraud: Public and private law responses in the EU, Italy, and the Netherlands. In A. Gray & R. Hinch (Eds.), *A handbook of food crime: Immoral and illegal practices in the food industry and what to do about them* (pp. 159–174). Bristol: Policy Press.

Cornell Law School. (2020). *21 U.S. Code Section 321. Definitions; Generally*. Retrieved from https://www.law.cornell.edu/uscode/text/21/321#fn002004

Corsica, J., & Pelchat, M. (2010). Food addiction: True or false? *Current Opinion in Gastroenterology*, 26(2), 165–169.

Corwin, R., & Grigson, P. (2009). Symposium overview—Food addiction: Fact or fiction? *Journal of Nutrition*, 139(3), 617–619.

Croall, H. (2007). Food crime. In P. Beirne & N. South (Eds.), *Issues in green criminology: Confronting harms against environments, human and other animals*. Portland, OR: Willan.

Croall, H. (2013). Food crime: A green criminology perspective. In N. South & A. Brisman (Eds.), *Routledge international handbook of green criminology* (pp. 167–183). New York: Routledge.

Crocq, M. A. (2007). Historical and cultural aspects of man's relationship with addictive drugs. *Dialogues in Clinical Neuroscience*, 9(4), 355–361.

Crosta, P. (2008). Researchers verify link between type 2 diabetes and diet. *Medical News Today*. Retrieved from www.medicalnewstoday.com/articles/116513.php

Culp, C. (2005, January 13). Monsanto assault on US farmers detailed in new report. *Center for Food Safety*. Retrieved from www.centerforfoodsafety.org

Davis, C., Curtis, C., Levitan, R., Carter, J., Kaplan, A., & Kennedy, J. (2011). Evidence that "food addiction" is a valid phenotype of obesity. *Appetite*, 57, 711–717.

Davis, C., Levitan, R., Kaplan, A., Kennedy, J, & Carter, J. (2014). Food cravings, appetite, and snack-food consumption in response to a psychomotor stimulant drug: The moderating effect of "food-addiction." *Frontiers in Psychology*, 5, 403.

Davis, C., & Saltos, E. (2022). Dietary recommendations and how they have changed over time. Retrieved from www.ers.usda.gov/webdocs/publications/42215/5831_aib750b_1_.pdf

De Ridder, D., Manning, P., Leong, S., Ross, S., Sutherland, W., Horwath, C., & Venneste, S. (2016). The brain, addiction, and obesity: An EEG neuroimaging study. *Scientific Reports*, 6, 34122.

De Rosa, M., Trabalzi, F., & Pagnani, T. (2019). The social construction of illegality within local food systems. In A. Gray & R. Hinch (Eds.), *A handbook of food crime: Immoral and illegal practices in the food industry and what to do about them* (pp. 43–58). Bristol: Policy Press.

De Vogli, R., Kouvonen, A., & Gimeno, D. (2014). The influence of market deregulation on fast food consumption and body mass index: A cross-national time series analysis. *Bulletin of the World Health Organization*, 92, 99–107.

Del Canto, S., & Engler-Stringer, R. (2019). Prohibitive property practices: The impact of restrictive covenants on the built food environment. In A. Gray & R. Hinch (Eds.), *A handbook of food crime: Immoral and illegal practices in the food industry and what to do about them* (pp. 141–156). Bristol: Policy Press.

Del Prado-Lu, J. (2019). Impact of hazardous substances and pesticides on farmers and farming communities. In A. Gray & R. Hinch (Eds.), *A handbook of food crime: Immoral and illegal practices in the food industry and what to do about them* (pp. 93–108). Bristol: Policy Press.

Dietary Guidelines for Americans. (2022). Retrieved from www.dietaryguidelines.gov/

Doggett, T., & Egan, A. (2016). Non-ideal food choices. In A. Chignell, T. Cuneo, & M. Halteman (Eds.), *Philosophy comes to dinner: Arguments on the ethics of eating*. New York: Routledge.

Dorling, D., Gordon, D., Hillyard, P., Pantazis, C., Pemberton, S., & Tombs, S. (2008). *Criminal obsessions: Why harm matters more than crime*. London: Center for Crime and Justice Studies.

Dresden, D. (2017, May 17). *Effects of diabetes on the body and organs*. Retrieved from www.medicalnewstoday.com/articles/317483.php

Drewnowski, A., & Almiron-Roig, E. (2010). *Human perceptions and preferences for fat-rich foods. National Center for Biotechnical Information*. Retrieved from www.ncbi.nlm.nih.gov/books/NBK53528/

E360. (2022). *Most Global Food Brands Continue to have a Dismal Record on Beef and Deforestation*. Retrieved from https://e360.yale.edu/digest/most-global-food-brands-continue-to-have-a-dismal-record-on-beef-and-deforestation

Eames-Sheavly, M., & Wilkins, J. (2006). A primer on community food systems: Linking food, nutrition and agriculture. In *Discovering the food system: An experiential learning program for young and inquiring minds*. Cornell University, Department of Nutritional Sciences and Department of Horticulture. Retrieved from www.discoverfoodsys.cornell.edu/primer.html

EcoNexus. (2015). *Agropoloy: A handful of corporations control world food production*. Retrieved from www.econexus.info/files/Agropoly_Econexus_BerneDeclaration.pdf

Edie. (2022). *World's biggest food companies accused of failing on climate action, human rights protections*. Retrieved from www.edie.net/worlds-biggest-food-companies-accused-of-failing-on-climate-action-human-rights-protections/

Elizabeth, L., Machado, P., Zinocker, M., Baker, P., & Lawrence, M. (2020). Ultra-processed foods and health outcomes: A narrative review. *Nutrients, 12*, 1–33.

Environmental Protection Agency. (2018). *Agency details*. Retrieved from www.usa.gov/federal-agencies/environmental-protection-agency#:~:text=The%20Environmental%20Protection%20Agency%20protects,develops%20and%20enforces%20environmental%20regulations.

Environmental Working Group. (2014). *Kids' cereals pack 40 percent more sugar*. Retrieved from www.ewg.org/release/kids-cereals-pack-40-percent-more-sugar#.Wly-eCPMyqA

Eördögh, E., Hoyer, M., & Szeleczky, G. (2016). Food addiction as a new behavioral addiction. *Psychiatria Hungarica, 31*(3), 248–255.

Fairley, P. (1999). Farmers sue seed farms. *Chemical Week, 35*, 13.

Fatka, J. (2013). Organic group sues over GM patents: Organic groups attempt to pre-emptively sue Monsanto to protect from being accused of patent infringement. *Feedstuffs, 85*(38), 23.

Federal Bureau of Investigation. (2013). *Crime in the United States 2012*. Retrieved from https://ucr.fbi.gov/crime-in-the-u.s/2012/crime-in-the-u.s.-2012/violent-crime/murder

Federal Bureau of Investigation. (2019). *2015 crime in the United States*. Retrieved from https://ucr.fbi.gov/crime-in-the-u.s/2015/crime-in-the-u.s.-2015/tables/table-1

Finlayson, G. (2017). Food addiction and obesity: Unnecessary medicalization of hedonic overeating. *Nature Reviews Endocrinology, 13*(8), 493–498.

Fitzgerald, A. (2010). The "underdog as "ideal victim"? The attribution of victimhood in the 2007 pet food recall. *International Review of Victimology, 17*(2).

Fletcher, P., & Kenny, P. (2018). Food addiction: A valid concept? *Neuropsychopharmacology, 43*(13), 2506–2513.

Fonseca, N., Molle, R., Costa, M., Goncalves, F., Silva, A., Rodrigues, Y., Price, M., Silveira, P., & Manfro, G. (2020). Impulsivity influences of food intake in women with generalized anxiety disorder. *Revista Brasileira de Psiquiatria, 42*, 382–388.

FoodPrint. (2021). *Factory farming and animal life cycles.* Retrieved from https://foodprint.org/issues/factory-farming-and-animal-life-cycles/

Fortuna, J. (2012). The obesity epidemic and food addiction: Clinical similarities to drug dependence. *Journal of Psychoactive Drugs, 44*(1), 56–63.

Frieden, J. (2020). *Global capitalism.* New York: W.W. Norton.

Friedman, M. (1970). *Capitalism and freedom.* Chicago, IL: Chicago Distribution Center.

Friedrichs, D. O., & Rothe, D. L. (2014). State-corporate crime and major financial institutions: Interrogating an absence. *State Crime, 3*(2), 146–162.

Fuhrman, J. (2017). *Fast food genocide: How processed food is killing us and what we can do about it.* New York: Harper One.

Garber, A., & Lustig, R. (2011). Is fast food addictive? *Current Drug Abuse Review, 4*(3), 146–162.

Gearhardt, A. (2021). *Foods high in added fats and refined carbs are like cigarettes—addictive and unhealthy.* Retrieved from https://theconversation.com/foods-high-in-added-fats-and-refined-carbs-are-like-cigarettes-addictive-and-unhealthy-165441

Gearhardt, A., Grilo, C., DiLeone, R., Brownell, K., & Potenza, M. (2011). Can food be addictive? Public health and policy implications. *Addiction, 106*(7), 1208–1212.

Gearhardt, A., Yokum, S., Orr, P., Stice, E., Corbin, W., & Brownell, K. (2011). Neural correlates of food addiction. *Archives of General Psychiatry, 68*(8), 808–816.

Geiss, L., Wang, J., Cheng, Y., Thompson, T., Barker, L., Li, Y., & Gregg, E. (2014). Prevalence and incidence trends for diagnosed diabetes among adults aged 20 to 79 years, United States, 1980–2012. *JAMA, 312*(12), 1218–1226.

General Mills Inc. (2018). *Brand Heritage.* Retrieved from https://www.generalmills.com/

Gewin, V. (2018). *How the food industry uses big tobacco tactics to manipulate the public.* Retrieved from www.eater.com/2018/11/28/18116273/ucsf-food-industry-documents-archive-coca-cola-policy-big-tobacco

Gillon, S. (2019). Food and the environment. In J. Konefal & M. Hatanaka (Eds.), *Twenty lessons in the sociology of food and agriculture.* New York: Oxford University Press.

Glantz, S. (2019). Cigarette giants bought food companies, used cartoon characters, colors, flavors to boost sales of sweetened beverages. *UCSF Center for Tobacco Control Research and Education.* Retrieved from https://tobacco.ucsf.edu/cigarette-giants-bought-food-companies-used-cartoon-characters-colors-flavors-boost-sales-sweetened-beverages

Glantz, S., Slade, J., Bero, L., Hanauer, P., & Barnes, D. (1997). *The cigarette papers.* Berkeley, CA: University of California Press.

Glenna, L., & Tobin, D. (2019). Science, technology, and agriculture. In J. Konefal & M. Hatanaka (Eds.), *Twenty lessons in the sociology of food and agriculture.* New York: Oxford University Press.

Goldberg, R. (2018). *Food citizenship: Food system advocates in an era of distrust.* New York: Oxford University Press.

Gomez-Donoso, C., Sanchez-Villegas, A., Martinex-Gonzalez, M., Gea Mendonca, R., Lahortiga-Ramos, F., & Nes-Rastrollo, M. (2019). Ultra-processed food consumption and the incidence of depression in a Mediterranean cohort: The SUN Project. *European Journal of Nutrition, 59*, 1093–1103.

Goodman, B. (2011). *Pesticide exposure in womb linked to lower IQ.* Retrieved from www.webmd.com

Gordon, E., Ariel-Donges, A., Bauman, V., & Merlo, L. (2018). What is the evidence for "food addiction?" A systematic review. *Nutrients, 10*(4), 477.

Gottlieb, R., & Joshi, A. (2010). *Food justice.* Cambridge, MA: MIT Press.

Gray, A. (2019a). A food crime perspective. In A. Gray & R. Hinch (Eds.), *A handbook of food crime: Immoral and illegal practices in the food industry and what to do about them* (pp. 11–26). Bristol: Policy Press.

Gray, A., & Hinch, R. (2015). Agribusiness, governments and food crime: A critical perspective. In R. Sollund (Ed.), *Green harms and crimes: Critical criminology in a changing world*. New York: St. Martin's Press.

Gray, A., & Hinch, R. (2018). *A handbook of food crime: Immoral and illegal practices in the food industry and what to do about them*. Bristol: Policy Press.

Gray, M. (2019b). Food and labor. In J. Konefal & M. Hatanaka (Eds.), *Twenty lessons in the sociology of food and agriculture*. New York: Oxford University Press.

Griffin, O., & Spillane, J. (2016). Confounding the process: Forgotten actors and factors in the state-corporate crime paradigm. *Crime, Law and Social Change*, 66(4), 421–437.

Hansen, L. (2009). Corporate financial crime: Social diagnosis and treatment. *Journal of Financial Crime*, 16(1), 28–40.

Harding, A., Wareham, N., Bingham, S. Khaw, K., Luben, R., Welch, A., & Forouhi, N. (2008). Plasma Vitamin C level, fruit and vegetable consumption, and the risk of new-onset type 2 Diabetes Mellitus: The European prospective investigation of cancer-Norfolk prospective study. *Archives of Internal Medicine*, 168(14), 1493–1499.

Hardy, R., Fani, N., Jovanovic, T., & Michopoulos, V. (2018). Food addiction and substance addiction in women: Common clinical characteristics. *Appetite*, 120, 367–373.

Harfy, R., Fani, N., Jovanovic, T., & Michopoulos, V. (2018). *Food addiction and substance addiction in women: Common clinical characteristics*. Retrieved from www.ncbi.nlm.nih.gov/pmc/articles/PMC5680129/

Hartney, E. (2019, September 5). DSM5 criteria for substance use disorders. *Verywellmind*. Retrieved from www.verywellmind.com/dsm-5-criteria-for-substance-use-disorders-21926

Hatanaka, M. (2019). Consuming food. In J. Konefal & M. Hatanaka (Eds.), *Twenty lessons in the sociology of food and agriculture*. New York: Oxford University Press.

Hatanaka, M., & Konefal, J. (2019). Conclusion: Toward more sustainable food and agriculture. In J. Konefal & M. Hatanaka (Eds.), *Twenty lessons in the sociology of food and agriculture*. New York: Oxford University Press.

Hauter, W. (2012). *Foodopoly: The battle over the future of food and farming in America*. New York: The New Press.

Hebebrand, J., Albayrak, Ö., Adan, R., Antel, J., Dieguez, C., de Jong, J., Leng, G., Menzies, J., Mercer, J., Murphy, M., van der Plasse, G., & Dickson, S. (2014). "Eating addiction", rather than "food addiction", better captures addictive-like eating behavior. *Neuroscience & Biobehavioral Reviews*, 47, 295–306.

Heidemann, C., Hoffmann, K., Spranger, J., Klipstein-Grobusch, K., Mohlig, M., Pfeiffer A., & Boeing, H. (2005) A dietary pattern protective against type 2 diabetes in the European Prospective Investigation into Cancer and Nutrition (EPIC)-Potsdam Study cohort. *Diabetologia*, 48, 1126–1134.

Hendrickson, M., & James, H. (2005). The ethics of constrained choice: How the industrialization of agriculture impacts farming and farmer behaviour. *Journal of Agricultural and Environmental Ethics*, 18, 269–291.

Herz, R. (2018). *Why you eat what you eat: The science behind our relationship with food*. New York: W.W. Norton Company.

Hilborn, R., Banobi, J., Hall, S., Pucylowski, T., & Walsworth, T. (2018). The environmental cost of animal source foods. *Frontiers in Ecology and the Environment*, 16(6), 329–335.

Hinch, R. (2018). Chocolate, slavery, forced labour, child labour and the state. In A. Gray & R. Hinch (Eds.), *A handbook of food crime: Immoral and illegal practices in the food industry and what to do about them* (pp. 77–92). Bristol: Policy Press.

Hinch, R., & Gray, A. (2019). Introduction. In A. Gray & R. Hinch (Eds.), *A handbook of food crime: Immoral and illegal practices in the food industry and what to do about them* (pp. 1–7). Bristol: Policy Press.

Holsen, L., Zarcone, J., Thompson, T., Brooks, W., Anderson, M., Ahluwalia, J., Nollen, N., & Savage, C. (2005). Neural mechanisms underlying food motivation in children and adolescents. *Neuroimage*, 27(3), 669–676.

Holt-Gimenez, E. (2019). Capitalism, food, and social movements: The political economy of food system transformation. *Journal of Agriculture, Food Systems, and Community Development*, 9(Suppl. 1), 23–35. https://www.foodsystemsjournal.org/index.php/fsj/article/view/723

Hormel Foods Corp. (2018). *Our Company*. Retrieved from https://www.hormelfoods.com/about/our-company/, and *Explore Our Brands*. Retrieved from https://www.hormelfoods.com/brands/?group=all

Howard, P. (2019). Increasing corporate control: From supermarkets to seeds. In J. Konefal & M. Hatanaka (Eds.), *Twenty lessons in the sociology of food and agriculture*. New York: Oxford University Press.

Humane League. (2021). *Dairy Cows: How long do dairy cattle live? Do they suffer?* Retrieved from https://thehumaneleague.org/article/dairy-cows

Hunt, M. (2020). What causes food addiction and what are the signs? *VirtuaHealth*. Retrieved from www.virtua.org/articles/what-causes-food-addiction-and-what-are-the-signs

Hyde, R., & Savage, A. (2019). Coming together to combat food fraud: Regulatory networks in the EU. In A. Gray & R. Hinch (Eds.), *A handbook of food crime: Immoral and illegal practices in the food industry and what to do about them* (pp. 229–244). Bristol: Policy Press.

Ifland, J., Preuss, H., Marcus, M., Rourke, K., Taylor, W., Burau, K., Jacobs, W., Kadish, W., & Manso, G. (2008). Refined food addiction: A classic substance use disorder. *Medical Hypotheses*, 72, 518–526.

Ifland, J., Preuss, H., Marcus, M., Rourke, K., Taylor, W., & Wright, T. (2015). Clearing the confusion around processed food addiction. *Journal of the American College of Nutrition*, 34, 240–243.

Inderbitzin, M., Bates, K., & Gainey, R. (2020). *Deviance and social control: A sociological perspective*. Thousand Oaks, CA: Sage.

Institute of Medicine and National Research Council. Committee to Ensure Safe Food from Production to Consumption (1998). *Ensuring safe food: From production to consumption*. Retrieved from www.ncbi.nlm.nih.gov/books/NBK209115/

Integrate. (2018). *Introduction to food access, food security, and food-insecure conditions*. Retrieved from https://serc.carleton.edu/integrate/teaching_materials/food_supply/student_materials/1063

Jabr, F. (2016, January 1). How sugar and fat trick the brain into wanting more food. *Scientific American*. Retrieved from www.scientificamerican.com/article/how-sugar-and-fat-trick-the-brain-into-wanting-more-food/

James, H. (2019). Ethical challenges facing farm managers. In A. Gray & R. Hinch (Eds.), *A handbook of food crime: Immoral and illegal practices in the food industry and what to do about them* (pp. 61–76). Bristol: Policy Press.

Johns Hopkins Center for a Livable Future. (2021). *Industrial food animal production*. Retrieved from www.foodsystemprimer.org/food-production/industrial-food-animal-production/

Jusko, J. (2017, August 9). 2017 IW 500: Meet the top US food manufacturers. *Industry Week*. Retrieved from www.industryweek.com/industryweek-us-500/2017-iw-500-meet-top-us-food-manufacturers

Kang, J. (2012). *Nutrition and metabolism in sports, exercise, and health*. New York: Routledge.

Kaplan, S. (2016, May 20). How much added sugar are you eating? You'll soon know. *STAT*. Retrieved from www.statnews.com/2016/05/20/fda-labels-added-sugar/

Kellogg Co. (2018). *Our Brand Portfolio*. Retrieved from https://www.kelloggcompany.com/en_US/brandportfolio.html

Kessler, D. (2009). *The end of overeating: Taking control of the insatiable American appetite*. Emmaus, PA: Rodale Books.

Killgore, W., Weber, M., Schwab, Z., Kipman, M., DelDonno, S., Webb, C., & Rauch, R. (2013). Cortico-limbic responsiveness to high-calorie food images predicts weight status among women. *International Journal of Obesity*, 37, 1435–1442.

Kimura, A. (2019). Food and nutrition. In J. Konefal & M. Hatanaka (Eds.), *Twenty lessons in the sociology of food and agriculture*. New York: Oxford University Press.

Kinsey, J. (2001). The new food economy: Consumers, farms, pharms, and science. *American Journal of Agricultural Economics*, 83(5), 1113–1130.

Klein, J. (2021). *PepsiCo Danone, and Kellogg's share ingredients for global food system transformation*. Retrieved from www.greenbiz.com/article/pepsico-danone-and-kelloggs-share-ingredients-global-food-system-transformation

Kluger, R. (1997). *Ashes to ashes: America's hundred-year cigarette war, the public health, and the unabashed triumph of Philip Morris*. New York: Vintage Books.

Konefal, J., & Hatanaka, M. (2019). *Twenty lessons in the sociology of food and agriculture*. New York: Oxford University Press.

Koopman, R., Mainous III, A., Diaz, V., & Geesey, M. (2005). Changes in age at diagnosis of type 2 Diabetes Mellitus in the United States, 1988 to 2000. *The Annals of Family Medicine*, 3(1), 60–63.

Kraft-Heinz Co. (2018). *Beloved Global Brands*. Retrieved from http://www.kraftheinzcompany.com/

Kramer, R. (1990, March 22). From white-collar to state-corporate crime. Paper presented to the North Central Sociological Association, Louisville.

Kramer, R. (1994). State violence and violent crime. *Peace Review*, 6(2), 171–175.

Kramer, R., & Michalowski, R. (1991). State-corporate crime. Paper presented to the American Society of Criminology, Baltimore, MD, November 7–12, 1990.

Kramer, R., Michalowski, R., & Kauzlarich, D. (2002). The origins and development of the concept and theory of state-corporate crime. *Crime & Delinquency*, 48(2), 263–282.

Kukla, R. (2018). Shame, seduction, and character in food messaging. In A. Barnhill, M. Budolfson, & T. Doggett (Eds.), *Oxford handbook of food ethics*. New York: Oxford University Press.

Lakhani, N., Uteuova, A., & Chang, A. (2021). Revealed: The true extent of America's food monopolies, and who pays the price. *The Guardian*. Retrieved from www.theguardian.com/environment/ng-interactive/2021/jul/14/food-monopoly-meals-profits-data-investigation

Lawrence, F. (2004). *Not on the label: What really goes into the food on your plate*. London: Penguin.

Lawrence, F. (2008). *Eat your heart out: Why the food business is bad for the planet and your health*. London: Penguin Books.

Leavitt, T. (1958). The dangers of social responsibility. *Harvard Business Review*, September–October (1958). Reprinted in 1979. http://57ef850e78feaed47e42-3eada556f2c82b951c467be415f62411.r9.cf2.rackcdn.com/Levitt-1958-TheDangersofSR.pdf

Lee, P., & Dixon, J. (2017). Food for thought: Reward mechanisms and hedonic overeating in obesity. *Current Obesity Reports*, 6(4), 353–361.

Legal Information Institute. (2021). *Mens rea*. Retrieved from www.law.cornell.edu/wex/mens_rea

Leigh, S., & Morris, M. (2018). The role of reward circuitry and food addiction in the obesity epidemic: An update. *Biological Psychology*, 131, 31–42.

Leighton, P. (2019). Mass Samonella poisoning by the Peanut Corporation of America: Lessons in state-corporate food crime. In A. Gray & R. Hinch (Eds.), *A handbook of food crime: Immoral and illegal practices in the food industry and what to do about them* (pp. 175–192). Bristol: Policy Press.

LeMotte, S. (2022, September 1). Ultra-processed foods linked to cancer and early death, studies find. *CNN*. Retrieved from www.wishtv.com/news/national/ultra-processed-foods-linked-to-cancer-and-early-death-studies-find/

Lennerz, B., & Lennerz, J. (2018). Food addiction, high-glycemix-index carbohydrates, and obesity. *Clinical Chemistry*, 64(1), 64–71.

Lent, M., & Swencionis, C. (2012). Addictive personality and maladaptive eating behaviors in adults seeking bariatric surgery. *Eating Behavior*, 13, 67–70.

Leon, K., & Ken, I. (2019). Legitimized fraud and the state-corporate criminology of food—a Spectrum-based theory. *Crime, Law & Social Change*, 71, 25–46.

Lerma-Cabrera, J., Carvajal, F., & Lopez-Legarrea, P. (2016). Food addiction as a new piece of the obesity framework. *Nutrition Journal*, 15, 5.

Liese, A., Weis, K., Schultz, M., & Tooze, J. (2009). Food intake patterns associated with incident type 2 diabetes: The Insulin Resistance Atherosclerosis Study. *Diabetes Care*, 32(2), 263–268.

Lima, J. (2018). Corporate practices and health: A framework and mechanisms. *Globalization and Health*, 14, 21.

Linardon, J. (2018). The relationship between dietary restraint and binge eating: Examining eating-related self-efficacy as a moderator. *Appetite, 127,* 126–129.

Lindgren, E., Gray, K., Miller, G., Tyler, R., Wiers, C., Volkow, N., & Wang, G. (2018). Food addiction: A common neurobiological mechanism with drug abuse. *Frontiers in Bioscience, 23,* 811–836.

Linnekin, B. (2016). *America's slaughterhouse mess.* Retrieved from https://thecounter.org/americas-slaughterhouse-mess/

Long, C., Blundell, J., & Finlayson, G. (2015). A systematic review of the application and correlates of YFAS-diagnosed "food addiction" in humans: Are eating-related "addictions" a cause for concern or empty concepts? *Obesity Facts, 8*(6), 386–401.

Long, M., & Lynch, M. (2018). Food waste (non-regulation). In A. Gray & R. Hinch (Eds.), *A handbook of food crime: Immoral and illegal practices in the food industry and what to do about them* (pp. 331–346). Bristol: Policy Press.

Lorr, B. (2020). *The secret life of groceries: The dark miracle of the American supermarket.* New York: Avery.

Lutter, M., & Nestler, E. (2009). Homeostatic and hedonic signals interact in the regulation of food intake. *Nutrition, 139,* 629–632.

Lyman, M. D. (2016). *Drugs in society.* New York: Routledge.

Lynch, M., & Michalowski, R. (2006). *Primer in radical criminology: Critical perspectives on crime, power and identity.* Boulder, CO: Lynne Rienner Publishers.

Malik, V., Popkin, B., Bray, G., Després, J., Willett, W., & Hu, F. (2010). Sugar-sweetened beverages and risk of metabolic syndrome and type 2 diabetes. *Diabetes Care.*

Manning, L., & Soon, J. (2019). The value of product sampling in mitigating food adulteration. In A. Gray & R. Hinch (Eds.), *A handbook of food crime: Immoral and illegal practices in the food industry and what to do about them* (pp. 127–140). Bristol: Policy Press.

Mardirossian, N., & Muiuri, K. (2021). *At UN food systems summit, did business show it is serious about addressing the crises facing global food systems.* Retrieved from https://news.climate.columbia.edu/2021/09/30/at-un-food-systems-summit-did-business-show-it-is-serious-about-addressing-the-crises-facing-global-food-systems/

Marion, N. (2017). *Soda Politics: Taking on big soda (and winning).* New York: Oxford University Press.

Marion, N. (2018). *Unsavory truth: How food companies skew the science of what we eat.* New York: Basic Books.

Markus, C., Rogers, P., Brouns, F., & Schepers, R. (2017). Eating dependence and weight gain: No human evidence for a "sugar-addiction" model of overweight. *Appetite, 114,* 64–72.

Martin, A. (2016). Factory farming and consumer complicity. In A. Chignell, T. Cuneo, & M. Halteman (Eds.), *Philosophy comes to dinner: Arguments on the ethics of eating.* New York: Routledge.

Matthes, E., & Matthes, J. (2018). The clean plate club? Food waste and individual responsibility. In A. Barnhill, M. Budolfson, & T. Doggett (Eds.), *Oxford handbook of food ethics.* New York: Oxford University Press.

Maxwell, A., Gardiner, E., & Loxton, N. (2020). Investigating the relationship between reward sensitivity, impulsivity, and food addiction: A systematic review. *European Eating Disorders Review, 28,* 368–384.

Mayer, C. (2005). McDonald's makes Ronald a health ambassador. *Washington Post,* January 28.

Mayer-Davis, E., Lawrence, J., Dabelea, D., Divers, J., Isom, S., Dolan, L., & Wagenknecht, L. (2017). Incidence trends of Type 1 and type 2 diabetes among youths, 2002–2012. *New England Journal of Medicine, 377*(3), 301.

Mayo Clinic. (2018). *Binge-eating disorder.* Retrieved from www.mayoclinic.org/diseases-conditions/binge-eating-disorder/symptoms-causes/syc-20353627

McDermott, A. (2016, August 11). Diabetes and amputation: Why it's done and how to prevent it. *Healthline.* Retrieved from www.healthline.com/health/diabetes/diabetes-amputation

McGrath, M. (2017, May 24). World's largest food and beverage companies 2017: Nestle, Pepsi and Coca-Cola dominate the field. *Forbes.* Retrieved from www.forbes.com/sites/

maggiemcgrath/2017/05/24/worlds-largest-food-and-beverage-companies-2017-nestle-pepsi-and-coca-cola-dominate-the-landscape/#1ed283863a69

McMahon, M., & Glatt, K. (2019). Food crime without criminals: Agri-food safety governance as a protection racket for dominant political and economic interests. In A. Gray & R. Hinch (Eds.), *A handbook of food crime: Immoral and illegal practices in the food industry and what to do about them* (pp. 27–42). Bristol: Policy Press.

Merriam-Webster. (2022a). *Definition for industrialization*. Retrieved from www.merriam-webster.com/dictionary/industrialization

Merriam-Webster. (2022b). *Definition for monopoly*. Retrieved from www.merriam-webster.com/dictionary/monopoly

Meule, A. (2019). A critical examination of the practical implications derived from the food addiction concept. *Current Obesity Reports*, 8(1), 11–17.

Meule, A., & Gearhardt, A. (2014). Food addiction in the light of DSM-5. *Nutrients*, 6, 3653–3671.

Meyersohn, N. (2023, January 26). McDonald's, in-n-out, and Chipotle are spending millions to block raises for their workers. *CNN*.

Micha, R., Penalvo, J., Cudhea, F., Imamura, F., Rehm, C., & Mozaffarian, D. (2017). Association between dietary factors and mortality from heart disease, stroke, and type 2 diabetes in the United States. *JAMA*, 317(9), 912–924. Retrieved from https://jamanetwork.com/journals/jama/fullarticle/2608221

Michalowski, R. & Kramer, R. C. (2006). *State-corporate crime: Wrongdoing at the intersection of business and government*. Chicago, IL: Rutgers University Press.

Minger, D. (2013). *Death by food pyramid: How shoddy science, sketchy politics and shady special interests have ruined our health*. Ocoee, FL: Primal Nutrition, Inc.

Mollenkamp, C., Menn, J., & Levy, A. (1998). *The people vs. big tobacco: How the states took on the cigarette giants*. New York: Bloomberg Press.

Mondelez International Inc. (2018). *Global Brand Montage*. Retrieved from http://www.mondelezinternational.com/en

Monteiro, C., Cannon, G., Lawrence, M., Louzada, M., & Machada, P. (2019). *Ultra-processed foods, diet quality, and health using the NOVA classification system*. Rome: Food and Agriculture Organization of the United Nations, FAO.

Montonen, J., Knekt, P., Härkänen, T., Järvinen, R., Heliövaara, M., Aromaa, A., & Reunanen, A. (2005). Dietary patterns and the incidence of type 2 diabetes. *American Journal of Epidemiology*, 161 (3), 219–227.

Morgan, D., & Goh, G. (2004). Genetically modified food labelling and the WTO agreements. *Review of European Community and International Environmental Law*, 13(3), 306–319.

Morland, K., Wing, S., Diez Roux, A., & Poole, C. (2002). Neighborhood characteristics associated with the location of food stores and food service places. *American Journal of Preventive Medicine*, 22(1), 23–29.

Morris, L., Voon, V., & Leggio, L. (2018, May 24). Stress, motivation, and the gut-brain axis: A focus on the ghrelin system and alcohol use disorder. *Alcoholism: Clinical and Experimental Research*. https://www.ncbi.nlm.nih.gov/pmc/articles/PMC6252147/

Moss, M. (2013). *Salt sugar fat: How the food giants hooked us*. New York: Random House.

Moss, M. (2021). *Hooked: Food, free will, and how the food giants exploit our addictions*. New York: Random House.

Moulesong, A. (2021, September 26). The problem with the energy balance model of weight loss. *Medium*. Retrieved from https://medium.com/in-fitness-and-in-health/the-problem-with-the-energy-balance-model-of-obesity-142cb51506b4

Mourdoukoutos, P. (2018, July 14). Pepsi beats Coke. *Forbes*. Retrieved from www.forbes.com/sites/panosmourdoukoutas/2018/07/14/pepsi-beats-coke/#60c5f9ce11d0

Mugni, H., Demetrio, P., Paracampo, A., Pardi, M., Bulus, G., & Bonetto, C. (2012). Toxicity persistence in runoff water and soil in experimental soybean plots following chlorpyrifos application. *Bulletin of Environmental Contamination and Toxicology*, 89(1), 208–212.

Myers, J. (2019). Food and hunger. In J. Konefal & M. Hatanaka (Eds.), *Twenty lessons in the sociology of food and agriculture*. New York: Oxford University Press.

Nally, D. (2011). The biopolitics of food provisioning. *Transactions of the Institute of British Geographies, 36*(1), 37–53.

Narayan, K., Boyle, J., Thompson, T., Sorenson, S., & Williamson, D. (2003). Lifetime risk for Diabetes Mellitus in the United States. *JAMA, 290*(14), 1884–1890.

National Institute on Drug Abuse. (2018, August). *Overdose death rates*. Retrieved from www.drugabuse.gov/related-topics/trends-statistics/overdose-death-rates

National Institute on Drug Abuse. (2020). *The science of drug use and addiction: The basics*. Retrieved from www.drugabuse.gov/publications/media-guide/science-drug-use-addiction-basics

Nesheim, M., Oria, M., & Yih, P. (2015). *A framework for assessing effects of the food system*. Committee on a Framework for Assessing the Health, Environmental, and Social Effects of the Food System; Food and Nutrition Board; Board on Agriculture and Natural Resources; Institute of Medicine; National Research Council.

Nestle, M. (2002). *Food politics*. London: University of California Press.

Nestle, M. (2013). *Eat, drink, vote: An illustrated guide to food politics*. Emmaus, PA: Rodale Books.

Nestle, M. (2018). *Unsavory truth: How food companies skew the science of what we eat*. New York: Basic Books.

Newman, K., Leon, J., & Newman, L. (2015). Estimating occupational illness, injury, and mortality in food production in the United States: A farm-to-table analysis. *Journal of Occupational and Environmental Medicine, 57*(7), 718–725.

Nguyen, K., Glantz, S., Palmer, C., & Schmidt, L. (2019). *Tobacco industry involvement in children's sugary drinks market*. Retrieved from https://pubmed.ncbi.nlm.nih.gov/30872273/

Nolan, J., & Jenkins, S. (2019). Food addiction is associated with irrational beliefs via trait anxiety and emotional eating. *Nutrition, 11*, 1711.

Nottingham, S. (2003). *Eat your genes: How genetically modified food is entering your diet*. New York: Zed Books.

Novelle, M., & Diéguez, C. (2018). Food addiction and binge eating: Lessons learned from animal models. *Nutrients, 10*(1), 71.

Nunez, X. (2021, August 11). If you think kids are eating mostly junk food, a new study finds you're right. *NPR*.

O'Connor, A. (2022). *What are ultra-processed foods? What should I eat instead?* Retrieved from www.washingtonpost.com/wellness/2022/09/27/ultraprocessed-foods/

Office of Disease Prevention and Health Promotion. (2017). *Dietary guidelines for Americans. Chapter 2. Shifts needed to align with healthy eating patterns*. Retrieved from https://health.gov/dietaryguidelines/2015/guidelines/chapter-2/current-eating-patterns-in-the-united-states/

Olive Oil Times. (2021). *Researchers identify main types of olive oil fraud, propose solutions*. Retrieved from www.oliveoiltimes.com/grades/main-types-of-olive-oil-fraud-revealed/91799

Open Secrets. (2022). *List of industries, 2021 data*. Retrieved from www.opensecrets.org/federal-lobbying/alphabetical-list?type=s

Oskam, A., Backus, G., Kinsey, J., & Frewer, L. (2010). *E.U. policy for agriculture, food and rural area*. Wageningen: Wageningen Academic Publishers.

Oxfam. (2016). *Behind the Brand*. Retrieved from https://www.oxfamamerica.org/explore/issues/humanitarian-response-and-leaders/hunger-and-famine/behind-the-brands/

Paarlberg, R. (2021). *Resetting the table: Straight talk about the food we grow and eat*. New York: Borzoi Books.

Packer, M., & Guthman, J. (2019). Food and obesity. In J. Konefal & M. Hatanaka (Eds.), *Twenty lessons in the sociology of food and agriculture*. New York: Oxford University Press.

Paddison, L. (2022, December 27). Beef burger or fish sandwich? These menu labels encourage people to eat less read meat, study shows. *CNN*.

Palmer, J., Boggs, D., Krishnan, S., Hu, F., Singer, M., & Rosenberg, L. (2008). Sugar-sweetened beverages and incidence of type 2 Diabetes Mellitus in African American women. *Archives of Internal Medicine*, 168(14), 1487–1492.

Pardue, A., Robinson, M., & Arrigo, B. (2013a). Psychopathy and corporate crime: A preliminary examination, part one. *Journal of Forensic Psychology Practice*, 13(2), 116–144.

Pardue, A., Robinson, M., & Arrigo, B. (2013b). Psychopathy and corporate crime: A preliminary examination, part two. *Journal of Forensic Psychology Practice*, 13(2), 145–169.

Passas, N. (2005). Lawful but awful: "Legal corporate crimes. *The Journal of Socio-Economics*, 34(6), 771–786.

Pelchat, M. (2009). Food addiction in humans. *Journal of Nutrition*, 139, 620–622.

People for the Ethical Treatment of Animals. (2021). *Animals used for food*. Retrieved from www.peta.org/issues/animals-used-for-food/

People for the Ethical Treatment of Animals. (2022). *Victory! Global food industry ditches deadly animal tests*. Retrieved from www.peta.org/features/victories-food-drink-companies-refuse-animal-tests/

Petersen, M. (2013). Economic costs of diabetes in the U.S. in 2012. *Diabetes Care*, 36(4), 1033–1046.

Pew Research Center. (2016). *What's on your table? How America's diet has changed over the decades*. Retrieved from www.pewresearch.org/fact-tank/2016/12/13/whats-on-your-table-how-americas-diet-has-changed-over-the-decades/

Pinto-Duschinsky, M., Postnikov, A., Nadeau, C., & Dahl, R. (1999). *Campaign finance in foreign countries: Legal regulation and political practices (a comparative legal survey and analysis. International Foundation for Election Systems*. Retrieved from https://ifes.org/sites/default/files/campaign_finance.pdf

Pollan, M. (2008). *In defense of food*. London: Penguin.

Pomeranz, J., & Roberto, C. (2014). The impact of "food addiction" on food policy. *Current Addiction Reports*, 1, 102–108.

Pontell, H. (2005). White-collar crime or just risky business? The role of fraud in major financial debacles. *Crime, Law and Social Change*, 42(4–5), 309–324.

Potato Pro. (2022). *SNAC International (formerly Snack Food Association)*. Retrieved from www.potatopro.com/companies/snac-international-formerly-snack-food-association

Powers, M. (2018). Food, fairness, and global markets. In Barnhill, A., Budolfson, N., & Doggett, T. (Eds.), *The Oxford handbook of food ethics*. Downloaded from: https://www.fewresources.org/uploads/1/0/5/2/10529860/powers_food_fairness_global_markets_-_final.pdf

Puhl, R., Moss-Racusin, C., Schwartz, M., & Brownell, K. (2008). Weight stigmatization and bias reduction: Perspectives of overweight and obese adults. *Health Education Research*, 23(2), 347–358.

Pursey, K., Davis, C., & Burrows, T. (2017). Nutritional aspects of food addiction. *Current Addiction Reports*, 4, 142–150.

Pursey, K., Stanwell, P., Gearhardt, A., Collins, C., & Burrows, T. (2014). The prevalence of food addiction as assessed by the Yale Food Addiction Scale: A systematic review. *Nutrients*, 6. Retrieved from www.mdpi.com/journal/nutrients

Randolph, T. (1956). The descriptive features of food addiction: Addictive eating and drinking. *Quarterly Studies on Alcohol*, 17(2), 198–224.

Ransom, E. (2019). Governing agriculture: Public policy and private governance. In J. Konefal & M. Hatanaka (Eds.), *Twenty lessons in the sociology of food and agriculture*. New York: Oxford University Press.

Rao, M., Afshin, A., Singh, G., & Mozaffarian, D. (2013). Do healthier foods and diet patterns cost more than less healthy options? A systematic review and meta-analysis. *BMJ Open*, 3(12).

Reilly, L. (2022, September 28). The FDA announces a new definition of what's healthy. *The Washington Post*.

Reiman, J. (2020). *The rich get richer and the poor get prison: Thinking critically about class and crime*. New York: Routledge.

Ritzer, G. (2000). *The McDonaldization of society*. Thousand Oaks, CA: Sage.

Robinson, M. (2015). *Criminal injustice: How politics and ideology distort American ideals*. Durham, NC: Carolina Academic Press.

Robinson, M. (2017). Food crime: An introduction to deviance in the food industries. Paper presented to the American Society of Criminology.

Robinson, M. (2020). *Criminal injustice*. Durham, NC: Carolina Academic Press.

Robinson, M. (2022). Eating ourselves to death: How food is a drug and what food abuse costs. *Drug Science, Policy and Law, 8*, 1–21.

Robinson, M., & Beaver, K. (2020). *Why crime?* Durham, NC: Carolina Academic Press.

Robinson, M., & Murphy, D. (2008). *Greed is good: Maximization and elite deviance in America*. Lanham, MD: Rowman & Littlefield.

Robinson, M., & Rogers, J. (2018). Applying contextual anomie and strain theory to recent acts of corporate deviance. *Journal of Theoretical & Philosophical Criminology, 10*, 71–92.

Robinson, M., & Tauscher, A. (2019). Big foods: Big benefits, big problems, or both? Presented to the annual conference of the North Carolina Criminal Justice Association.

Robinson, M., & Turner, C. (2019). Incidence and prevalence of type 2 diabetes in America: Is there culpability in the food industry? *State Crime, 8*(2), 175–218.

Robinson, W. (2014). *Global capitalism and the crisis of humanity*. Cambridge, MA: Cambridge University Press.

Rodin, J., Mancuso, J., Granger, J., & Nelbach, E., (1991). Food cravings in relation to body mass index, restraint, and estradiol levels: A repeated measures study in healthy women. *Appetite, 17*, 177–185.

Ross, J. (1998). *Cutting the edge: Current perspectives in Radical/Critical Criminology and criminal justice*. Westport, CT: Praeger.

Ross, J. (2017). Protecting democracy: A parsimonious, dynamic and heuristic model of controlling crimes by the powerful. *Criminal Justice Studies, 30*(3), 289–306.

Ross, M. (2013). *Salt, sugar, fat*. New York: Random House.

Ross, S. (2022). *4 countries that produce the most food*. Retrieved from www.investopedia.com/articles/investing/10061/4-countries-produce-most-food.asp

Rowland, J. (2017). *Biologic regulation of physical activity*. East Peoria, IL: Versa Press.

Rowland, W. (2005). *Greed, Inc.: Why corporations rule our world and how we let it happen*. Toronto: Dundurn Group.

Ruddock, H., Christiansen, P., Halford, J., & Hardman, C. (2017). The development and validation of the Addiction-like Eating Behaviour Scale. *International Journal of Obesity, 41*(11), 1710–1717.

Ryan, K. (2017, June 26). This infographic shows how only 10 companies own all the world's food brands. *Good Money*. Retrieved from https://money.good.is/articles/food-brands-owners

Safety of Aspartame. (2023). *The Science of the Safety of Aspartame*. Retrieved from https://www.safetyofaspartame.com/?gclid=Cj0KCQjwocShBhCOARIsAFVYq0gMlzW9jgdrCHiFq672FEcGzMlvQbFm-SB6lyPKMkkMKcs93lYd9rIaAuAVEALw_wcB

Saljoughian, M. (2016, February 16). Cardioetabolic syndrome: A global health issue. *US Pharmacist*. Retrieved from www.uspharmacist.com/article/cardiometabolicsyndrome-a-global-health-issue

Samaha, J. (2016). *Criminal law*. Boston, MA: Cengage Learning.

Sami, W., Ansari, T., Butt, N., & Hamid, M. (2017). Effect of diet on type 2 diabetes mellitus: A review. *International Journal of Health Sciences, 11*(2), 65–71.

Satre, L. (2005). *Chocolate on trial Slavery and the ethics of business*. Athens, OH: University Press.

Saunders, R. (2001). Compulsive eating and gastric bypass surgery: What does hunger have to do with it? *Obesity Surgery, 11*, 757–761.

Scholtz, S., Miras, A., Chhina, N., Prechtl, C., Sleeth, M., Daud, N., Ismail, N., Durighel, G., Ahmed, A., Olbers, T., Vincent, R., Alaghband-Zadeh, J., Ghatei, M., Waldman, A., Frost, G., Bell, J., le Roux, C., & Goldstone, A. (2014). Obese patients after gastric bypass surgery have lower brain-hedonic responses to food than after gastric banding. *Gut, 63*(6), 891–902.

Schrage, E., & Ewing, A. (2005). The cocoa industry and child labour. *The Journal of Corporate Citizenship, 18*, 19–112.

Schrempf-Stirling, J., & Phillips, R. (2019). Agency and responsibility: The case of the food industry and obesity. In A. Gray & R. Hinch (Eds.), *A handbook of food crime: Immoral and illegal practices in the food industry and what to do about them* (pp. 111–126). Bristol: Policy Press.

Schulte, E., Avena, N., & Gearhardt, A. (2015). Which foods may be addictive? The roles of processing, fat content, and glycemic load. *PLoS One, 10.*

Schulte, E., & Gearhardt, A. (2018). Associations of food addiction in a sample recruited to be nationally representative of the United States. *European Eating Disorders Review, 26,* 112–119.

Schulte, E., Jacques-Tiura, A., Gearhardt, A., & Naar, S. (2018). Food addiction prevalence and concurrent validity in African American adolescents with obesity. *Psychology of Addictive Behaviors, 32*(2), 187–196.

Schulte, E., Potenza, M., & Gearhardt, A. (2017). A commentary on the "eating addiction" versus "food addiction" perspectives on addictive-like food consumption. *Appetite, 115,* 9–15.

Schulze, M., Hoffmann K., Manson J., Willett W., Meigs J., Weikert C., Heidemann C., Colditz G., Hu, F. (2005). Dietary pattern, inflammation, and incidence of type 2 diabetes in women. *American Journal of Clinical Nutrition, 82,* 675–684.

Sebo, J. (2018). Multi-issue food activism. In A. Barnhill, M. Budolfson & T. Doggett (Eds.), *Oxford handbook of food ethics.* New York: Oxford University Press.

Senauer, B., & Venturini, L. (2005). The globalization of food systems: A concept framework and empirical patterns. In E. Defrancesco, L. Galletto, & M. Thiene (Eds.), *Food agriculture and the environment.* Milan: The Food Industry Center.

Sharma, A. (2014). *Why the energy balance equation result in flawed approaches to obesity prevention and management.* Retrieved from www.drsharma.ca/why-the-energy-balance-equation-results-in-flawed-approaches-to-obesity-prevention-and-management

Sharma, L., Teret, S., & Brownwell, K. (2010). The food industry and self-regulation: Standards to promote success and to avoid public health failures. *American Journal of Public Health, 100*(2), 240–246.

Silbergeld, E. (2016). *Chickenizing farms & food: How industrial meat production endangers workers, animals, and consumers.* Baltimore, MD: Johns Hopkins University Press.

Simester, A. (2021). *Fundamentals of Criminal Law: Responsibility, Culpability, and Wrongdoing.* New York: Oxford University Press.

Simon, M. (2006). *Appetite for profit: How the food industry undermines our health and how to fight back.* New York: Nation Books.

Skyler, J., & Oddo, C. (2002). Diabetes trends in the USA. *Diabetes/Metabolism Research and Reviews, 18*(S3), S21-S23.

Small, D., Zatorre, R., Dagher, A, Evans, A., & Jones-Gotman, M. (2001). Changes in brain activity related to eating chocolate: From pleasure to aversion. *Brain, 124*(9), 1720–1733.

Smart Label. (2022). *Kellog's Froot Loops Cereal.* Retrieved from https://smartlabel.kelloggs.com/Product/Index/00038000937675

Smith, R., Manning, L., & McElwee, G. (2017). Critiquing the inter-disciplinary literature on food fraud. *International Journal of Rural Criminology, 3*(2).

Soto-Escageda, J., Vidal, B., Vidal-Victoria, C., Chavez, A., Sierra-Beltran, M., & Bourges-Rodriquez, H. (2016). Does salt addiction exist? *Salud Mental, 39*(3). Retrieved from www.revistasaludmental.mx/index.php/salud_mental/article/view/SM.0185-3325.2016.016/2995

Source Magazine. (2017). Global food companies ranked on water risk management. *Source Magazine.* Retrieved from www.thesourcemagazine.come/global-food-companies-ranked-water-risk-management/

SourceWatch. (2022). *American council for fitness and nutrition.* Retrieved from www.sourcewatch.org/index.php/American_Council_for_Fitness_and_Nutrition

Spettigue, W., Obeid, N., Santos, A., Norris, M., Hamati, R., Hadjiyannakis, S., & Buchholz, A. (2019). Binge eating and social anxiety in treatment—seeking adolescents with eating disorders or severe obesity. *Eating and Weight Disorders – Studies on Anorexia, Bulimia and Obesity, 25,* 787–793.

Spring, B., Schneider, K., Smith, M., Kendzor, D., Appelhans, B., Hedeker, D., & Pagoto, D. (2008). Abuse potential of carbohydrates for overweight carbohydrate cravers. *Psychopharmacology, 197,* 637–647.

Stanish, J. (2010). The obesity epidemic in America and the responsibility of big food manufacturers. *Inquiries*, 2(11), 1.

Statistics About Diabetes. (2017, July 19). Retrieved from www.diabetes.org/diabetes-basics/statistics/

Stice, E., Spoor, S., Bohon, C., Veldhuizen, M., & Small, D. (2008). Relation of reward from food intake and anticipated food intake to obesity: A functional magnetic resonance imaging study. *Journal of Abnormal Psychology*, 117(4), 924–935.

Stice, E., Yokum, S., Burger, K., Epstein, L., & Small, D. (2011). Youth at risk for obesity show greater activation of striatal and somatosensory regions to food. *The Journal of Neuroscience*, 31(12), 4360–4366.

Sugar Science. (2022). *How much is too much?* Retrieved from http://sugarscience.ucsf.edu/the-growing-concern-of-overconsumption/#.WllnySPMyqA

Sustainable Brands. (2015). *Global investors urging food, beverage companies to better manage water risks*. Retrieved from http://sustainablebrands.com/read/behavior-change/global-investors-urging-food-beverage-companies-to-better-manage-water-risks

Tappan, P. (1947). Who is the criminal? *American Sociological Review*, 12(1), 96–102.

Taubes, G. (2021, September 13). How a "fatally, tragically flawed" paradigm has derailed the science of obesity. *STAT*. Retrieved from www.statnews.com/2021/09/13/how-a-fatally-tragically-flawed-paradigm-has-derailed-the-science-of-obesity/

The Guardian. (2022). *Revealed: The True Extent of America's Food Monopolies, and Who Pays the Price*. Retrieved from https://www.theguardian.com/environment/ng-interactive/2021/jul/14/food-monopoly-meals-profits-data-investigation

The Journal. (2022). *Nearly 3,000 Thoroughbred Horses Slaughtered for Meat in Ireland Since 2020*. Retrieved from www.thejournal.ie/horse-meat-ireland-5764684-May2022/

Thompson, N. (2012, March 5). International campaign finance: How do countries compare? *CNN*.

Tinker, L., Bonds, D., Margolis, K., Manson, J., Howard, B., Larson, J., Perri, M., Beresford, A., Robinson, J., Rodríguez, B., Safford, M., Wenger, N., Stevens, V., & Parker, L. (2008). Low-fat dietary pattern and risk of treated Diabetes Mellitus in postmenopausal women: The Women's Health Initiative randomized controlled dietary modification trial. *Archives of Internal Medicine*, 168(14), 1500–1511.

Tombs, S. (2012). State-corporate symbiosis in the production of crime and harm. *State Crime 1*(2), 170–195.

Tomiyama, A., Dallman, M., & Epel, E. (2011). Comfort food is comforting to those most stressed: Evidence of the chronic stress response network in high stress women. *Psychoneuroendocrinology*, 36, 1513–1519.

Truman, J., & Morgan, R. (2018). *Criminal victimization, 2015*. Retrieved from www.bjs.gov/content/pub/pdf/cv15.pdf

Tuomisto, T., Hetherington, M., Morris, M., Tuomisto, M., Turjanmaa, V., & Lappalainen, R. (1999). Psychological and physiological characteristics of sweet food "addiction." *International Journal of Eating Disorders*, 25, 169–175.

Tyson. (2018). *Who We Are*. Retrieved from https://www.tysonfoods.com/who-we-are

United Nations Office of Drugs and Crime. (2019). *Expert group meeting. Transparency in political finance*. Retrieved from www.unodc.org/documents/corruption/PragueEGM2019/Report_EGM_Transparency_in_Political_Finance_Prague.pdf

US Census Bureau. (2018). *Economic census: Industry snapshots*. Retrieved from https://census.gov/econ/snapshots/

US Centers for Disease Control and Prevention (CDC). (2016). *Division of Nutrition, Physical Activity, and Obesity. Adult obesity facts*. Retrieved from www.cdc.gov/obesity/data/adult.html

US Centers for Disease Control and Prevention (CDC). (2017a). *Diabetes Basics*. Retrieved from www.cdc.gov/diabetes/basics/index.html

US Centers for Disease Control and Prevention. (2017b). *More than 100 Million Americans have Diabetes or Prediabetes*. Retrieved from www.cdc.gov/media/releases/2017/p0718-diabetes-report.html

US Centers for Disease Control and Prevention. (2017c). *Rates of New Diagnosed Cases of Type 1 and Type 2 Diabetes on the Rise Among Children, Teens*. Retrieved from www.cdc.gov/media/releases/2017/p0412-diabtes-rates.html

US Centers for Disease Control and Prevention. (2018a). *Defining Adult Overweight and Obesity*. Retrieved from www.cdc.gov/obesity/adult/defining.html

US Centers for Disease Control and Prevention. (2018b). *Physical Inactivity*. Retrieved from www.cdc.gov/healthcommunication/toolstemplates/entertainmented/tips/PhysicalInactivity.html

US Centers for Disease Control and Prevention. (2018c). *Foodborne Germs and Illnesses*. Retrieved from https://www.cdc.gov/foodsafety/foodborne-germs.html

US Department of Agriculture. (2018). *Agencies*. Retrieved from www.usda.gov/our-agency/agencies

US Department of Agriculture. (2022). *Agriculture, food, and related industries contributed more than $1 trillion to U.S. GDP in 2017*. Retrieved from www.ers.usda.gov/data-products/chart-gallery/gallery/chart-detail/?chartId=92983

US Food and Drug Administration. (2018). *CVM vision and mission*. Retrieved from www.fda.gov/AboutFDA/CentersOffices/OfficeofFoods/CVM/CVMVisionandMission/default.htm

Valenti, C. (2001, June 13). Kraft's Philip Morris connection. *ABC News*. Retrieved from https://abcnews.go.com/Business/story?id=88088&page=1

Van Nieuwkoop, M. (2019). Do the costs of the global food system outweigh its monetary value? *World Bank Blogs*. Retrieved from https://blogs.worldbank.org/voices/do-costs-global-food-system-outweigh-its-monetary-value

Waldman, P. (2014). *How our campaign finance system compares to other countries*. Retrieved from https://prospect.org/power/campaign-finance-system-compares-countries/

Walker, K. (2019). *The grand food bargain and the mindless drive for more*. Chicago, IL: Island Press.

Walters, R. (2006a). Crime, bio-agriculture and the exploitation of hunger. British Journal of *Criminology*, 46(1), 26–45.

Walters, R. (2008). Criminology and genetically modified food. *The British Journal of Criminology*, 44(2), 151–167.

Walters, R. (2019). Food, genetics and knowledge politics. In A. Gray & R. Hinch (Eds.), *A handbook of food crime: Immoral and illegal practices in the food industry and what to do about them* (pp. 265–280). Bristol: Policy Press.

Warburton, D., Nicol, C., & Bredin, S. (2006). Health benefits of physical activity: The evidence. *Canadian Medical Association Journal*, 174(6), 801–809.

WebMD. (2017). *Diabetes guide*. Retrieved from www.webmd.com/diabetes/guide/default.htm

WebMD. (2020). *Food addiction*. Retrieved from www.webmd.com/mental-health/eating-disorders/binge-eating-disorder/mental-health-food-addiction#1

Wen, H., Gris, D., Lei, Y., Jha, S., Zhang, L., Huang, M., Brickey, W., & Ting, J. (2011). Fatty acid—induced NLRP3-ASC inflammasome activation interferes with insulin signaling. *Nature Immunology*, 12(5), 408–415.

Westwater, M., Fletcher, P., & Ziauddeen, H. (2016). Sugar addiction: The state of the science. *European Journal of Nutrition*, 55, 55–69.

White, R. (2012). Land theft as rural eco-crime. *International Journal of Rural Criminology*, 1(2), 203.

White, R. (2014). Environmental instability and fortress mentality. *International Affairs*, 90(4), 835–851.

White, R., & Yeates, J. (2018). Farming and climate change. In A. Gray & R. Hinch (Eds.), *A handbook of food crime: Immoral and illegal practices in the food industry and what to do about them* (pp. 315–330). Bristol: Policy Press.

Whyte, D. (2018). Regimes of permission and state-corporate crime. *State Crime*, 3(2), 237–246.

Winson, A. (2013). *The industrial diet: The degradation of food and the struggle for healthy eating*. New York: New York University Press.

Winson, A., & Choi, J. (2019). Food, diets, and industrialization. In J. Konefal & M. Hatanaka (Eds.), *Twenty lessons in the sociology of food and agriculture*. New York: Oxford University Press.

Wise, T. (2019). *Eating Tomorrow: The Battle for the Future of Food*. Retrieved from https://tawise01.medium.com/eating-tomorrow-the-battle-for-the-future-of-food-93b14eeb5496

Wiss, D., & Brewerton, T. (2020). Separating the signal from the noise: How psychiatric diagnoses can help discern food addiction from dietary restraint. *Nutrients, 12*(10), 2937.

Wittman, H. (2019). Getting to food sovereignty (locally?) in a globalized world. In J. Honefsal & M. Hatanaka (Eds.), *Twenty lessons in the sociology of food and agriculture*. New York: Oxford University Press.

Wolf, S. (2018). The ethics of being a foodie. In A. Barnhill, M. Budolfson, & T. Doggett (Eds.), *Oxford handbook of food ethics*. New York: Oxford University Press.

World Population Review. (2022). *Tar Heel, North Carolina Population 2022*. Retrieved from https://worldpopulationreview.com/us-cities/tar-heel-nc-population

Yokum, S., Gearhardt, A., Harris, J., Brownell, K., & Stice, E. (2014). Individual differences in striatum activity to food commercials predict weight gain in adolescents. *Obesity, 22*(12), 2544–2551.

Young, C., & Quinn, E. (2015, April 15). Food safety scientists have ties to big tobacco. *The Center for Public Integrity*. Retrieved from https://publicintegrity.org/politics/food-safety-scientists-have-ties-to-big-tobacco/

Zegart, D. (2001). *Civil warriors: The legal siege on the tobacco industry*. New York: Delta.

Zhang, Y., M. von Deneen, K., Tian, J., S. Gold, M., & Liu, Y. (2011). Food addiction and neuroimaging. *Current Pharmaceutical Design, 17*(12), 1149–1157.

Zhuo, X., Zhang, P., & Hoerger, T. J. (2013). Lifetime direct medical costs of treating type 2 diabetes and diabetic complications. *American Journal of Preventive Medicine, 45*(3).

Ziauddeen, H., & Fletcher, P. (2013). Is food addiction a valid and useful concept? *Obesity Reviews, 14*, 19–28.

# INDEX

Note: Page numbers in *italics* indicate figures; page numbers in **bold** indicate tables.

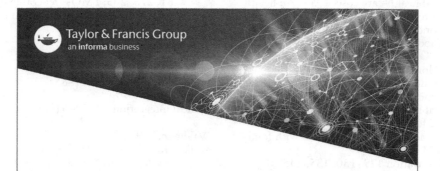